Liquid Crystals and Ordered Fluids

Volume 3

Liquid Crystals and Ordered Fluids

A Continuation Order Plan is available for this series. A continuation order will bring
delivery of each new volume immediately upon publication. Volumes are billed only upon
actual shipment. For further information please contact the publisher.

Liquid Crystals and Ordered Fluids

Volume 3

Edited by
JULIAN F. JOHNSON
Institute of Materials Science
University of Connecticut
Storrs, Connecticut

and

ROGER S. PORTER
Materials Research Laboratory
Department of Polymer Science and Engineering
University of Massachusetts
Amherst, Massachusetts

PLENUM PRESS • NEW YORK-LONDON

The Library of Congress cataloged the second volume of this title as follows:

Main entry under title:

Liquid crystals and ordered fluids.

   Papers, from a symposium of the Division of Colloid and Surface Chemistry held in Chicago during the national meeting of the American Chemical Society, August 1973, of the 3d of a series of meetings; papers of the 1st are entered under the title: Ordered fluids and liquid crystals; papers of the 2d are entered under: Symposium on Ordered Fluids and Liquid Crystals, 2d, New York, 1969.
   1. Liquid crystals – Congresses. I. Johnson, Julian Frank, 1923-   ed. II. Porter, Roger Stephen, 1928-   ed. III. American Chemical Society. Division of Colloid and Surface Chemistry.
QD923.L56   548'.9   74-1269
ISBN 0-306-35182-X (v. 2)

Library of Congress Card Catalog Number 74-1269
ISBN 0-306-35183-8

Selected papers from a symposium of the Division of Colloid and
Surface Chemistry held in Chicago, August 30–September 3, 1977

© 1978 Plenum Press, New York
A Division of Plenum Publishing Corporation
227 West 17th Street, New York, N.Y. 10011

PREFACE

This volume represents a collection of selected papers from a
symposium of the Division of Colloid and Surface Chemistry held during the national meeting of the American Chemical Society in Chicago,
August 1977.

A variety of experimental techniques has been used in these
studies including dynamic calorimetry, equilibrium calorimetry,
Raman spectroscopy, NMR, depolarized fluorescence, acoustical/opti-
cal effects, EPR, and photochemical methods. The range of systems
studied is similarly broad, including both lyotropic and thermotropic materials ranging from biological systems through synthetic
organic liquid crystals.

The variety of study modes ranges from applied studies to
highly theoretical treatments.

Historically, the amount of research in the area of liquid
crystals has fluctuated widely, with peaks of activity around 1900
and in the 1930's. For the period that began in 1960 liquid crystal
research appears to have peaked and has now settled down to be a
stable field with many investigators involved but lacking both the
periods of rapid acceleration and deceleration of effort that characterized it earlier.

Julian F. Johnson
Institute of Materials Science
University of Connecticut
Storrs, Conn. 06268

Roger S. Porter
Polymer Science & Engineering
University of Massachusetts
Amherst, Mass. 01002

v

CONTENTS

RAMAN SPECTROSCOPIC AND CALORIMETRIC INVESTIGATION OF THE MULTIPLE

SMECTIC PHASE MATERIAL TBBA

J. M. Schnur and J. P. Sheridan

Naval Research Laboratory

Washington, D. C. 20375

INTRODUCTION

Smectic mesophase formation and structure have been the subject of vigorous scientific inquiry in recent years.[1-6] Among those phases investigated most intensively, the B and H phases are thought to possess the highest degree of local three-dimensional order. As determined by X-ray crystallography [1-5] both phases exhibit long-range two dimensional pseudo-hexagonal order within each layer; in addition there appears to be some degree of correlation between planes.[1,5] In the H phase the director lies at an angle to the smectic planes, while in the B phase it is normal to the planes. The nature of the degrees of freedom available and the intermolecular ordering in these phases has been the subject of a number of theoretical and experimental studies.[3,7,8] Since terephthal-bis-butylaniline (TBBA) exhibits a smectic H phase as well as at least eight other phases (see schematic below), it has been extensively studied by various experimental techniques including X-ray,[4] NMR,[9] DSC,[10] neutron scattering,[11] infrared[12] and Raman[10,13-14] spectroscopy. This paper provides new information on the nature of the various smectic phases observed in TBBA. We present new spectroscopic and thermodynamic data about TBBA, and two of its deuterated isotopes, obtained in its various solid and smectic phases. Both the Raman and calorimetric techniques have provided new insights into the nature of these smectic phases and the mechanisms of the transition.

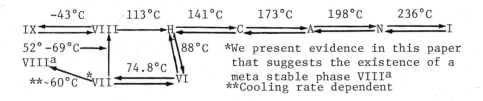

Changes in the Raman spectra have been observed between a number of the phases, the most dramatic differences being seen between the solid and H phases. Specifically, there is clear evidence that substantial "melting" of the butyl hydrocarbon endgroups occurs upon entering the H phase. There are also changes in other spectral regions which suggest that one or two of the aromatic rings begin to rotate with respect to each other in the H phase. These points are consistent with the NMR,[9] and neutron scattering[11] studies which support the model of the alkyl end groups being fairly mobile and liquid-like in the H-phase, and the possible existence of a correlated motion of the molecules around their long axes, which also can explain the pseudo-hexagonal symmetry observed in the X-ray[1] experiments. Furthermore, the Raman data suggest that phase VI lies intermediate in structure between the smectic H   phase and the solid phase VIII while phase VII is very similar on a local scale to the molecular structure of phase VIII.

In addition, while the spectroscopic data obtained upon heating a virgin crystal compared quite well with previously published results,[13] it was found that recycling the samples produced unusual and, at first, not easily reproducible results.

Specifically, a sharp low frequency band (~20 cm^{-1}) appeared in some cases but not in others, and the solid-smectic H phase transition temperature appeared to vary. Consequently a comprehensive DSC study of the effects of thermal history on TBBA was initiated. This study demonstrated that there are three solid phases, one that exists below −43°C, the other two existing above that temperature. One of these phases may be metastable with respect to the other or the effect may be a kinetic one. As a result of this study, we have learned how to control the sample history so as to yield consistent reproducible results. Recently, Raman spectra have been published[14] which did not seem to agree with our previously reported results.[13] By appropriately varying sample history and quenching we can now reproduce both Dvorjetski's[14] work and our earlier spectra.

EXPERIMENTAL

Raman

The experimental arrangement consisted of a Spex 14018 double monochromator with an RCA 31034 cooled photomultiplier tube and a photon counting system. The spectra were enhanced by the use of zero suppression and scale expansion techniques. The excitation source was a Coherent Krypton 500K laser providing 80 mW of 7525Å radiation at the sample. This near infrared line of the Krypton laser provided the best spectra, since it minimized both photo-degradation of the sample and background fluorescence. The samples were contained in 1 mm glass capillary tubes. The laser beam was focused upon the sample to approximately a 50 μm diameter, and 90° scattering from the illuminated region was then sampled. Both the exciting and scattered radiation were scrambled to minimize the spurious birefringent contributions possible in anisotropic meso-morphic systems. The temperature was continuously monitored and was accurate to within $\pm$ 0.5°C; fluctuations were less than $\pm$0.2°C.

At higher temperatures the spectra were obtained by running several samples in different but overlapping spectral regions. TBBA and two of its isotopes were studied in all observed phases except the isotropic. The structures of these materials are described elsewhere.[10] It was found that the material with the deuterated centers (TBBA-DC) was of highest initial purity, and consequently provided the best spectra at highest temperatures. TBBA and TBBA deuterated tails (TBBA-DT) were obtained as a gift from Jean Charvolin, Laboratoire des Solides, Universite de Paris, Orsay, France and TBBA-DC was a gift from Sol. Meiboom, Bell Laboratories, Murray Hill, New Jersey.

Calorimetry

A Perkin-Elmer scanning microcalorimeter model #DSC-2 was used to obtain the calorimetric data. The machine was calibrated for use to 100°K. The samples were placed in hermetically sealed alum-inum containers in a rigorously dry argon atmosphere. Scanning rates varied between 1.25°/mm and 5°/mm.

RESULTS

Spectroscopic

The most dramatic differences between the H and solid phase occur in the 300 cm^{-1} and 1500 cm^{-1} region. In order to attribute these changes to molecular conformations and motions, TBBA-DT (deuterated tail) and TBBA-DC (deuterated centers) were studied. Figure 1 shows the Raman spectra for TBBA, TBBA-DC, and TBBA-DT,

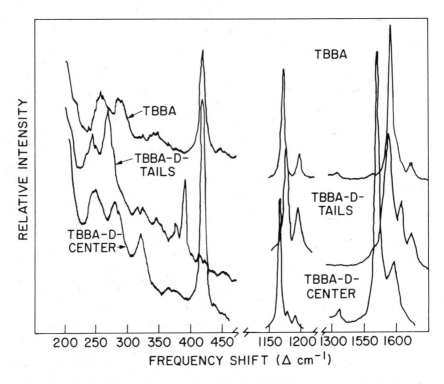

FIGURE 1 - Raman Spectra of TBBA, TBBA-DC, and TBBA-DT in the
200-450, 1150-1200, and 1500-1650 cm^{-1} regions.
Spectral resolution is 3 cm^{-1}.

at 30°C for the 200 cm^{-1}-450 cm^{-1}, 1150 cm^{-1}-1200 cm^{-1}, 1500 cm^{-1}-1650 cm^{-1} regions. The deuteration of the tails has apparently reversed order of the peaks in the 250-300 cm^{-1} region, the smaller of the two appears to have shifted from ~285 cm^{-1} to ~245 cm^{-1}. The more intense band at ~418 cm^{-1} has also shifted slightly lower in frequency. The shift upon deuteration, the location of the non-deuterated band, coupled with the Raman studies of alkanes,[18] alkylbenzenes,[19] and alkoxy azoxybenzenes[20] strongly suggest that the 285 cm^{-1} band is associated with a group motion of the butyl side groups, most probably the all trans accordian band.

Previous work on Schiff bases[21] has indicated that the bands

in the 1550 cm^{-1}-1650 cm^{-1} region are associated with the R-C = N
$\phantom{in the 1550 cm^{-1}-1650 cm^{-1} region are associated with the R-C}$ H

linkage. Upon deuteration of the centers TBBA-DC - the bands observed in this region (both solid and H phase) shift downward approximately 30 cm^{-1} thus confirming that these bands provide information on the central portion of the TBBA molecule.

Raman spectra for TBBA solid and H phases in the 200-600, 1100-1250, and 1500-1700 cm^{-1} region are shown in Figure 2. The most dramatic spectral change is observed in the 1500-1700 cm^{-1} region. In the solid phase only a trace of a band exists at about 1560 cm^{-1}. Upon entering the H phase this band dramatically increases a factor of 40 when compared to the 1170 cm^{-1} band. The 1170 cm^{-1} band does not change appreciably as a function of temperature. Figure 3 shows a plot of the intensity of the 1560 cm^{-1} band relative to the 1170 cm^{-1} band as a function of temperature through the solid-smectic H phase transition region. There is no evidence of pretransition effects and the sharpness of the intensity change at the phase transition temperature is indicative of a highly cooperative event.

Another important set of spectral changes occur in the 280-350 cm^{-1} region. The band at 285 cm^{-1} in the solid phase decreases in a more or less sigmoidal fashion upon nearing and entering the smectic H phase. Simultaneously a group of bands at somewhat higher frequencies (310-350 cm^{-1}) begins to increase in intensity in the phase transition region and becomes quite dominant in the H phase. The temperature dependence of the relative intensities (compared to the 1170 cm^{-1} band) of the 285 cm^{-1} band and the envelope of bands at 310-350 cm^{-1} is shown in Fig. 4. As can be seen from the figure, the intensity changes are less sharp at the phase transition temperature than that displayed by the 1560 cm^{-1} band and there is evidence of some pretransition changes. However both curves display the sigmoidal behavior expected from a cooperative first-order phase change. Upon heating through the smectic phases C and A and on into the nematic phase only small increases in the relative intesities of the 310-350 cm^{-1} envelope occur accompanied by a slight decrease in intensity of the 285 cm^{-1} band. Similarly, there is little or no change in the relative intensity of the 1560cm^{-1} as a function of phase upon heating.

However, upon cooling the sample, we observed some further interesting changes in these spectral regions. Figure 5(a) and (b) show the spectra obtained in the 200-450 cm^{-1} region for Phases VI and VII respectively. In Phase VI it appears that the 285 cm^{-1} has regained considerable intensity while some of the higher frequency bands (notably at ~315 cm^{-1}) persist. In Phase VIII most of the 310-350 cm^{-1} envelope has disappeared and the 285 cm^{-1} is even stronger. Furthermore, the strong quasi-elastic scatter

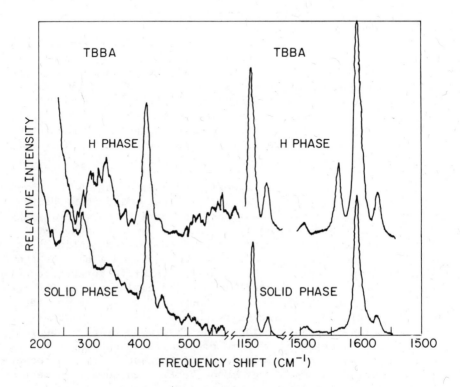

FIGURE 2 - Raman spectra of TBBA solid and H phase in the
200-600, 1100-1250, and 1500-1700 cm^{-1} region.
Spectral resolution is 3 cm^{-1}.

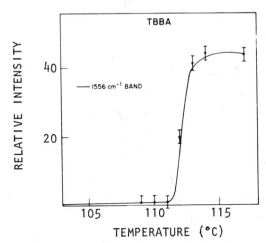

FIGURE 3 – Temperature dependence of the intensity of the 1560 cm^{-1} band relative to the 1170 cm^{-1} band through the solid-smectic H transition region.

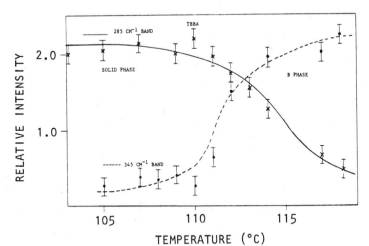

FIGURE 4 – The temperature dependence of the relative intensities (compared to the 1170 cm^{-1} band) of the 285 cm^{-1} band and the envelope of bands at 310–350 cm^{-1}.

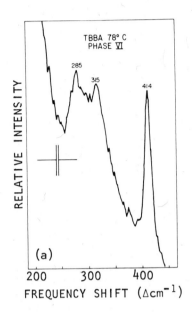

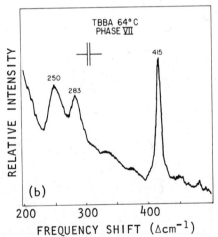

FIGURE 5(a) and (b) – Raman spectra of TBBA obtained in the
              200–450 cm^{-1} region for phases VI and
              VII respectively.

observed in Phase VI and the smectic H phase has considerably diminished and the 250 cm^{-1} band is now observable, resulting in a spectrum remarkably similar to that observed in the solid phase VIII for the region 200-450 cm^{-1}. Figure 6 shows the spectral range 1100 cm^{-1}-1650 cm^{-1} for phase VI and VII respectively. In phase VI we see that the 1560 cm^{-1} band has shifted slightly in frequency but persists at about the same relative intensity as it does in the H phase. On the other hand, in phase VII there is a discontinuous drop in intensity of the 1560 cm^{-1} band to a value commensurate with that of the solid VIII phase. Thus, in this spectral region the Raman spectrum of phase VII is remarkably similar to that of solid phase VIII.

Turning now to the low frequency regime, Fig. 7 illustrates the spectra obtained in the range 0-25 cm^{-1} for the solid and smectic H phases obtained on heating a virgin crystal of TBBA as well as the phases VI and VII obtained upon cooling. The previously observed band at ~20 cm^{-1} in the solid is seen to persist into the smectic H phase corroborating earlier observations of Schnur and Fontana.[13] Furthermore, it is also observed in both of the cooled phases VI and VII. In the smectic H and VI phases the band is partially obscured by strong quasi-elastic scattering and the frequency appears to have shifted to a lower value. However, careful work by Fontana and Bini[22] on single crystal mono-domain samples of TBBA has revealed that the 20 cm^{-1} band is relatively temperature insensitive and that its frequency remains constant over a wide temperature range including the smectic H phase. The band does disappear discontinuously at the smectic H-smectic C phase transition - again in agreement with the previously reported observations of Schnur and Fontana.

At this point an interesting observation can be made which may throw some light upon the variance of some of the observations reported here from those of Dvorjetski et al.[14] The low frequency Raman data obtained upon heating from a virgin crystal of TBBA solid phase VIII to the smectic H modification agrees well with similar spectral data obtained from other virgin TBBA crystals. However, we have found that temperature recycling of the samples produces unusual and not easily reproducible results. Specifically, the 20 cm^{-1} band appeared in some cases but not in others and the solid-smectic H transition temperature seemed to vary with thermal history of the sample. Consequently, a comprehensive DSC study of the effects of thermal history upon TBBA was undertaken and correlated with the Raman spectral observations.

<u>Calorimetric</u>

Figure 8 illustrates a calorimetric scan for a virgin crystal of TBBA in the temperature range ~230-390°K. It clearly shows the

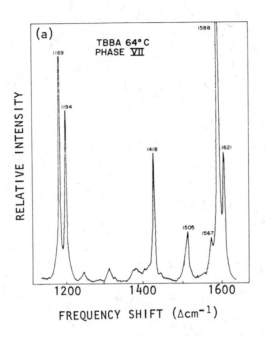

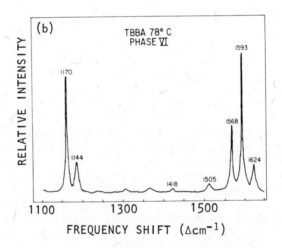

FIGURE 6(a) and (b) – Raman spectra of TBBA in the 1100–1650 cm⁻¹
region for Phase VI and VII respectively.

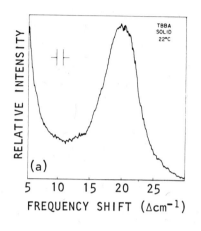
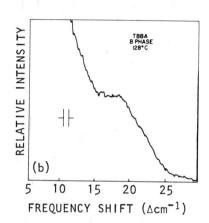
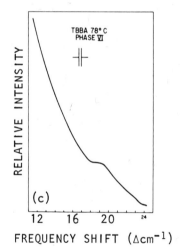
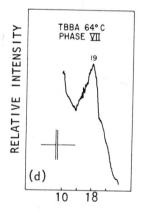

FIGURE 7(a), (b), (c), and (d) – Raman spectra of TBBA in
0–25 cm^{-1} region in Solid, H   phase, phase VI, and
phase VII respectively. Spectra resolution is
0.1 cm^{-1}.

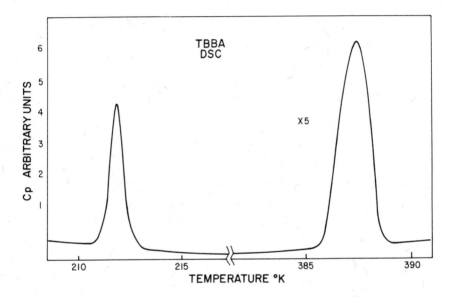

FIGURE 8 – Tracing of a DSC calorimetric scan for a virgin
crystal of TBBA in the temperature range
210–390°K.

existence of the solid (IX)-solid (VIII) phase transition at ~230°K
and the solid VIII-smectic H phase transition at ~387°K. However, if that sample were now cooled below 243°K (but not to liquid
N_2 temperature) and recycled up in temperature, three observations
can be made: (1) no solid IX-solid VIII transition is observed;
(2) the apparent solid-smectic H transition begins at a variable
temperature about 4 degrees below the previously observed ~113°C;
(3) a doublet is observed in the melting curve with a separation
of 2-3 degrees. The relative magnitudes of the two peaks in this
doublet are dependent upon the sample's previous thermal history.
Figure 9 shows scans of this temperature region corresponding to
three different thermal histories on the same sample of TBBA. The
lower trace corresponds to a second melting curve after cycling the
initially melted sample to ~200°K for a few minutes in the calorimeter until apparent equilibration has occurred. The middle trace
corresponds to a melting curve obtained after a half-hour equilibration at liquid N_2 temperatures. Subsequent Raman scans were
conducted on samples similarly treated. In each case, the onset
of the solid-smectic H phase transition (as measured by the sharp
increase in the 1560 cm^{-1} band) occurs at a temperature several
degrees below that observed for the virgin crystal. Concomitantly,
the 20 cm^{-1} band seems to disappear at the transition while the
quasi-elastic scatter increases enormously.

In contrast to these results, if the sample is maintained at
liquid N_2 temperature for several hours, complete equilibration to
solid phase IX appears to occur. A calorimetric scan on this
equilibration to solid phase IX appears to occur. A calorimetric
scan on this equilibrated sample now reveals the reappearance of
the solid IX-solid VIII transition and a solid VIII-smectic H phase
transition occurring at ~387°K accompanied by a single melting
curve. Raman temperature scans of sample similarily equilibrated
reveals the persistence of the low frequency 20 cm^{-1} band into the
smectic H phase.

DISCUSSION AND CONCLUSION

Our calorimetric data therefore appear to explain the thermal
cycling problems experienced by Venugapalan[12] as well as the apparent differences betwee the published data of Schnur and Fontana,[13]
Fontana and Bini,[22] and Dvorjetski.[14] Clearly TBBA is a complicated material and it is now quite apparent that the thermal history is extremely important in determining which phases can be
reached. The thermal data give clear evidence of a metastable
solid phase [VIIIa] which melts at a slightly lower temperature
than the phase VIII-H phase transition. Phase VIII is easily obtained by crystallization from a solvent or upon heating of phase
IX. Phase VIIIa is only obtained by cooling phase VII (pressure
effects not withstanding). The data presented in Fig. 9 suggests

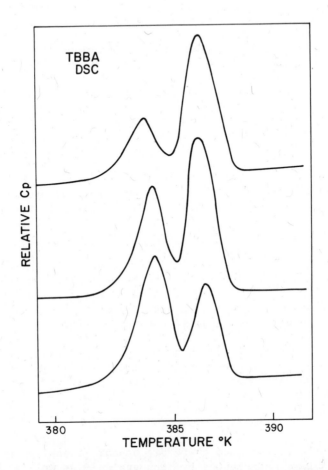

FIGURE 9 – Tracing of DSC calorimetric scans of TBBA in the
380–390°K temperature range as a function of
thermal history.

that both phase VIII and the metastable phase VIII[a] are formed
upon cooling from phase VII with the relative proportion of each
phase dependent upon the cooling rate. The structures of phase
VIII, VIII[a], and VII are quite similar according to our Raman
data and it is quite possible that phase VIII[a] is a glass formed
by cooling phase VII rapidly.

It is now possible to explain the disappearance of the 20 cm^{-1}
band observed by Dvorjetski[14] as due to masking by strong light
scattering obtained in a mixed phase region of VIII crystal, VIII[a]
crystal and H phase.

The earlier triple monochromator Raman data of Schnur and
Fontana[13] was always acquired after cooling to phase IX and thus
only a simple phase transition was observed with much decreased
light scattering from the sample. This then made possible the
observation of the 19 cm^{-1} band on both sides of the phase transition.

The Raman data presented in this paper coupled with the
earlier spectroscopic data provide an understanding of the molecular basis of some of the observed phases and phase transitions.
The changes observed in the 200-400 cm^{-1} region between the solid
VIII phase and the H phase upon heating are most probably caused
by the melting of the butyl tails. This interpretation is consistent with the previous NMR,[9] neutron,[11] and infrared[12] results.
The behavior of the 1560 cm^{-1} band suggests that some intramolecular rotation of the benzene rings with respect to each
other is allowed in the H phase.

The results upon cooling suggest that phase VI is somewhat
more conformationally ordered that the H phase. The Raman spectral information in the 200 cm^{-1}-400 cm^{-1} show that the all trans
conformation of the butyl tails is more strongly preferred relative to the gauche conformers than in phase H. The internal rotation of the aromatic rings is still allowed in this phase.

Phase VII is quite similar conformationally to phase VIII
according to our Raman data. The butyl tails are now primarily
in the all trans conformation and the internal rotation is now
substantially restricted.

Thus a general picture of these highly ordered smectic phases
now emerges which suggests that there is a high degree of intermolecular two dimensional order in all of these phases (VII, VI,
H) with some lower degree of three dimensional correlations.

The primary differences between phase VII, VI, and H phase
would thus appear to be in their intramolecular order with each

higher temperature phase being more conformationally disordered.

ACKNOWLEDGEMENTS

We gratefully acknowledge the advice and long discussions of this subject with Prof. M. Fontana, Dr. R. Priest, and Dr. P. Schoen.

REFERENCES

1. A. Levelut and M. Lambert, C. R. Acad. Sci. (Paris) 272, 1018 (1971).

2. A. de Vries and D. L. Fishel, Mol. Cryst. 16, 211 (1972).

3. J. Doucet, A. M. Levelut, and M. Lambert, Phys. Rev. Lett. 32, 201 (1974).

4. A. de Vries, Chem. Phys. Lett. 28, 252 (1974).

5. J. Doucet, A. M. Levelut, M. Lambert, L. Liebert, and L. Strzelecke, J. de Phys. 36, Supp. 3 (1975).

6. J. Billard, C. R. Hebd, Seances Acad. Sci. (Paris) B280, 573 (1975).

7. R. J. Meyer and W. L. McMillan, Phys. Rev. A9, 899 (1974).

8. R. J. Meyer, Phys. Rev. A12, 1066 (1975).

9. Z. Luz and S. Meiboom, J. Chem Phys. 59, 275 (1973).

10. J. Schnur, J. P. Sheridan and M. Fontana, Int. Liq. Cryst. Conf. Bangalore (1973); Pramana Sup. #1, P.175 (1975).

11. H. Hervet, F. Volino, A. J. Dianoux, and R. E. Lechner, J. de Phys. L35, 151 (1974).

12. S. Venugopalan et al. Mol. Crystl and Liq. Cryst. 40, 149-161 (1977).

13. J. M. Schnur and M. Fontana, J. de Phys. L35, 53 (1975).

14. D. Dvorjetski, V. Volterra and E. Wiener-Avnear, Phys. Rev. A 12, 681 (1975).

15. J. M. Schnur, Mol. Cryst. Liquid Cryst. 23, 155 (1973).

16. B. J. Bulkin and F. T. Prochaska, J. Chem. Phys. 54, 635 (1971).

17. S. J. Borer, S. S. Mitra and C. W. Brown, Phys. Rev. Lett. 27, 379 (1971).

18. R. F. Schaufele, J. Chem. Phys. 49, 4168 (1968).

19. F. Behroozi, R. Priest, and J. Schnur, J. Raman Spectroscopy 4, 379 (1976).

20. J. M. Schnur, Phys. Rev. Lett. 29, 1141 (1971).

21. G. Vergoten, Dissertation, University of Lille (1973).

22. M. Fontana and S. Bini, Phys. Rev. A 14, 1555 (1976).

THERMAL PROPERTIES OF A NEW SERIES OF LIQUID CRYSTAL FORMING
MATERIALS: p,p'-DI-n-ALKYLDIPHENYL AND p,p'-DI-n-ALKOXY-
DIPHENYLDIACETYLENES

E. M. Barrall, II, Barbara Grant, Annie R. Gregges

IBM Research Laboratory

San Jose, California 95193

It is impossible to predict the mesomorphic behavior (if any)
of a compound using modern liquid crystal theory. When the
compound belongs to an already recognized liquid crystalline
homologous series, it is safe to assume that some liquid crystal
behavior will be exhibited. Given the heats and temperatures of
transition of neighboring homologs, the transition entropy of
the new homolog may be estimated.[1] However, if the compound in
question belongs to an entirely new homologus series, only
intuition and previous experience furnish any guide to possible
mesomorphic properties.

It is known that compounds with a linear rigid central axis
tend to be liquid crystalline. Para-substitution on this core
further adds to the likelihood of liquid crystallinity as does
the existence of a dipole moment.[2,3] Indeed, the most divergent
groups of materials have been found to be liquid crystalline when
they satisfy all of these conditions.[4]

The 1,4-disubstituted diacetylenes have attracted significant
attention as monomeric matrices for solid (crystalline) phase
polymerizations.[5] In many cases the intermolecular distances
favor the production of highly regular polymer structures upon
irradiation or annealing.[6] In a recent study diacetylenes of
the types:

R—⬡—C≡C–C≡C—⬡—R (p,p'-di-n-alkyldiphenyldiacetylene)

I

19

and

R-O—⟨O⟩—C≡C–C≡C—⟨O⟩—OR (p,p'-di-n-alkoxydiphenyldiacetylene)

II

R = n-alkane

were prepared.[7] The authors have chosen to name these compounds
in terms of diacetylene, since there is no generally accepted
shorter nomenclature for the parent

—⟨O⟩—C≡C–C≡C—⟨O⟩—

as there is for the symmetrical diphenylacetylene, i.e., tolane.
The name 1,4-bis(p-alkyl or alkyloxyphenyl)1,3-butadiyne would
also be acceptable, but slightly more cumbersome.

The central portion of this molecule, benzene rings included,
is quite rigid and linear. By analogy to the tolanes which
contain only one acetylene linkage, the diphenyldiacetylenes may
be expected to be liquid crystal-forming materials. Hot stage
microscopy and differential scanning calorimetry (DSC) have
supported this expectation. Indeed, compounds of types I and II
have proven to be stable mesophase forming materials. Thus, they
constitute a previously unrecognized series of nematic and smectic
mesogens. This paper is concerned with the description of the
new series of liquid crystals in terms of calorimetry and optical
microscopy. The synthesis and purification are given in detail
elsewhere.[7]

EXPERIMENTAL

Microscopy

The transition characteristics of four of the dialkyl,
diphenyldiacetylene I, and eight dialkoxydiphenyldiacetylene,
II, compounds were surveyed with a Zeiss Photomicroscope III
equipped with a Mettler FP-5 hotstage and liquid nitrogen cooler.
All samples were viewed between crossed polarizers. In this way
it was possible to determine all transition temperatures and to
identify solid, smectic, and nematic mesophases. Special care
was used to determine the existence of monotropic mesophases.
None were found. Uniformly, both compounds of types I and II
produced one (nematic) or more (various smectics and nematic)

mesophases on heating. These transition temperatures were used
as a guide in the differential scanning calorimetry (DSC) studies.

Calorimetry

The DSC data were obtained with a Perkin-Elmer Model 2
differential scanning calorimeter connected to an IBM System/7
computer. The digital data were reduced using a Tektronix 4013
interactive display and Tektronix model 4631 hard copy unit
supported by an IBM 360/195 host computer. This system has been
defined fully elsewhere.[9,10] At least three heating runs were
obtained on all compounds. These were added together digitally
to produce thermograms with an improved signal to noise ratio
(1.7 times) before selecting regions for integration.

The samples were degassed under nitrogen for one hour prior
to weighing and encapsulating in standard aluminum Perkin-Elmer
volatile sample sealers. The sample was compressed between the
pan bottom and an additional metal plate to improve the system
thermal conductivity and temperature accuracy as described
previously.[11] All samples were under 2 mg and were both heated
and cooled at 5°C/min.

RESULTS

p,p'-Di-n-Alkyldiphenyldiacetylenes

The DSC scans of these four compounds are roughly similar.
Figure 1 shows the complete scan for the octyl homolog as an
example. The large peak is the solid to nematic transition and
is relatively narrow, indicating a purity of better than 99.8%.
The small peak is the nematic to isotropic liquid transition,
see Figure 2 for a detail of this region. There is evidence of
some pretransitional heat capacity change in the nematic mesophase
similar to that observed for p-azoxyanisole.[12]

The thermal properties for this series on heating are
summarized in Table I, and the transition temperatures are shown
in Figure 3. Results are quoted in the tables in both calories
and joules for convenience in comparing with the older and newer
literature. Both the solid to nematic and nematic to isotropic
liquid temperatures decrease sharply with the butyl substitution
and then level off in the paraffinic melting range. This is
indicative of a significant change in molecular interaction as
the alkyl substituent increases in length beyond four. The same
trend is exhibited in the transition entropy, see Figure 4. The
entropy sharply decreases from propyl to butyl and then increases

TABLE I

Thermal Properties of a Series of p,p'-Di-n-alkyldiphenyldiacetylenes

(I) Derived from DSC Measurements.   R ⬡ C≡C—C≡C ⬡ R

R =	Transition[*]	Transition Temperature[**]		Transition Heat, ΔH		Transition Entropy, ΔS	
		°K	°C	Cal/mole	Joules/mole	Cal/mole/°K	Joules/mole/°K
$CH_3(CH_2)_2-$	Solid→Nematic	381.05	107.76	5620	23510	14.75	61.71
	Nematic→I.L.	404.21	130.91	330.3	1382	0.8171	3.419
$CH_3(CH_2)_3-$	Solid→Nematic	347.07	73.77	3442	14400	9.918	41.49
	Nematic→I.L.	371.60	98.30	226.7	948.5	0.6101	2.553
$CH_3(CH_2)_5-$	Solid→Nematic	329.35	56.05	3514	14700	10.67	44.64
	Nematic→I.L.	356.45	83.15	253.2	1059	0.710	2.972
$CH_3(CH_2)_7-$	Solid→Nematic	337.38	64.08	4212.3	17620	12.48	52.24
	Nematic→I.L.	350.92	77.62	298.5	1249	0.8506	3.559

[*] Identified by hotstage microscopy; I.L. = isotropic liquid.
[**] Temperature at endothermal minimum.

TABLE II

Thermal Properties of a Series of p,p'-di-n-alkoxydiphenyldiacetylenes

(II) Derived from DSC Measurements.   RO ⬡ C≡C—C≡C ⬡ OR

RO =	Transition[*]	Transition Temperature[**]		Transition Heat, ΔH		Transition Entropy, ΔS	
		°K	°C	Cal/mole	Joules/mole	Cal/mole/°K	Joules/mole/°K
CH_3CH_2O-	Solid→Nematic	461.44	188.14	8402	35150	18.20	76.18
	Nematic→I.L.	482.39	209.09	581.2	2441	1.205	5.039
$CH_3(CH_2)_2O-$	Solid→Nematic	413.36	140.06	6525	27300	15.79	66.04
	Nematic→I.L.	449.02	175.72	458.7	1919	1.021	4.274
$CH_3(CH_2)_3O-$	Solid→Nematic	426.64	153.33	10400	43510	24.61	102.0
	Nematic→I.L.	449.14	175.84	536.3	2244	1.194	4.996
$CH_3(CH_2)_5O-$	Solid→Nematic	393.33	120.03	10250	42890	26.07	109.0
	Nematic→I.L.	423.80	150.50	585.4	2449	1.381	5.779
$CH_3(CH_2)_7O-$	SolidI→SolidII	346.71	73.41	2817	11790	8.125	33.99
	SolidII→Nematic	381.53	108.23	6012	25150	15.76	65.93
	Nematic→I.L.	407.19	133.89	464.3	1943	1.140	4.771
$CH_3(CH_2)_9O-$	SolidI→SolidII	366.21	92.91	2179	9117	5.949	24.90
	SolidII→Nematic	372.40	99.10	6293	26330	16.90	70.70
	Nematic→I.L.	398.09	124.79	485.0	2029	1.218	5.097
$CH_3(CH_2)_{13}O-$	SolidI→SolidII	354.69	81.39	1749	7318	4.932	20.63
	SolidII→SmI	368.55	95.25	9423	39430	25.57	107.0
	SmI→SmII	377.60	104.30	25.82	108	0.0684	0.286
	SmII→Nematic	381.86	108.56	1031	4314	2.701	11.30
	Nematic→I.L.	386.86	113.56	576.8	2413	1.491	6.238
$CH_3(CH_2)_{14}O-$	SolidI→SolidII	365.47	92.17	2864	11980	7.837	32.79
	SolidII→SmI	366.65	91.35	1004	4200	27.39	11.46
	SmI→SmII	378.80	105.50	--	--	--	--
	SmII→Nematic	382.27	108.97	1851	7745	4.843	20.26
	Nematic→I.L.	384.19	110.89	1071	4481	2.788	11.66

[*] Sm = Smectic, I.L. = Isotropic Liquid. Phases were identified by optical microscopy.
[**] Taken at endothermal minimum.

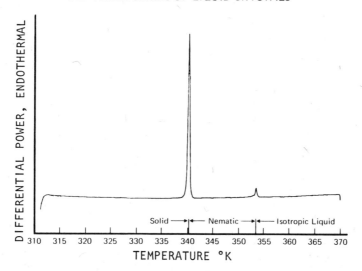

Figure 1. Complete DSC scan for p,p'-Di-n-octyldiphenyldiacetylene at 5°C/min.

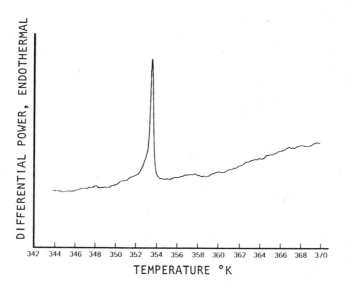

Figure 2. Detail of the Nematic to Isotropic Liquid Transition of p,p'-Di-n-octyldiphenyldiacetylene at 5°C/min.

in a regular order for both the solid and nematic transitions as
carbon units are added. This is additional evidence for
significant changes in molecular interactions when the substituent
chain reaches four carbons in length. From the transition
temperature trend, apparently the propyl solid and nematic phases
have a higher degree of order (fewer degrees of freedom) and
greater molecular interaction than the higher homologs. The
alternative explanation for the decrease in ΔS, greater
interaction in the liquid phase, does not fit with the decreasing
transition temperature trend.

p,p'-Di-n-Alkoxydiphenyldiacetylenes

This series exhibits a significantly more complicated pattern
of transitions and entropy changes than the alkyl series. Indeed,
each member of the series appears to have unique properties. In
general, the temperatures of transition would be expected to be
higher due to the introduction of the alkoxyphenyl ether by
analogy to other series.[13,14] Beyond agreement with that
prediction, this series is unique. The thermal properties are
summarized in Table II and general trends in temperature and
transition entropy are given in Figures 5, 6, and 7.

The p,p'-di-n-ethyloxydiphenyldiacetylene is thermally
unstable. After the first heating the nematic to isotropic liquid
transition endotherm vanishes, and the solid to isotropic liquid
endotherm becomes broader indicating additional decomposition.
Usual diacetylene polymerization appears unlikely, since the
formation of a crosslinked insoluble polymer similar to the type
formed by irradiation of other diacetylenes would not affect
phase transitions in the unreacted material. The crosslinked
product would be insoluble in all phases and would act as an
inert component, decreasing the magnitude of the transition
endotherms but not broadening them. Of course, there is no reason
to expect thermal polymerization to produce the same product as
radiation. The p,p'-di-n-propyloxydiphenyldiacetylene is also
thermally unstable, but to a lesser extent than the ethoxy
material. The transition temperatures are lower than those of
the ethoxy material as is the total transition entropy. The
nematic to isotropic liquid transition entropy is lower in the
propyloxy case than in the ethoxy, but not by a large amount,
0.18 cal/mole/°K.

The p,p'-di-n-butyloxy- and hexyloxydiphenyldiacetylenes are
both thermally stable and may be heated repeatedly through the
nematic to isotropic liquid transition with no reduction in the
size of the transition endotherm. Although the addition of one
carbon unit to the propyloxy compound results in a large entropy
of transition increase for the solid to nematic transition in

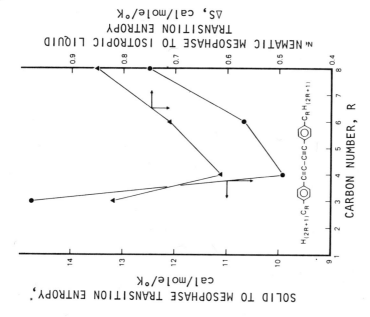

Figure 4. Solid to Nematic to Isotropic Liquid Transition Entropies for a Series of p,p'-n-Di-n-alkyldiphenyldiacetylenes.

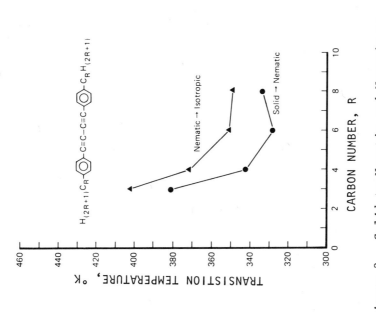

Figure 3. Solid to Nematic and Nematic to Isotropic Liquid Transition Temperatures for a Series of p,p'-Di-n-alkyldiphenyl-diacetylenes.

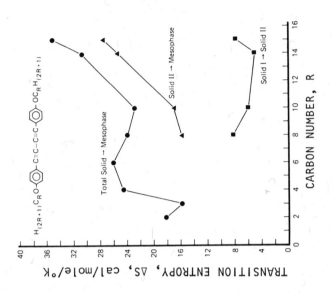

Figure 6. Solid to Solid to Mesophase
Transition Entropies for a Series of
P,p'-Di-n-alkoxydiphenyldiacetylenes.

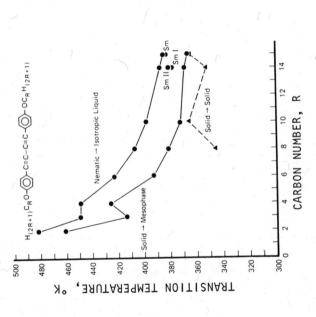

Figure 5. Transition Temperatures for a
Series of P,p'-Di-n-alcoxydiphenyldiacetyl-
enes.

p,p'-di-n-butyloxydiphenyldiacetylene, little increase is noted in the nematic to isotropic liquid transition. The next higher homolog, hexyloxyphenyl, shows little increase in the solid to nematic transition entropy over the level established by the butyloxyphenyl material.

The DSC scans, Figure 8, of the p,p'-di-n-octyloxydiphenyl-diacetylene indicate two solid phase transitions. Figure 9 shows the first solid transition in detail. Solid phase polymorphism is not uncommon in mesogenic compounds, but it may be of two types:

A. $\quad$ Solid I $\underset{\longleftarrow}{\overset{T_1}{\rightleftharpoons}}$ Mesophase $\overset{T_3}{\rightleftharpoons}$ Isotropic Liquid

$$\Big\updownarrow T_2$$

Solid II

where $T_2 > T_1$

B. Solid I $\overset{T_1}{\rightleftharpoons}$ Solid II $\overset{T_2}{\rightleftharpoons}$ Mesophase $\overset{T_3}{\rightleftharpoons}$ Isotropic Liquid

Case A implies no conversion between solids without passing through the mesophase. p-n-Hexyloxy-p'-cyanobiphenyl is a good example of this case.[13] Case B implies a true solid to solid phase transition with equilibrium between solid phases. In Case A either solid may form below T_2 depending on conditions of nucleation as a mixed crystal phase. The percentage of either solid formed is variable and irreproducible, and the endotherms of both solid A and B will obey van't Hoff's equation of melting. This will give both endotherms a lower temperature tail. In Case B the solid to solid transition endotherm will have a shape determined by the rate at which the lattice can transform, and the amount of heat required for each endotherm will be constant. The endotherm between solid II and the mesophase will have a shape determined by the van't Hoff equation.

It is clear in Figures 8 and 9 that the first endotherm for p,p'-di-n-octyloxydiphenyldiacetylene has a very sharp onset, and the second endotherm has a typical melting low temperature endothermal tail. Also, in six runs the areas of the Solid I→Solid II and Solid II→mesophase endotherms remained constant on a ratio basis. Thus, the transition path for the octyloxydiphenyl material is

$$\text{Solid I} \underset{\longleftarrow}{\overset{346.71°K}{\rightleftharpoons}} \text{Solid II} \underset{\longleftarrow}{\overset{381.53°K}{\rightleftharpoons}} \text{Nematic} \underset{\longleftarrow}{\overset{407.19°K}{\rightleftharpoons}} \text{Isotropic Liquid.}$$

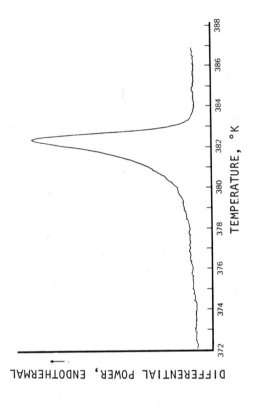
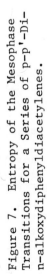

Figure 8. Detail of the Higher Temperature Solid Phase Transition in p,p'-Di-n-octyloxydiphenyl-diacetylene.

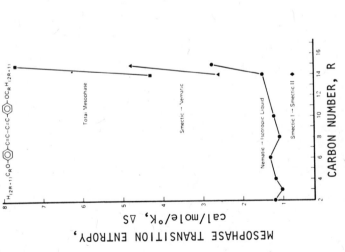

Figure 7. Entropy of the Mesophase Transitions for a Series of p-p'-Di-n-alkoxydiphenyldiacetylenes.

For the purposes of comparison between homologs and with other mesogenic series, the entropy of the solid to mesophase transition is the sum of the Solid I and Solid II transition entropies, or 23.89 cal/mole/°K.

The solid-solid phase transition also appears in p,p'-di-n-decyloxydiphenyldiacetylene, see Figure 10. It is of the same type as above. The compound produces a nematic mesophase with a large premelting change in heat capacity. This change has been eliminated as completely as possible from the data shown in Table II for the heat and entropy of nematic→isotropic liquid transition. When this change in heat capacity is included in the heat measurement, the entropy of the transition increases to 1.82 cal/mole°K from 1.218 cal/mole/°K, or about 50%. On cooling of the isotropic liquid the nematic mesophase reforms sharply at 398°K, and the Solid II reforms at 367.8°K. However, as cooling proceeds, a new solid form, Solid IIA, crystallizes at 363.8°K requiring -203.8 cal/mole. This is an entropy change of 0.560 cal/mole/°K. On further cooling, Solid I crystallizes at 358.37°K. The DSC scan for this crystallization is shown in Figures 11 and 12. Solid IIA appeared on cooling four times and in three separate samples. Thus, it is not an artifact. The small entropy change could be caused by a change in the tilt of the molecules in the crystal. Changes of similar magnitude have been reported previously[15] and can have profound effects on crystal electronic structure.

The smectic mesophase is first exhibited in this series by p,p'-di-n-tetradecyloxydiphenyldiacetylene. Also, two solid phases are evident. The phase path of this compound is significantly more complicated than that of the lower members. From the averaged thermogram shown in Figure 13 the phase transitions are:

$$\text{Solid I} \underset{354.69}{\rightleftharpoons} \text{Solid II} \underset{368.55}{\rightleftharpoons} \text{Smectic I} \underset{377.60}{\rightleftharpoons} \text{Smectic II}$$

$$\underset{381.86}{\rightleftharpoons} \text{Nematic} \underset{386.86}{\rightleftharpoons} \text{Isotropic Liquid}$$

The Solid I→Solid II transition, from microscopy, is accompanied by a large change in specific volume. On cooling Solid II the volume change causes the preparation to shatter abruptly as Solid I is formed. Solid II converts to a clearly smectic mesophase at 368.55°K. From detailed microscopy this phase appears to be a tilted smectic B ($S_{B\angle}$ in conventional notation). This phase has been characterized for N_1N'-terephthalylidene-bis(4-n-butyl-aniline) (TBBA):

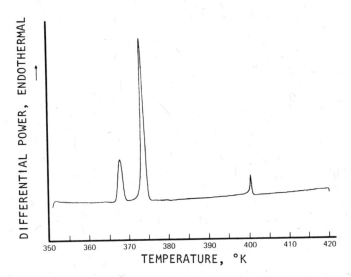

Figure 10. Complete Heating DSC Scan on p,p'-Di-n-decyloxydiphenyldiacetylene at 5°C/min.

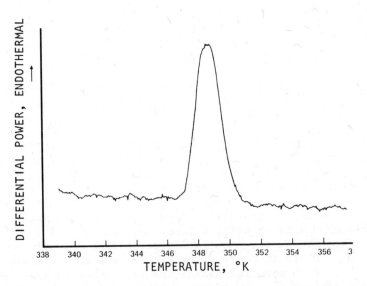

Figure 9. Detail of the Lower Temperature Solid Transition in p,p'-Di-n-octyloxydiphenyldiacetylene.

$$C_4H_9-\bigcirc-N{=}CH-\bigcirc-CH{=}N-\bigcirc-C_4H_9 \; .$$

X-ray studies on single $S_{B\angle}$ crystals of this compound demonstrated
that the molecules are hexagonally close packed and tilted at
32° to the normal.[16]   According to Coates and Gray, the $S_{B\angle}$ phase
is followed on heating by either a smectic C or a smectic A
mesophase.[17] The smectic II observed at 377.60°K is obviously
not a smectic A by microscopy, but has the mottled structure
reported for smectic C, D, and F. Since these latter two are
not reported to follow a $S_{B\angle}$, it seems safe to conclude that
smectic II is identical to S_C. The S_C phase is a tilted variant
of the smectic A. That is, the layer thickness is less than one
molecular length and there is no order in the plane of the
smectic. Thus, we have a transition from a hexagonal close pack
tilted layer structure to a randomly spaced tilted layer
structure.

The thermal events surrounding the $S_{B\angle} \to S_C$ transition are most
unusual. The total summed thermogram for all transitions is
shown in Figure 13 and a detail of the mesophase transitions is
shown in Figure 14. The transition is only a small sigmoid change
in the base line slope, similar to that reported for the $S_C \to S_A$
transition in TBBA.[18] This transition has been variously reported
to be second order, but definitive volume change measurements
have not been reported.[19] In the tetradecyloxy material the
$S_{B\angle} \to S_C$ transition shows no thermal hystersis, and the area between
arbitrary limits of temperature is heating rate and cooling rate
invariant. This is not usual in second order processes, which
must have a large time constant. The authors prefer to view the
$S_{B\angle} \to S_C$ transition as a small first order thermal transition with
large pretransitional change in heat capacity. A transition heat
is reported in Table II. The other transitions, smectic→nematic
and nematic→isotropic liquid, are relatively straightforward.
The nematic mesophase has a higher heat capacity than either the
preceding smectic or the following liquid, see Figure 13. This
is characteristic of the phase and has been observed
previously.[12,20]

The phase transitions in p,p'-n-di-n-pentadecyloxydiphenyl-
diacetylene are significantly more compact than the
tetradecyloxydiphenyl material, see Figure 15. The two solid
phase transitions, although reproducible, overlap significantly.
The heats for Solid I→Solid II transition given in Table II were
obtained by arbitrarily constructing a vertical line dropped from
the minimum between the two solid transition endotherms. This
probably resulted in a positive error in evaluating the heat of
the solid I→solid II transition. However, the total heat and
entropy of transition should not be affected. The material also

exhibits <u>two</u> smectic mesophases, which are observable by hot
stage microscopy but overlap so closely as not to be resolved by
DSC. The cooling curves are significantly clearer, see Figure
16, but are not clear enough to evaluate the "second order-like"
smectic C→smecticB$_{\angle}$in any way other than by transition
temperature. The transition path for the pentadecyloxy material
is as follows:

$$
\text{Solid I} \xrightarrow{365.47} \text{Solid II} \xrightarrow{366.65} S_{B\angle} \xrightarrow{378.80} S_C \xrightarrow{382.27} N \xrightarrow{384.19} IL
$$

This is illustrated on Figure 15 for the mesophase range.

DISCUSSION

The thermodynamic approach to crystal and mesophase structure
determination is the only viable route for the study of the
diacetylene compounds, since all these materials are potentially
unstable in an x-ray beam. Indeed, ionizing radiation is the
usual method for polymerizing these materials in the solid state.[5]
Thus, comparisons, which in other homologous series would be
better and more precisely served by powder and single crystal
x-ray diffraction studies, are useful in diacetylenes.

In the p,p'-di-n-alkyldiphenyl series the melting point and
entropy trend with increasing alkyl tail length indicates that
there is a profound change in crystal organization between the
propyl and butyl materials, see Figures 3 and 4. For butyl and
above the melting point and entropy of transition trend for the
solid to mesophase transition follows that of the n-alkanes,
although lower on an entropy basis. This indicates that the
tails must be arranged in a crystal order similar to the
n-alkanes. Below butyl the order is significantly different and
is probably determined by the diphenyldiacetylene portion of the
molecules. This predicts that the methyl, ethyl, and propyl
homologs may be more similar in structure to previously reported
diacetylenes than the butyl and higher homologs, which have
paraffinic packing characteristics.

The entropy change and transition temperature for the nematic
to isotropic liquid in the p,p'-di-n-alkyldiphenyl materials
follows the same trend as the crystal to nematic transition.
This indicates a high degree of interaction in the nematic region
between alkyl tails. In many homologous series the alkyl tails
do not contribute to the order of the nematic mesophase, but are
essentially as free to rotate as in the isotropic liquid.[21,22]
Thus, the nematic mesophase of the alkyldiphenyl homologs have
a type or order missing in many known liquid crystals as well as
in the lower p,p'-di-n-alkoxydiphenyldiacetylenes. The leveling

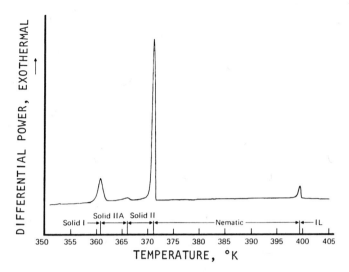

Figure 11.   Complete Cooling Scan on p,p'-Di-n-decyloxydiphenyldiacet-
ylene.

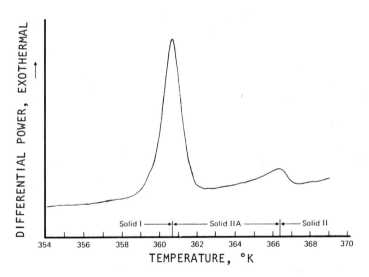

Figure 12.   Detail of the Solid II → Solid IIA and Solid IIA → Solid I
Transitions for p,p'-Di-n-decyloxydiphenyldiacetylene.

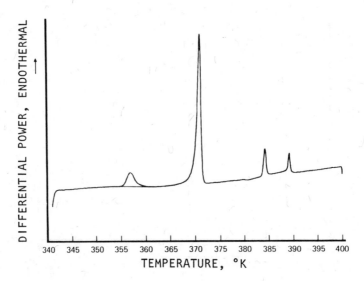

Figure 13. Averaged DSC Scan of p,p'-Di-n-tetradecyloxydiphenyl-diacetylene.

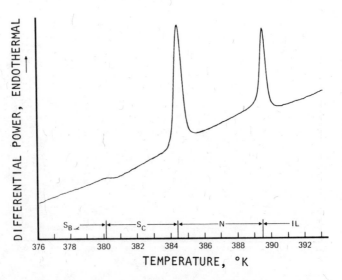

Figure 14. Detail of the Mesophase Transition Regions of p,p'-Di-n-tetradecyloxydiacetylene.

off of the nematic to isotropic liquid transition temperature
above p,p'-di-n-hexyldiphenyldiacetylene suggests that the alkyl
chains are extended in paraffinic fashion in the nematic
mesophase, see Figure 3.

The p,p'-di-n-alkoxydiphenyldiacetylenes are somewhat more
complex in phase behavior. This is generally true for
alkoxyphenyl ethers compared to alkylphenyl homologs for many
series.[13,21,22] The interposition of the oxygen not only bends
the alkyl tail out of the plane determined by the "stiff" portion
of the molecule, but also permits additional molecular interaction
via van der Waals forces. This has a twofold effect: increased
transition temperatures and/or entropies compared to alkylphenyl
homologs and increased opportunity for solid and mesophase
polymorphy.

For the p,p'-di-n-alkoxyphenyldiacetylenes polymorphism is
first noted at octyloxydiphenyl and appears to be characteristic
of higher members. The temperature of the solid to mesophase
transition decreases with additional carbon units, but not as
abruptly as with the alkyldiphenyl series. Indeed, there appears
to be no clear cut split in crystal order between the lower and
higher members of the alkoxydiphenyl series. The total entropy
of the solid to mesophase transition is little affected by alkyl
tail length up to p,p'-di-n-decyloxydiphenyldiacetylene, see
Figure 6. This is a good indication that the crystal order is
not determined by the packing of the hydrocarbon tails. This is
in sharp contrast with the alkyldiphenyl series. For all
materials which show a solid I to solid II transition,
alkoxydiphenyl length is not a packing controlling factor. This
transition is also of relatively low entropy and by analogy to
the triglycerides is probably due to a change of base plane angle
in the hydrocarbon tails.[23] This would result in an alteration
of the intermolecular spacing of the diacetylene bonds at the
transition temperature.

The nematic mesophase transition entropy up to
tetradecyloxydiphenyl is chain length insensitive, see Figure 7.
Therefore, in the nematic mesophase these tails should have a
degree of freedom essentially equivalent to the isotropic liquid
state.

The increment in total mesophase order at C_{14} is the result
of the appearance of a hexagonally close packed tilted smectic
which converts to the tilted disordered smectic C. The order or
entropy change increment is borne primarily on the smectic C to
nematic transition. Only in the pentadecyloxydiphenyl materials
are the chains restricted in rotation in the nematic phase. By
analogy to other series, smectic mesomorphism should start with
the dodecyloxydiphenyl material and nematic mesomorphism should

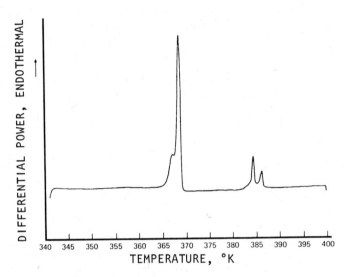

Figure 15.　Complete DSC Scan of p,p'-Di-n-pentadecyloxydiphenyldi-
acetylene.

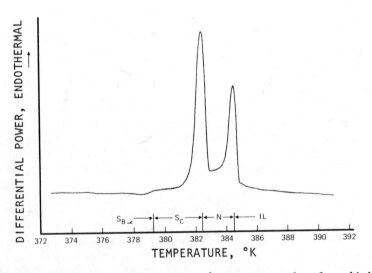

Figure 16.　Mesophase Range of p,p'-Di-n-pentadecyloxydiphenyl-
diacetylene.

vanish in favor of a complex smectic system at the hexa- or heptadecyloxydiphenyl compound.

The apparent second order nature of the $S_{B \angle} \rightarrow S_C$ transition may be due to relatively large post transitional changes in the smectic C phase after a conventional first order transition. Coates and Gray[17] have reported and the present authors have verified that there is a significant change in the birefringence of the smectic C mesophase with temperature. It has been suggested that this continuous change is due to the change of the molecular tilt with temperature.[17] Such changes in tilt alter the heat capacity of the phase with temperature. There is nothing particularly unusual about tilt-temperature sensitivity, for it is a well known and well understood phenomenon in the cholesteric mesophase.[24,25] It is more noticeable in the cholesteric mesophase, since the screw axis dislocation in the cholesteric mesophase is regular and diffracts light of different frequencies as a function of temperature. As stated earlier, the smectic C is not regular in spacing.

The type of smectic mesomorphism, i.e., a highly tilted structure, may indicate a tilted structure in the solid phase. The solid I to solid II transition has already been discussed in terms of tilt. The known hexagonal close pack structure for the smectic $B \angle$ mesophase suggests the same or easily derivable crystal structure for the solid II. There is no evidence for intercalation noted previously for the p,p'-n-alkyl and alkoxycyanobiphenyls.[13]

CONCLUSIONS

The p,p'-di-n-alkyl and alkoxydiphenyldiacetylenes form two homologous series of liquid crystal forming materials. The alkyldiphenyl materials form a series which shows a sharp change in crystal structure between the propyl and butyl homologs. The shorter alkyl tail homologs probably have a structure dictated by the diphenyldiacetylene portion of the molecule, while materials above butyl have a paraffinic crystal structure of lower order. This series exhibits a nematic mesophase at least up to p,p'-di-n-octyldiphenyldiacetylene, which is usually associated with significant order (rotation restriction) in the alkyl chains. The propyl material is much more highly associated in the nematic mesophase than the higher members.

The p,p'-di-n-alkyloxydiphenyldiacetylenes exhibit a solid phase in which the molecular order is little affected by the alkoxy chain length up to decyloxy. The high melting temperatures indicate significant van der Waals interaction in the solid phase. This series is also characterized by solid phase polymorphism from the octyloxydiphenyl homolog. The nematic mesophase

transition entropy is little affected by alkoxy tail length up
to pentadecyloxy. The appearance of the tilted hexagonally close
packed smectic B mesophase in the tetradecyloxy homolog indicates
that the high temperature crystal phase may be similarly packed
with the diacetylene groups along one crystal axis in a diagonal
array. Above the tetradecyloxydiphenyl homolog, paraffinic order
becomes dominant.

The stability of both series of diacetylenes towards x-rays
is now under investigation. If the rates of polymerization or
degradation are low enough, it may be possible to verify these
conclusions by more conventional direct x-ray measurements.

REFERENCES

1. Bondi, A., (1967) Chem. Rev. 67, 565.

2. Brown, G. H. and Shaw, W. G. (1957) Chem. Rev. 57, 1049.

3. McMillan, W. L. (1974) in Liquid Crystals and Ordered Fluids,
        J. F. Johnson and R. S. Porter, eds., Vol. 2, Plenum Press,
        N.Y., p. 141.

4. Verbit, L. and Tuggey, R. L., ibid., p. 307.

5. Wegner, G. (1972) Die Makromol. Chem. 154, 35.

6. Ibid. (1971) 145, 85.

7. Grant, B., to be submitted or J. of Organic Chemistry.

9. Doelman, A. (1975) San Francisco ACS Meeting Preprints,
        Symposium on Analytical Calorimetry, August 31, ANAL77.

10. Doelman, A., Gregges, A. R., and Barrall, E. M., II, in
        Analytical Calorimetry, R. S. Porter and J. F. Johnson,
        eds., Vol. 4, Plenum Press, N.Y., in press.

11. Barrall, E. M., II, and Diller, R. D. (1970) Thermochim. Acta
        1, 509.

12. Barrall, E. M., II, Porter, R. S., and Johnson, J. F. (1967)
        J. Phys. Chem. 71, 895.

13. Barrall, E. M., II, Cox, R. J., Doelman, A., Clecak, N.,
        Logan, J. A., and Gregges, A. R., in Analytical
        Calorimetry, op. cit., in press.

14. Barrall, E. M. II, and Johnson, J. F. (1974) in Liquid Crystals

and *Plastic Crystals*, G. W. Gray and P. A. Winsor, eds., John Wiley and Sons, N.Y., p. 268.

15. Chow, L. C. and Martire, D. E. (1969) J. Phys. Chem. *73*, 1127.

16. Levelut, A. M. and Lambert, M. (1971) C.r.hebd. Seanc. Acad. Sci. Paris *272*, 1018.

17. Coates, D. and Gray, G. W. (1976) Microscope *24*, 117.

18. Taylor, T. R., Arora, S. L., and Ferguson, J. L. (1970) Phys. Rev. Lett. *25*, 722.

19. Flick, J. R., Marshall, A. S., and Petrie, S. E. B., in *Analytical Calorimetry*, op. cit., p. 97.

20. Arnold, H. (1964) Z. Phys. Chem. *226*, 146.

21. Herbert, A. J. (1967) Trans. Faraday Soc. *63*, 555.

22. Arnold H. (1964) Z. Phys. Chem. *225*, 45.

23. Barrall, E. M., II, and Guffy, J. C. (1967) in *Advances in Chemistry*, *63*, R. S. Porter and J. F. Johnson, eds., American Chemical Society, Washington, D.C., p. 1.

24. Stein, R. S., Rhodes, M. B., and Porter, R. S. (1968) Interface Sci. *27*, 336.

25. Baessler, H., Labes, M. M. (1970) Molecular Cryst. and Liquid Cryst. *6*, 419.

THE PREPARATION AND THERMODYNAMIC PROPERTIES OF

SOME CHIRAL 4-ALKYL-4'-CYANOSTILBENES AND TOLANS

Robert J. Cox, Nicholas J. Clecak, and Julian F. Johnson[*]

IBM Research Laboratory

5600 Cottle Road, San Jose, California 95193

INTRODUCTION

It is well established that optically active nematogens exhibit cholesteric mesophases. The term "chiral nematic" was coined by Dolphin[1] to describe this general class of liquid crystals. Examples of chiral nematic liquid crystals have been synthesized based on Schiff bases[1,2] aromatic esters,[3] cinnamic esters[4] and biphenyls.[5] We have recently prepared two new series of liquid crystals with the trans stilbene and tolan structures,[6,7] and have now obtained the optically active analogs in these series. This paper will describe the compound preparation and list the thermodynamic properties that were obtained.

The starting point for most of the syntheses of chiral nematics has been the commercially available ℓ-2-methylbutanol. We have also chosen this starting point and have synthesized several members of both the 4-alkyl-4'-cyanostilbene and the 4-alkyl-4'-cyanotolan series. In each reaction care was taken not to involve the chiral center and thus lose the original

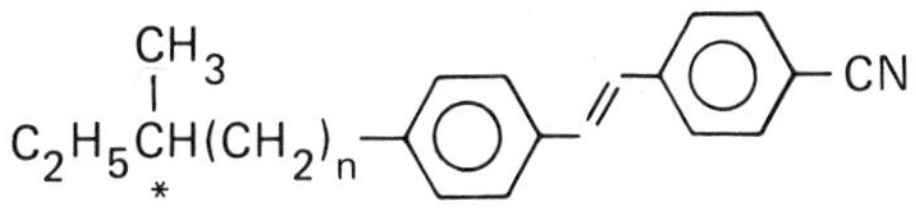

Figure 1. Structure of Chiral 4-Alkyl-4'-cyanostilbenes

[*] University of Connecticut, Storrs, Connecticut.

41

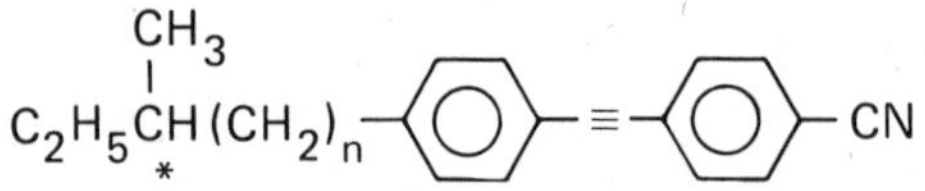

$$CH_3$$
$$C_2H_5\overset{|}{\underset{*}{CH}}(CH_2)_n\text{—}\langle\bigcirc\rangle\text{—}\equiv\text{—}\langle\bigcirc\rangle\text{—}CN$$

Figure 2.   Structure of Chiral 4-Alkyl-4'-cyanotolans

optical activity.   The formula of the trans-4-alkyl-4'-
cyanostilbenes is shown in Figure 1 in which n is 2, 3, 4, or 5.
The structure of this series was thus varied in two ways
simultaneously.   The length of the alkyl chain was increased from
five to eight carbons, while at the same time the chiral center
was moved from the three to the six position.

The structure of the tolans is shown in Figure 2 where n is
varied from two to four.   In this case only three members of the
series were prepared; they were those in which the alkyl chain
was lengthened from five to seven carbons and the chiral center
moved from the three to the five position.

COMPOUND PREPARATION

The preparation of the stilbenes, outlined in Figure 3, was
through chloromethylation of alkylbenzenes using chloromethyl
ethyl ether to give 4-alkylbenzylchlorides.   These were treated
with triphenylphosphine, followed by butyl lithium to produce
the Wittig reagents in the normal manner.   Subsequent reaction
with p-cyanobenzaldehyde gave the required stilbene.   These were
always obtained as mixtures of the cis and trans isomers and the
trans could only be obtained pure by repeated recrystallization.

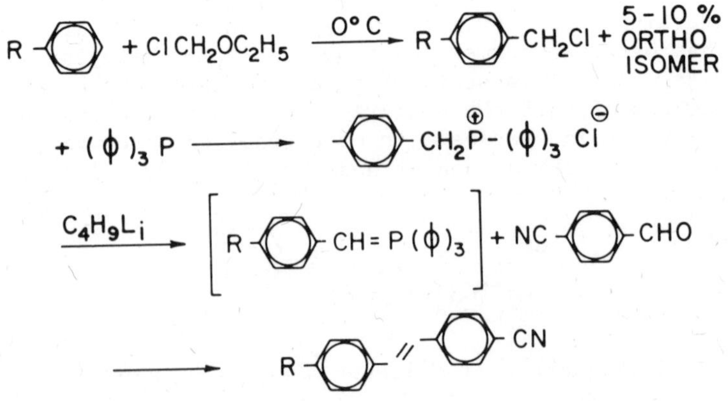

$$R\text{—}\langle\bigcirc\rangle + ClCH_2OC_2H_5 \xrightarrow{0°C} R\text{—}\langle\bigcirc\rangle\text{—}CH_2Cl + \begin{array}{l}5\text{-}10\,\%\\ \text{ORTHO}\\ \text{ISOMER}\end{array}$$

$$+ (\phi)_3 P \longrightarrow \langle\bigcirc\rangle\text{—}CH_2\overset{\oplus}{P}\text{-}(\phi)_3 \ \overset{\ominus}{Cl}$$

$$\xrightarrow{C_4H_9Li} \left[R\text{—}\langle\bigcirc\rangle\text{—}CH\text{=}P(\phi)_3 \right] + NC\text{—}\langle\bigcirc\rangle\text{—}CHO$$

$$\longrightarrow R\text{—}\langle\bigcirc\rangle\text{—}\text{//}\text{—}\langle\bigcirc\rangle\text{—}CN$$

Figure 3.   Synthesis of 4-Alkyl-4'-cyanostilbenes

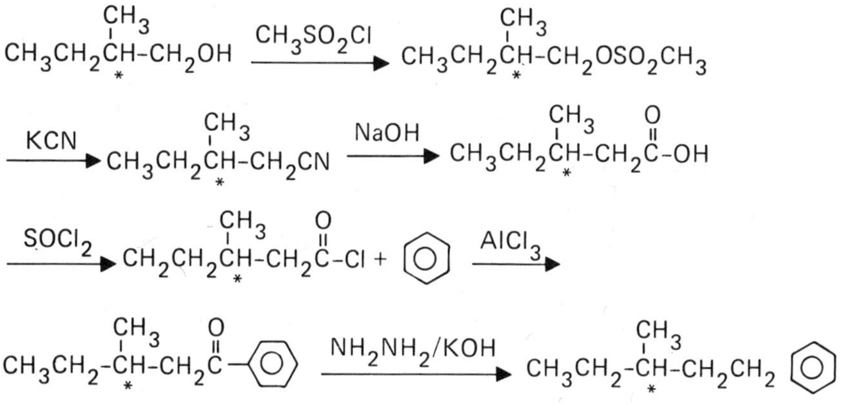

$$CH_3CH_2\underset{*}{CH}(CH_3)-CH_2OH \xrightarrow{CH_3SO_2Cl} CH_3CH_2\underset{*}{CH}(CH_3)-CH_2OSO_2CH_3$$

$$\xrightarrow{KCN} CH_3CH_2\underset{*}{CH}(CH_3)-CH_2CN \xrightarrow{NaOH} CH_3CH_2\underset{*}{CH}(CH_3)-CH_2\overset{O}{C}-OH$$

$$\xrightarrow{SOCl_2} CH_2CH_2\underset{*}{CH}(CH_3)-CH_2\overset{O}{C}-Cl + \bigcirc \xrightarrow{AlCl_3}$$

$$CH_3CH_2-\underset{*}{CH}(CH_3)-CH_2\overset{O}{C}-\bigcirc \xrightarrow{NH_2NH_2/KOH} CH_3CH_2-\underset{*}{CH}(CH_3)-CH_2CH_2\bigcirc$$

Figure 4. Synthesis of 3-Methylpentylbenzene

The fact that we were required to start from 2-methylbutanol in order to incorporate the optical activity into the molecule made the preparation of the alkylbenzenes, particularly those with the longer chain alkyl groups, a somewhat tedious process involving a number of synthetic steps.

Figure 4 shows the preparation of optically active 3-methylpentylbenzene. The alcohol was mesylated, then treated with potassium cyanide to replace the mesyl group with cyano. Hydrolysis to the acid, formation of the acid chloride, and a Friedel-Craft reaction with benzene gave the ketone which was reduced with alkaline hydrazine hydrate.

The preparation of 4-methylhexylbenzene, in Figure 5, is similar, and starting from 2-methylbutanol, the mesylate is treated with diethylmalonate in order to add two carbons to the chain. Subsequent formation of the acid chloride, the ketone, and finally the hydrocarbon is the same as for the pentyl compound.

Increasing the chain length by three carbons, shown in Figure 6, requires starting from 3-methylpentanoic acid, one of the intermediates used for the synthesis of 3-methylpentylbenzene. This, or its ester, is reduced to the alcohol with lithium aluminum hydride, is mesylated and then converted to 5-methylhepantoic acid with diethylmalonate. The acid chloride is prepared, reacted with benzene, and reduced in the usual way to 5-methylheptylbenzene.

Figure 7 outlines the method used for the preparation of 6-methyloctylbenzene. The route is the same as that employed for the heptyl compound except that the starting point is

4-methylhexanol. This is obtained from the reduction of
4-methylhexanoic acid, an intermediate prepared during the
synthesis of 4-methylhexylbenzene.

The corresponding tolans were prepared from the stilbenes by
the method shown in Figure 8, that is bromination of the double
bond using pyridine hydrobromide perbromide and subsequent
dehydrobromination with 1,5-diazobicyclo[5,4,0] undec-5-ene (DBU)
in refluxing DMF.

THERMAL PROPERTIES

Figure 9 summarizes the thermal transitions which are found
in the stilbene series. The two lower members in which n is 2
and 3 are both monotropic. They can be supercooled sufficiently
on a microscope slide to observe a mesophase. These were
identified as cholesteric both by the texture, observed
microscopically between crossed polarizers, and by the typical
colors obtained by shearing a thin layer of the compound.

In the case of the 3-methylpentyl compound no qualitative
heats could be measured as it could not be supercooled in the
D.S.C. far enough to observe the mesomorphic transition. The
enthalpy for the solid to isotropic liquid transition was found
to be 4.2 Kcal/mole with an entropy change of 12.0 eu.

The 4-methylhexyl compound, also monotropic, shows two solid
forms, one of which is obtained directly from the
recrystallization and melts at 57.2°C, the other is formed by
allowing the molten compound to crystallize, and transforms
directly to the first crystal at 45.6°C. In this case the
compound supercools enough in the D.S.C. to allow the measurement
of the enthalpy of the liquid crystal transition. This was found
to be .074 Kcal/mole with an entropy change of 0.227 eu.

The other two members of the series, the 5-methylheptyl and
6-methyloctyl compounds are enantiotropic, and the texture of
the liquid crystal phases, as seen between crossed polarizers,
shows them to be smectic in both cases. This is also indicated
by the high viscosity that is evident on shearing the microscopic
preparations. The appearance of only a smectic phase in the
heptyl compound was somewhat unexpected as the corresponding
straight chain stilbene with a seven carbon tail has both a
nematic and a smectic mesophase. A second unusual feature of
this compound is the very small enthalpy change obtained in going
from the mesophase to the isotropic liquid. This is only
0.48 Kcal/mole with an entropy change of 1.4 eu. The usual smectic
to isotropic transitions are accompanied by an entropy change
6-7 eu.[8] It should be pointed out also that the total entropy

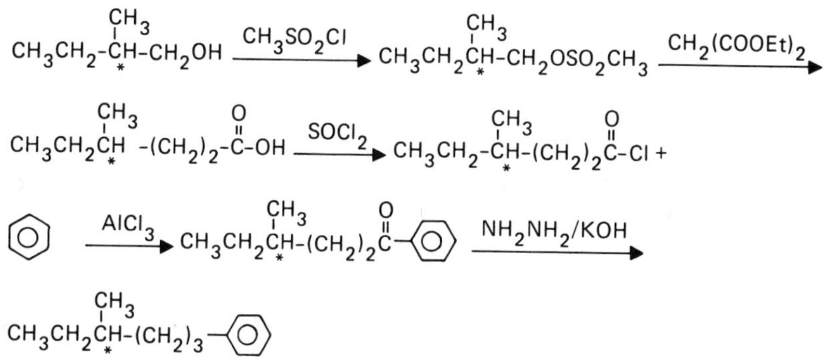

$$CH_3CH_2\underset{*}{\overset{\overset{\displaystyle CH_3}{|}}{CH}}-CH_2OH \xrightarrow{CH_3SO_2Cl} CH_3CH_2\underset{*}{\overset{\overset{\displaystyle CH_3}{|}}{CH}}-CH_2OSO_2CH_3 \xrightarrow{CH_2(COOEt)_2}$$

$$CH_3CH_2\underset{*}{\overset{\overset{\displaystyle CH_3}{|}}{CH}}-(CH_2)_2-\overset{\overset{\displaystyle O}{||}}{C}-OH \xrightarrow{SOCl_2} CH_3CH_2-\underset{*}{\overset{\overset{\displaystyle CH_3}{|}}{CH}}-(CH_2)_2\overset{\overset{\displaystyle O}{||}}{C}-Cl +$$

$$\bigcirc \xrightarrow{AlCl_3} CH_3CH_2\underset{*}{\overset{\overset{\displaystyle CH_3}{|}}{CH}}-(CH_2)_2\overset{\overset{\displaystyle O}{||}}{C}-\bigcirc \xrightarrow{NH_2NH_2/KOH}$$

$$CH_3CH_2\underset{*}{\overset{\overset{\displaystyle CH_3}{|}}{CH}}-(CH_2)_3-\bigcirc$$

Figure 5.   Synthesis of 4-Methylhexylbenzene

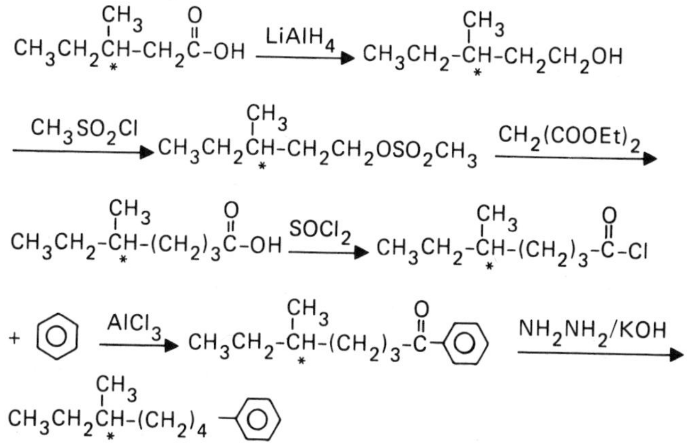

$$CH_3CH_2\underset{*}{\overset{\overset{\displaystyle CH_3}{|}}{CH}}-CH_2\overset{\overset{\displaystyle O}{||}}{C}-OH \xrightarrow{LiAlH_4} CH_3CH_2-\underset{*}{\overset{\overset{\displaystyle CH_3}{|}}{CH}}-CH_2CH_2OH$$

$$\xrightarrow{CH_3SO_2Cl} CH_3CH_2\underset{*}{\overset{\overset{\displaystyle CH_3}{|}}{CH}}-CH_2CH_2OSO_2CH_3 \xrightarrow{CH_2(COOEt)_2}$$

$$CH_3CH_2-\underset{*}{\overset{\overset{\displaystyle CH_3}{|}}{CH}}-(CH_2)_3\overset{\overset{\displaystyle O}{||}}{C}-OH \xrightarrow{SOCl_2} CH_3CH_2-\underset{*}{\overset{\overset{\displaystyle CH_3}{|}}{CH}}-(CH_2)_3-\overset{\overset{\displaystyle O}{||}}{C}-Cl$$

$$+ \bigcirc \xrightarrow{AlCl_3} CH_3CH_2-\underset{*}{\overset{\overset{\displaystyle CH_3}{|}}{CH}}-(CH_2)_3-\overset{\overset{\displaystyle O}{||}}{C}-\bigcirc \xrightarrow{NH_2NH_2/KOH}$$

$$CH_3CH_2\underset{*}{\overset{\overset{\displaystyle CH_3}{|}}{CH}}-(CH_2)_4-\bigcirc$$

Figure 6.   Synthesis of 5-Methylheptylbenzene

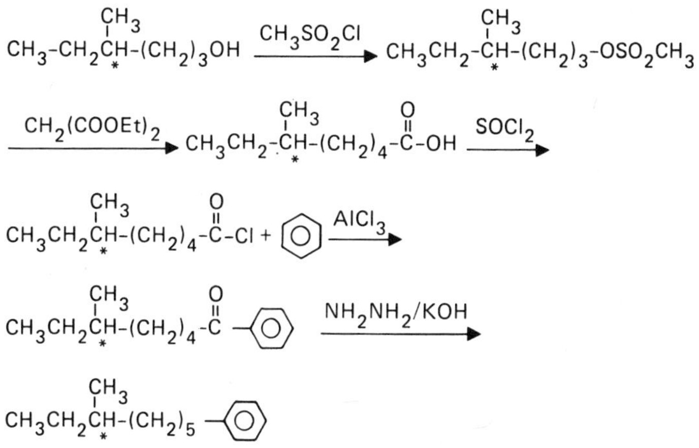

$$CH_3-CH_2\underset{*}{\overset{\overset{\displaystyle CH_3}{|}}{CH}}-(CH_2)_3OH \xrightarrow{CH_3SO_2Cl} CH_3CH_2-\underset{*}{\overset{\overset{\displaystyle CH_3}{|}}{CH}}-(CH_2)_3-OSO_2CH_3$$

$$\xrightarrow{CH_2(COOEt)_2} CH_3CH_2-\underset{*}{\overset{\overset{\displaystyle CH_3}{|}}{CH}}-(CH_2)_4-\overset{\overset{\displaystyle O}{||}}{C}-OH \xrightarrow{SOCl_2}$$

$$CH_3CH_2\underset{*}{\overset{\overset{\displaystyle CH_3}{|}}{CH}}-(CH_2)_4-\overset{\overset{\displaystyle O}{||}}{C}-Cl + \bigcirc \xrightarrow{AlCl_3}$$

$$CH_3CH_2\underset{*}{\overset{\overset{\displaystyle CH_3}{|}}{CH}}-(CH_2)_4-\overset{\overset{\displaystyle O}{||}}{C}-\bigcirc \xrightarrow{NH_2NH_2/KOH}$$

$$CH_3CH_2\underset{*}{\overset{\overset{\displaystyle CH_3}{|}}{CH}}-(CH_2)_5-\bigcirc$$

Figure 7.   Synthesis of 6-Methyloctylbenzene

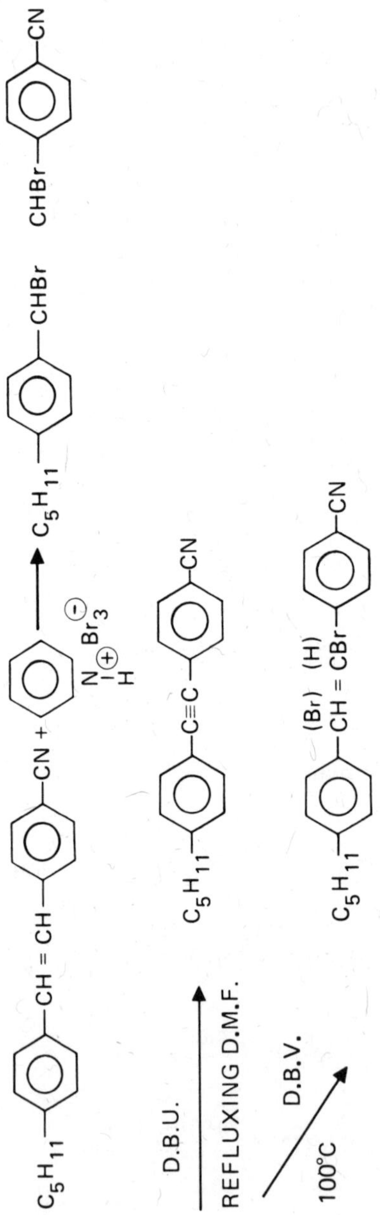

Figure 8. Synthesis of 4-Alkyl-4'-cyanotolans

change for the smectic and nematic phases to the isotropic in
the corresponding straight chain stilbene is only 0.29 eu and that
the smectic to nematic transition entropy is extremely small,
about 0.01 eu.

The 6-methyloctyl compound follows this same pattern, the
microscopic examination clearly indicating a smectic mesophase
with an entropic change to the isotropic liquid of only 1.4 eu.
This compound also shows solid polymorphism with the higher
melting form being obtained from a recrystallization solvent and
the lower from cooling the isotropic melt. The corresponding
eight carbon straight chain stilbene has both a smectic and a
nematic mesophase, and in this case also the total entropy change
from the smectic to the isotropic liquid is very small, 0.061 eu.
Clearly the thermodynamic changes in these seven and eight carbon
compounds needs a closer examination.

The tolans that were prepared are shown in Figure 10 with the
melting and transition temperatures that were observed. All
three members of this series were monotropic, and in each case,
samples run in the D.S.C. crystallized before the mesophase heats
could be determined. They could be observed, however, by cooling
a sample on a microscope slide between crossed polarizers,
although in each case the mesophase is quite unstable and the
solid grows rapidly into the sample.

When compared to the corresponding straight chain
derivatives,[7] the melting point of the crystals has not changed
a great deal, however, the mesophase transition temperature is
considerably reduced for the compounds in which the chain has a
branch. This latter is certainly not unexpected and seems to be
general for all of the cases in which the hydrocarbon group has
a branch. The lack of any such trend in the crystalline melting
temperatures is simply evidence for the complex and diverse
crystalline habits that these molecules display.

The stilbenes, not unexpectedly, were found to be quite
sensitive to light and could not be examined in the microscope
without adequate optical filtration. The tolans, on the other
hand, are quite light stable and an attempt was made to prepare
a binary tolan mixture which would have a stable mesophase in
order to study cholesteric properties such as the helical pitch.
Figure 11 shows the binary phase diagram for mixtures of the
5-methylheptyl and the 3-methylpentyltolans. Unfortunately no
combination of these two lowers the freezing point enough to
create a stable mesophase. The monotropic transition is shown
by the dotted line at the bottom of the figure. The solid line
is the calculated value obtained from the enthalpies and melting
points of the pure components, using the Schröder-Van Laar
equation. It can be seen that the experimental points do not

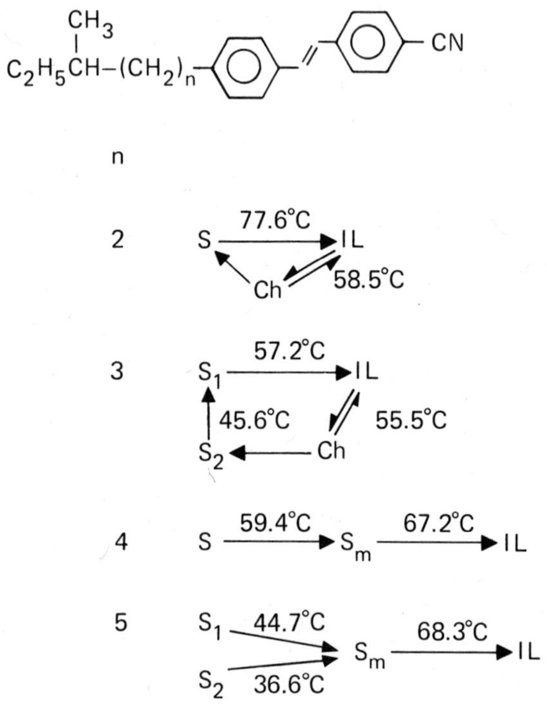

S = Crystal (Solid)
Sm = Smectic
IL = Isotropic Liquid
Ch = Cholesteric

Figure 9. Melting and Liquid Crystal Transition Temperatures of Chiral trans 4-Alkyl-4'-cyanostilbenes

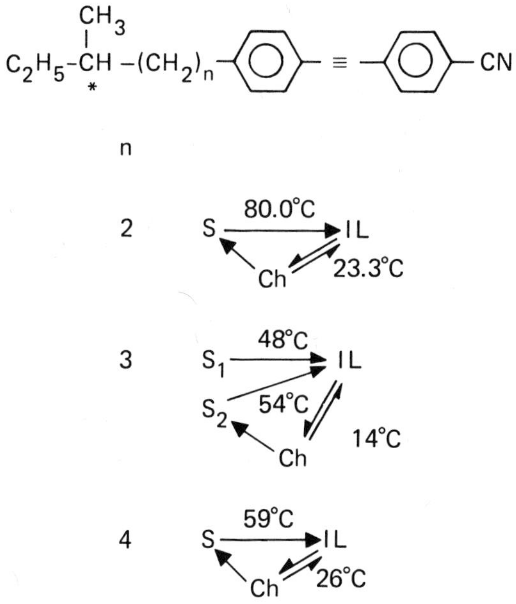

$$C_2H_5-\overset{\overset{\displaystyle CH_3}{|}}{\underset{*}{CH}}-(CH_2)_n-\bigcirc- \equiv -\bigcirc-CN$$

n

2   S $\xrightarrow{\text{80.0°C}}$ IL
Ch ⟋ 23.3°C

3   S$_1$ $\xrightarrow{\text{48°C}}$ IL
S$_2$ ⟋ 54°C
Ch   14°C

4   S $\xrightarrow{\text{59°C}}$ IL
Ch ⟋ 26°C

S   = Crystal (Solid)
Ch = Cholesteric
IL = Isotropic Liquid

Figure 10.   Melting and Liquid Crystal Transition Temperatures of Chiral 4-Alkyl-4'cyanotolans

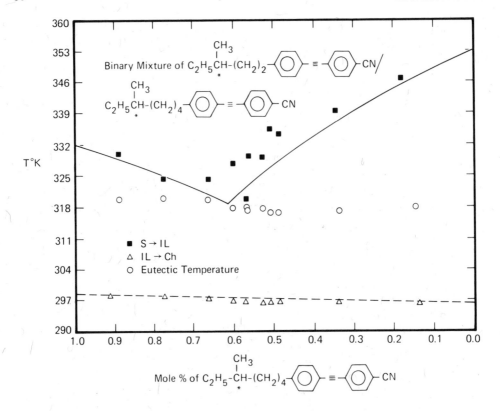

Figure 11. Binary Phase Diagram for 3–Methylpentyl and 5–Methylheptylcyanotolans

coincide particularly well with this theoretical curve. This is possibly due to some solid polymorphism in the pure materials which we did not detect, or more likely, to the formation of solid solution.

ACKNOWLEDGMENTS

We would like to thank Dr. E. M. Barrall II for doing the DSC measurements on these compounds.

REFERENCES

1. D. Dolphin and Z. Muljiani, J. Chem Phys. $\underline{58}$, 413 (1973).

2. J. A. Castellano, C. H. Oh, M. T. McCaffrey, Mol. Cryst. Liq. Cryst. $\underline{27}$, 417 (1973).

3. B. H. Klanderman, T. R. Criswell, J. Am. Chem. Soc. $\underline{97}$, 1585 (1975).

4. M. Leclercq, J. Billard, J. Jacques, Mol. Cryst. Liq. Cryst. $\underline{8}$, 367 (1969).

5. G. W. Gray and D. G. McDonnell, presented at the Sixth International Liquid Crystal Conference, Kent State University, Kent, Ohio, August 1976.

6. R. J. Cox and N. J. Clecak, Mol. Cryst. Liq. Cryst. $\underline{37}$, 263 (1976).

7. R. J. Cox and N. J. Clecak, Mol. Cryst. Liq. Cryst. $\underline{37}$, 241 (1976).

8. E. M. Barrall II and J. F. Johnson, Liquid Crystals and Plastic Crystals, G. W. Gray and P. A. Winsor, editors, Vol. II, E. Horwood, Chichester, England (1974), p. 254.

A SIMPLE AND INEXPENSIVE SYNTHETIC METHOD FOR ALKYLCYANOBIPHENYLS

Chan S. Oh

Beckman Instruments, Inc.

2500 Harbor Boulevard, Fullerton, California   92634

ABSTRACT

4'-substituted-4-cyanobiphenyl[1] has been widely used in the manufacture of twisted nematic display devices. This class of compounds offers several advantages, such as their moisture resistance, photo chemical stabilities, low-threshold voltages, and high speed of switching times. Until now, the availability of this material to small laboratories for testing purposes has been very limited, and it has not been amendable to synthesis by simple chemical reactions. As an alternative to the original cumbersome method, a very simple high yield synthetic method has been developed. This method utilizes very inexpensive starting materials and chemical reagents, and can be used for making gram quantities to multi kilo lots.

4-alkylbiphenyl is reacted with oxalylchloride in presence of aluminum chloride to yield pure 4-biphenylcarboxylic acid chloride. This acid chloride is converted to carboxamide with ammonia. The resultant amide can be readily converted to 4'-n-alkyl-4-cyanobiphenyl with a quantitative yield of high purity.

INTRODUCTION

The several advantageous properties of biphenyl derived nematogens in liquid crystal display applications motivated us to develop a simpler and more economical method for synthesizing these compounds. Although this class of nematic materials is readily available for commercial sources, it is limited to purchases

in bulk quantities of a fixed composition and may have inherent
impurity levels of various chemical species. There has been little
information on simple preparative methods for this class of compounds.

A method had been developed in our laboratories for some years
and it is intended in this paper to suggest an alternative method
to that of the original[1] synthetic route. It was noted, in the
original method, that the starting material was expensive and
some of the reagents, such as cuprous cyanide, may be objectionable
in terms of industrial safety.

The alternative method described here is economical, safer to
perform, and yet results in a high yield end product. The reaction
sequences are shown in Figure 1.

RESULTS AND DISCUSSIONS

As the starting material, the most readily available and
inexpensive chemical, biphenyl ($2.2/mole) was chosen. The alkyl
pendant group at the para-position (4'R' in Figure 2) of the
biphenyl was attached first by introducing an alkanoyl group, and
subsequent reduction. A Friedel-Craft reaction on the biphenyl
readily occurred with acid chloride and aluminum chloride. The
yield varied between 70% to 85% theoretical. Also, note that
virtually all the alkanoyl group attach at the desired 4'-position
without other isomers. Thus, the purification was relatively
easy. Simple vacuum distillation yielded pure compound.

The reduction of the ketone to 4'-alkylbiphenyl was readily
achieved via a Huang-Minlon reaction. Other less expensive reduc-

TABLE 1

THE YIELDS AND THE PHYSICAL CONSTANTS OF 4-SUBSTITUTED BIPHENYLS

	Ketones		Hydrocarbons	
Pendant Group	$n-C_4H_9CO-$	$n-C_6H_{13}CO-$	$n-C_5H_{11}$	$n-C_7H_{15}$
Yield (% Theor.)	79	71	65	65
Melting Point °C,	78.5-79.0	85.5-86.0		
Boiling Point °C, (0.1 mm Hg)			106-109	124-127

tive methods have not been explored. Commercial sources, such as custom synthetic services, could be used to secure this hydro-carbon at lower cost in larger quantities. Table 1 lists the yields and the physical constants of the ketones and the hydro-carbons.

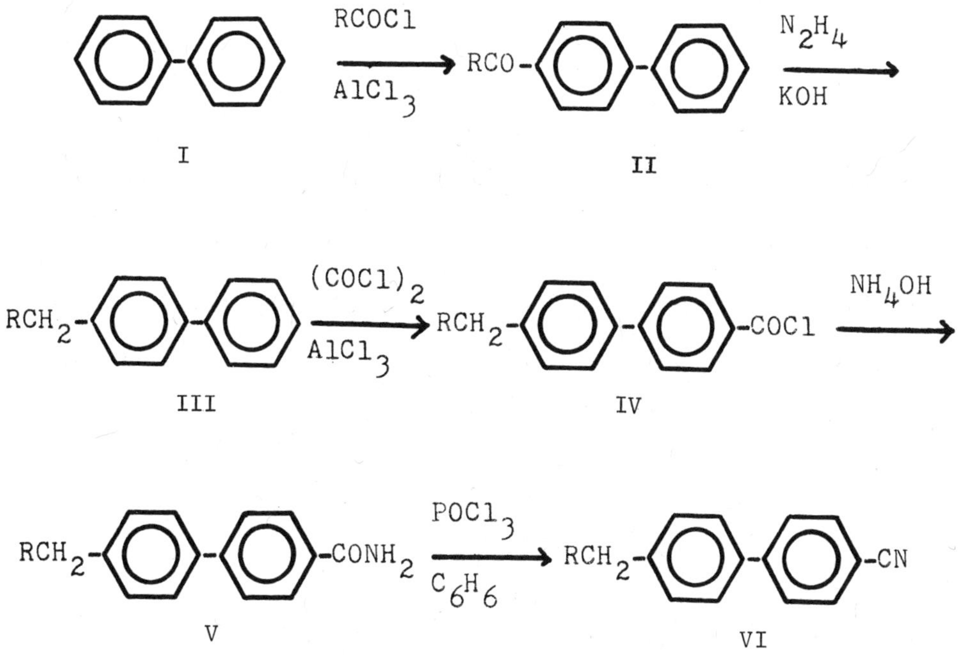

Figure 1

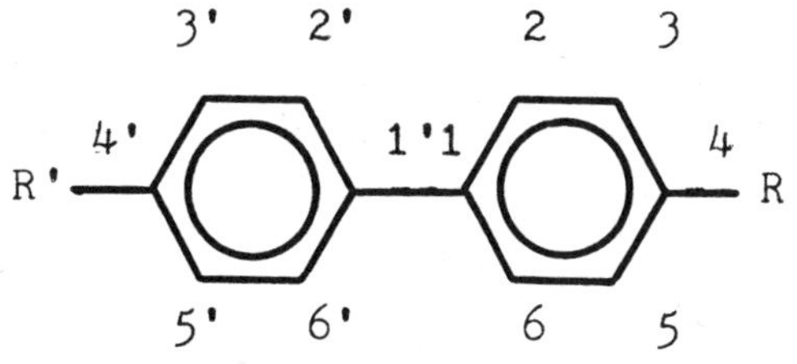

Figure 2

The next step was to introduce a functional group at the para-position of the second aromatic ring (4-R of Figure 2).   An electrophilic substitution reaction, such as a Friedel-Craft reaction, was considered.   Since acetylation had been performed on this substrate[2], it seemed feasible to perform the reaction with another electrophile.   Neubert, et. al.[3] reported a simple preparative method for 4-substituted benzoyl chlorides, under mild conditions, by reacting an appropriate aromatic compound and oxalyl chloride.   Thus 4-alkylbiphenyl was reacted with oxalyl chloride in presence of aluminum chloride.   Among the many reaction conditions tried in this study, the method described in the experimental section of this note turned out to be optimum.   As for the solvent, carbon disulfide gave the most satisfactory result. Other solvents, such as carbon tetrachloride, methylene chloride and nitrobenzene rendered poor yields; often an insoluble resinous material.   These biphenylcarboxylic acid chlorides are relatively stable and did not decompose under the work-up procedures and the crude product could readily be vacuum distilled to obtain a white solid at room temperature.   The shelf life has been good in a sealed container.

One of the unique features of this reaction is the absence of ortho substituted acid chloride.   It was suspected that the activat-ing alkylphenyl radical would encourage the formation of isomers with substituents at 2- and 5- positions to a certain extent.   The presence of these isomers would not only have interferred with the purification of the product, but also carried through the subsequent reactions and contaminated the final product, of which purification would have been yet more difficult.   Thus the purification of the acid chloride was achieved by simple vacuum distillation.   This enabled us to handle relatively large quantities of pure material, which helped the overall throughput.   Another method[4] of preparation of cyano group containing biphenyls or techphenyls has been described. In their procedures, an acetyl group was first introduced into the aromatic moiety, then oxidized to carboxyl acid, which in turn was converted to the acid chloride.   One step synthesis of this acid chloride, as described in this note, is a much simpler approach.

Once pure acid chlorides are prepared, it is an easy task to convert them into the respective carboxamides by reacting with ammonia.   One simple method is to pour the melt acid chloride in-to vigorously stirred ammonium hydroxide in a Warring Blender. Subsequent filtration yields white amide.   These amides are vir-tually insoluble in water, thus one or two washings with warm water eliminates the excess ammonia.

After drying, the amide can be reacted with any commonly used dehydrating agent, such as thionyl chloride, phosphorus pentoxide or phosphorus oxychloride.   The latter seemd to give the best yield.   These dehydrating agents can be used in excess,

which serves as both reagent and solvent at the same time. Or
alternatively, a separate solvent can be used. By merely refluxing in benzene with phosphorus oxychloride converted the
amide to nitrile in almost quantitative yield. The final product,
nitrile, then can be purified from the polar impurities, such as
amide, carboxylic acid, or phosphoric acids. Again this reaction
scheme was designed in such a manner that an easy work-up and
less troublesome purification can be used. This contrasts the
difficult purification procedure to separate the 4-bromo-4'-
alkylbiphenyl and 4-cyano-4'alkylbiphenyl as described in the
original method. Table II lists the yields and the physical constants of the acid chlorides and the carboxamides.

TABLE II

THE YIELDS AND THE PHYSICAL CONSTANTS OF 4'-SUBSTITUTED
4-BIPHENYLCARBOXYLIC ACID CHLORIDES AND CARBOXAMIDES.

	ACID CHLORIDES		CARBOXAMIDES	
4'-Pendant Group	$n\text{-}C_5H_{11}$	$n\text{-}C_7H_{13}$	$n\text{-}C_5H_{11}$	$n\text{-}C_7H_{13}$
Yields (% Theor.)	61	58	86	82
Melting Point °C			231-233	223-225
Boiling Point °C (0.1 mmHg)	150-155	160-165		

Although the nitriles obtained by the described synthesis and
purification method appeared chemically pure, their quality was not
good enough for liquid crystal display devices. Mainly, the product exhibited too low electrical resistivity and the required
homogeneous alignment would not withstand the accelerated life
testings of the devices. These two problems have been eliminated
by a further purification of the carboxamide and the nitrile. A
single recrystallization of the carboxamides from boiling p-
dioxane yielded white plates which is single component (TLC, Silica
Gel/Ethyl Acetate). This purified carboxamide was converted into
the nitrile, which subsequently distilled under vacuo and passed
through a short column of basic aluminum oxide (Woehlm, Basic
Alumina Act. I) using alcohol free chloroform as eluant.

Table III lists the intermediates and the final products along with the costs which take into account the raw material prices and typical reaction yields. Although the calculation is not thorough, the table does illustrate the comparative merit of this alternative synthesis method over the reported.

TABLE III

COSTS OF THE INTERMEDIATES AND THE FINAL PRODUCT

The Present Method		The Conventional Method	
Compounds	$/mole	Compounds	$/mole
Biphenyl	2.2	4-Bromobiphenyl	90.4
4-Acylbiphenyl	37.5	4'-Acyl-4-bromobiphenyl	137
4-Alkylbiphenyl	59.2	4'-Alkyl-4-bromobiphenyl	180
4'-Alkyl-4-biphenyl- carboxylic acid chloride	169		
4'-Alkyl-4-biphenyl- carboxamide	196		
4'-Alkyl-4-cyano- biphenyl	215	4'-Alkyl-4-cyano- biphenyl	321

EXPERIMENTAL

All the reagents and solvents used in this study were commercially purchased reagent grade chemicals and used without further purifications. The transition temperatures were determined with a Thomas-Hoover Melting Point Apparatus, a Differential Thermal Analyzer (duPont Model 990) and a polarizing microscope with Mettler Model FP5 Hot Stage. The intermediates were purified until the transition temperatures remained constant and the purity was confirmed by a thin layer chromatography. The identities of the functional groups were confirmed by inspection of the infrared absorption spectra.

Preparation of 4-n-pentanoylbiphenyl (II)

Into an 18-litre reaction flask, 1542 grams (10 moles) biphenyl, 1330 grams of anhydrous aluminum chloride and 6 litres of carbon disulfie are charged. With vigorous stirring, 1200 grams of n-pentanoyl chloride is added into the above mixture during 60 minutes. The reaction mixture is left stirring for another hour, then treated with cold dilute hydrochloric acid (250 ml conc. hydrochloric acid in 1500 ice water). The organic layer is separated and washed with water, 10% sodium hydroxide, brine and dried over anhydrous sodium sulfate. After filtration and evaporation of the solvent, the resultant white solids are recrystallized from isooctane. The yield is 1890 grams (79% Theor.).

Preparation of 4-n-pentylbiphenyl (III)

Into 18-litre reaction flask 1890 grams (7.9 moles) of II, 793 ml of 64% hydrazine, 1585 grams of potassium hydroxide and 2.2'-oxydiethanol are charged. The stirred mixture is heated to 160°C for a period of 60 minutes, then the temperature is allowed to rise 200°C. This temperature is maintained for another 60 minutes. After the reaction, the mixture is cooled to room temperature; one litre of water and one litre of bezene are added and stirred. The organic layer is separated and the lower layer is further extracted with 2x250 ml bezene. The combined organic layers are washed with water and brine, and dried over anhydrous sodium sulfate. After filtration and evaporation of the solvent, the residue is vacuum distilled, collecting the fraction at 106-109°C (0.1 m m Hg). The yield is 1147 grams (65% Theor.).

Preparation of 4'-n-pentyl-4-biphenylcarboxylic Acid Chloride (IV)

Into a 5-litre reaction flask, 707 grams (3.15 moles) of III, 420 grams (3.15 moles) of anhydrous aluminum chloride and 2 litres of carbon disulfide are placed and cooled to ca. 15°C on an ice bath. Oxalyl chloride (400 grams, 3.15 mole) is added to the above mixture in portions during ca. 30 minutes maintaining the temperature at 15°C. When the addition is completed, the ice bath is removed and the resultant dark-red colored reaction mixture is left stirring at room temperature for 30 minutes. This complex is treated with 4 litres of ice-water. The organic layer is separated and washed with cold dilute hydrochloric acid, cold water, brine, and dried over anhydrous sodium sulfate. After filtration and evaporation of the solvent, the residue is vacuum distilled, with the fraction collected at 150-160°C (0.1 mm Hg). The yield is 327 grams (61% Theor.).

Preparation of 4'-n-pentyl-4-biphenylcarboxamide (V)

The acid chloride (IV, 500 grams) is treated with 1.2 litre of ammonium hydroxide by pouring the melt acid chloride into the ammonia in a Warring Blender. The resultant amide is collected and washed with warm water. After drying at 120°C overnight, the amide

is recrystallized from boiling p-dioxane to yield colorless plates.
The yield is 400 grams (86% Theor.).

Preparation of 4'-n-pentyl-4-cyanobiphenyl (VI)

Into 5-litre reaction flask are placed 400 grams (1.5 moles)
of V, 250 ml of phosphorus oxchloride and 2 litres of bezene.
The mixture is heated to reflux for 4 hours with vigorous stirring. After cooling to room temperature, the reaction mixture is
treated with 1 litre of cold water. The organic layer is separated
and washed with water, 10% sodium hydroxide and brine. The solution is then dried over anhydrous sodium sulfate, filtered, and the
solvent evaporated to obtain the crude product. Upon vacuum distillation, the fraction distilled over at 150-160°C is collected. The
yield is 340 grams (92% Theor.).

ACKNOWLEDGEMENT

The author wishes to thank John C. Powers who performed many
of the reactions and purifications, and William T. Kelly who encouraged and supported this study.

REFERENCES

1. G.W. Gray, K.J. Harrison, and J.A. Nash, Liquid Crystal and
Ordered Fluids, Vol. 2., Ed., J.F. Johnson and R.S. Porter,
Plenum Press, New York, 1974, p.617.

2. D.J. Byron, G.W. Gray, and R.C. Wilson, J. Chem. Soc. (C),
1966 (9), 840-5.

3. M.E. Neubert, L.T. Carlino, Richard D'Sidocky, and D.L. Fishel,
Liquid Crystal and Ordered Fluids, Vol. 2., Ed., Johnson, J.F.,
Porter, R.S., Plenum Press, New York, 1974, p.293.

4. a) B.K. Sadashiva and G.S.R. Subba Rao, 6th International
Liquid Crystal Conference, Aug. 1976, Kent, Ohio, Abstract
#I-17. b) Inukai, Takashi; Sato, Hideo; Sugimori, Shigeru
and Ishibe, Tetsuya (Chisso Corp.), Ger. Offen., 2,545,121.
c) Fujii, Yasuyuki; and Matsumura, Naotake (Fugi Dyestuff
Co. LTD), Japan Kokai; 76 75050.

EFFECT OF MOLECULAR STRUCTURE ON MESOMORPHISM.   6.   NON-LINEAR

THERMAL BEHAVIOR IN SOME BINARY LIQUID CRYSTAL SYSTEMS

Anselm C. Griffin, Thomas R. Britt, Neal W. Buckley,
Richard F. Fisher, Stephen J. Havens

Department of Chemistry, University of Southern
Mississippi; Hattiesburg, Mississippi 39401 U.S.A.

and

D. Wayne Goodman

National Bureau of Standards
Gaithersburg, Maryland 20760 U.S.A.

Most binary mesophase systems exhibit mesophase-isotropic transition temperatures which are linearly dependent on composition.
There are however exceptions. Demus (1) reported non-linear behavior
for some mixtures when one component was a chloro mesogen. Dave (2)
found similar behavior with chloro and nitro liquid crystals and
postulated a structural rationale based on dipolar characteristics
of the constituent molecules. Schroeder and Schroeder (3) reported
in 1968 that non-mesomorphic nitro-terminated anils enhanced (4) the
smectic phase of di-n-hexyloxyazoxybenzene. These authors suggested
a loose dimeric association of the nitro anils as part of a molecular
explanation for this enhanced smectic phase.

In more recent times Labes (5,6) and Oh (7) have found that
cyano-terminated liquid crystals are often successful in promoting
enhanced mesomorphic behavior in mixed liquid crystals. Labes (5,6)
favors a charge transfer interaction as the reason for such non-ideal
solution behavior while Oh (7) prefers a lamellar structure of constituent molecules. One of us (8) has recently studied 4-nitrophenyl
4'-decyloxybenzoate by x-ray diffraction and postulated an explanation for the non-linear behavior of mixtures containing the compound
based on the bimolecular smectic structure (9) of this nitro

ester. The alkyl cyanobiphenyls studied by Labes (5,6) and Oh (7) also have bimolecular smectic phases. The purpose of this work was to further probe the relationship between smectic layer structure and non-linear thermal behavior in binary mesophase systems.

In order to test this theory (8) we decided to use four mesogens; x-ray data for each of them has been previously reported. We chose two unimolecular smectogens ("unimolecular" in the sense that the x-ray spacing d is equal to or less than the measured extended molecular length, 1); and two bimolecular smectogens("bimolecular" in the sense that d is greater than 1). Compounds 1 and 3 have unimolecular smectics; compounds 2 and 4 have bimolecular smectics. Values for d and 1 are shown below in figure 1 along with structures for these compounds. It was intended to form binary systems by admixture of each of these aforementioned compounds with three compounds which themselves differ only in the electronic nature of one terminal group. In this way it was possible to examine both the effect of smectic layer arrangement and also the electronic effects of terminal groups on non-linear thermal behavior of binary mesophases.

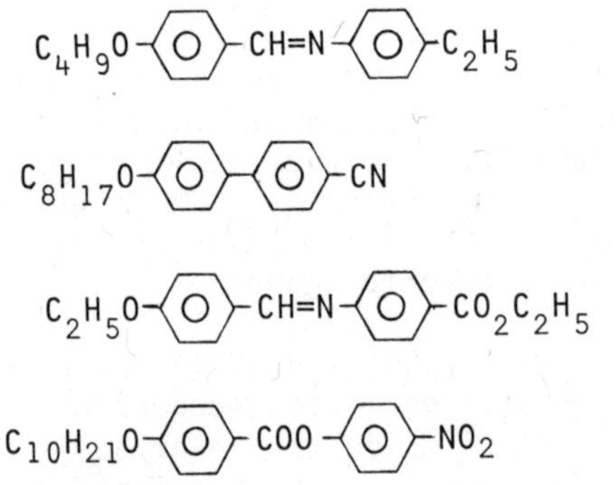

		$\underline{d}$ (Å)	$\underline{1}$ (Å)
1	C_4H_9O—⟨O⟩—$CH{=}N$—⟨O⟩—C_2H_5	17.3	21.5
2	$C_8H_{17}O$—⟨O⟩—⟨O⟩—CN	31	22
3	C_2H_5O—⟨O⟩—$CH{=}N$—⟨O⟩—$CO_2C_2H_5$	19.9	21.3
4	$C_{10}H_{21}O$—⟨O⟩—COO—⟨O⟩—NO_2	31.4	27.4

Figure 1. The four smectogens used in phase studies along with x-ray d values and molecular lengths 1 as measured from Dreiding stereomodels.

EXPERIMENTAL

Synthesis:

     Compounds 1 and 3 were obtained by reacting equimolar quantities
of the appropriate aldehyde and aniline in absolute ethanol. The
reaction solution was stirred at room temperature for 16 hours and
then cooled at 0° in an ice-water mixture. The resulting solid was
collected by vacuum filtration and crystallized from absolute ethanol
to constant transition temperatures. Agreement with literature data
when available was excellent. Compound 2 was obtained as a gift
from Dr. Robert Cox, IBM, San Jose, CA, and was used as received.
Compound 4 has been previously reported by us (8). Compounds A, B,
and C were synthesized using the method described above for the
preparation of 1 and 3. The homologous series of ethyl 4-n-alkoxy-
benzylidene-4'-aminobenzoates (10) were synthesized as 1 and 3.
Our transition temperatures are very close to those in the literature (10) and are the temperatures used in calculations described
later.

Mixture Preparation:

     Binary mixtures were prepared by the Kofler contact procedure
(11) and phase diagrams were constructed by polarized light microscopy of these mixtures using a hot stage. It should be noted that
50:50 mole percent compositions were examined for the phase diagrams
to insure a degree of accuracy for the composition coordinate. For
evaluation of the effect of alkoxy chain length, mixtures were examined in two ways: [1] by packing one component in a glass melting
point capillary tube; fusing; then packing the second component on
top of the first. Enhanced mesophases were observed at the interface of the two components in a Thomas-Hoover melting point apparatus [2] by conventional microscopy to determine the type of liquid
crystal phase which is enhanced. Regions of mixed composition were
seen in the phase diagrams, but were not included in the diagrams
in an effort to facilitate readability.

RESULTS AND DISCUSSION

Mixtures of A:

     The four phase diagrams for binary mixtures of compound A with
compounds 1-4 are shown below in figure 2.

     Compound A itself exhibits only an enantiotropic smectic A
phase, K59S93I. We find that mixtures of A with compounds 1 and 3
(hereafter referred to as A-1 and A-3, respectively) exhibit marked
non-linearity in the mesophase-isotropic transition temperatures as
one scans the composition coordinate. Mixture A-1 shows an enhanced

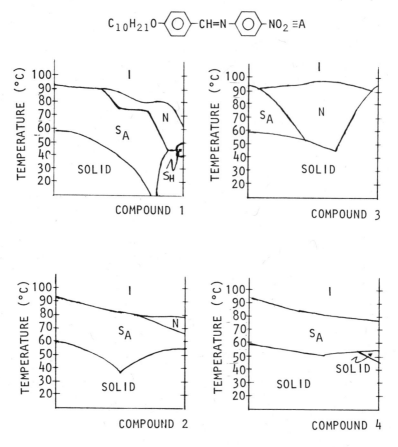

Figure 2. Set of phase diagrams for binary mixtures of A. Compounds 1 and 3 are unimolecular smectics; compounds 2 and 4 are bimolecular smectics.

nematic phase in the region of 50:50 mole precent composition, some 28° above the N→I temperature of the only nematogenic component, 1. There is also definite non-linearity in the S→N temperatures in the region of the diagram near pure 1. Mixture A-3 also shows a mixture-induced nematic phase of higher N→I temperature than either pure component even when neither pure component is enantiotropically nematogenic.

Mixtures of A with compounds 2 and 4, the bimolecular smectics, on the other hand, show ideal solution behavior. In summary, compound A exhibits non-linear behavior with unimolecular smectics and linear behavior with bimolecular smectics.

Mixtures of B:

Compound B is a most unusual smectogen. It has an enantiotropic smectic phase, K66S77.5I. Gray (12) has recently reported an
analogous compound, 4-n-hexyloxybenzylidene-4'-trifluormethylaniline,
and found it to possess no mesophase. We (13) have, however, prepared numerous members of this homologous series including the hexy-
loxy member and found them to be smectogenic. Evidence for a smectic
phase and other data for these compounds will be reported in the
near future. Although this smectic phase does flow it is extremely
viscous. It exhibits a most intriguing variety of microscopic textures (mosaic, homeotropic, and rodlike). We are here describing
this smectic as S? due to our current uncertainty as to its appropriate smectic classification.

In mixtures with compounds 1 and 3, compound B promotes and
enhances an enantiotropic smectic A (S_A) mesophase, even in mixtures
with compounds which themselves do not possess an enantiotropic S_A
phase. The usual linear behavior is found for mixtures B-2 and
B-4. For compound B we find non-linear behavior with unimolecular
smectics and linear behavior with bimolecular smectics.

Mixtures of C:

Compound C has both enantiotropic smectic and nematic phases,
K66S73N77I. Behavior of C in binary mixtures is completely the re-
verse of that of compounds A and B in mixtures. Mixtures C-2 and
C-3 show non-linear, non-ideal thermal behavior exhibiting S_A phases
having enhanced S→I transition temperatures. In mixtures C-1 and
C-3 one finds ideal, linear behavior. Thus for mixtures of compound
C with unimolecular smectics we find linear thermal behavior; for
mixtures of C with bimolecular smectics we find non-linear thermal
behavior.

From observation of these phase diagrams two points are pre-
iminent:
(1) a relationship exists between smectic layering of constituent
molecules and non-linear thermal behavior in binary liquid crystal
systems.
[2] the molecular explanation for observation (1) lies in an electronic phenomenon not a geometric one. This is apparent as the
only structural difference in molecules A, B, and C is in the nature
of one small, terminal substituent.
The first point is obvious from a general examination of the three
sets of phase diagrams. We feel that non-linear behavior (which
arises when Y-Z interactions are significantly different from interactions between the pure components Y-Y and Z-Z) is to be anticipated
when smectics of different layering arrangements are mixed. From
this postulate we predict that compounds A and B will exhibit x-ray
photographs in which d is greater than 1, i.e. a bimolecular smectic.
Likewise we predict C to have a d less than 1, i.e. a unimolecular
smectic phase.

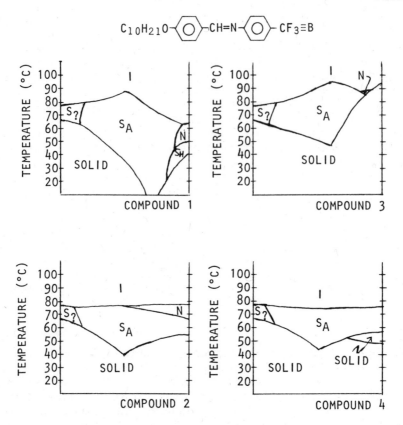

Figure 3. Set of phase diagrams for binary mixtures of B. Compounds 1 and 3 are unimolecular smectics; compounds 2 and 4 are bimolecular smectics.

From a consideration of terminal groups which promote non-linear thermal behavior, -CN, $-NO_2$, and $-CF_3$; we believe that it is the electron-withdrawing effect of these substituents which is responsible for this behavior. Bimolecular smectics are we feel a result of molecular complex formation (14,15) in the pure components. The analogy with Neubert's domino model (16) for nematics is noted. As an extension of this model, we feel that for the cyano and nitro compounds 2 and 4, the most satisfactory model is a bimolecular one such as that shown schematically below in figure 5. Cladis et al. (17) have recently proposed a similar model to account for re-entrant nematic phases. It should be noted that the driving force for molecular association in our model is π-molecular complexation, whereas they favor attraction between hydrocarbon chains as the dominant factor.

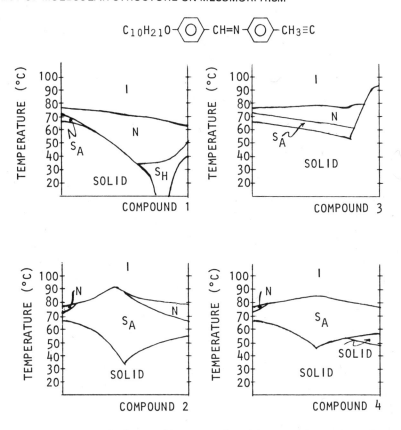

$$C_{10}H_{21}O-\langle O \rangle-CH=N-\langle O \rangle-CH_3\equiv C$$

Figure 4. Set of phase diagrams for binary mixtures of C. Compounds 1 and 3 are unimolecular smectics; compounds 2 and 4 are bimolecular smectics.

The schematic is proposed to explain the unusual d-spacings in compounds 2 and 4. It also provides a molecular rationale consistent with the electronic character of cyano and nitro groups and with Labes'(6) finding that donors such as p-aminobiphenyl (no hydrocarbon chain) can promote non-linear thermal behavior in mixed systems.

Binary mesophase systems in which a bimolecular smectic is mixed with one which significantly perturbs the molecular complexing of the pure bimolecular smectic will, we predict, exhibit non-linear thermal behavior, i.e. (non-ideal solution behavior). An example of such perturbation would be found in the addition of compounds (solute) with electron rich aromatic rings which can successfully compete for molecular complexation with an electron deficient ring of a solvent molecule. Thus Y-Z interactions are considerably different than Y-Y or Z-Z interactions, or the concentration weighted

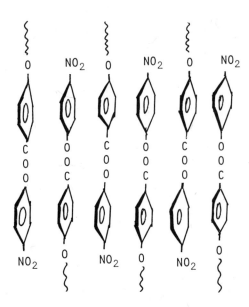

Figure 5. Model for bimolecular complex.

sum of Y-Y + Z-Z interactions. Indeed in such cases, system A-1 as
an example, non-linear thermal behavior is found. Likewise one
should expect that compounds with similar smectic layering characteristics, i.e. 2 and 4, should when mixed with each other exhibit
thermal behavior which is linear with concentration. This is
experimentally found to be the case (18).

We have also examined the effect of alkoxy chain length as it
relates to enhancement of the smectic A phase in mixtures. Figure
6 allows a comparison of trifluoromethyl and nitro as terminal
groups in elevating the $S_A \to I$ temperature in binary mesophases since
the only difference in the molecules is in the terminal group. The
alkoxy chain length of a homologous series of ethyl 4-n-alkoxyben-
zylidene-4'-aminobenzoates (10) is plotted against maximum meso-
phase-to-isotropic transition temperatures when the homologues are
each mixed separately with both the trifluormethyl and nitro compounds. These plotted temperatures represent values which are
higher than mesophase-isotropic temperatures of either pure component and are the direct result of mixing the two components. It is
easily seen that alkoxy chain length does have a significant influence on the $S_A \to I$ temperature. It is a perturbation on a larger
theme. The effect is different for the two types of terminal
groups, nitro and trifluoromethyl. For the nitro compound one finds
nematic behavior until the alkoxy chain length reaches five carbons
while the trifluoromethyl compound exhibits smectic behavior even
at low carbon chain number. In addition the curves differ radically
as chain length is increased.

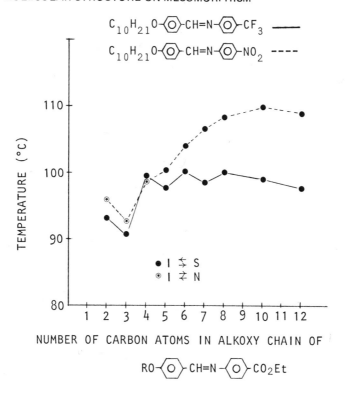

Figure 6. Comparison of effectiveness of trifluoromethyl and nitro as terminal groups in elevating the $S_A{\to}I$ temperature in binary meso-phase systems.

Using figure 7 below a comparison between ester vs Schiff's base linkages is possible. Again we have used a common homologous series to probe the differences. The two curves are remarkably similar indicating that the alkoxy chain length exerts a similar effect for both molecules being compared. Transition from enhanced nematic to enhanced smectic behavior occurs at identical carbon number. In both cases the maximum $S_A{\to}I$ occurs at an alkoxy chain length of ten carbons, an identical alkoxy length to that of the two nitro compounds being compared. The even-odd alternation is reminiscent of such alternation in stability constants for charge-transfer complexes of fluoranil: n-alkylbenzene mixtures (19).

Figure 8 below shows data for comparison of ester vs. Schiff's base linkage in terms of the degree of enhancement of the $S_A{\to}I$ temperature. For this comparison we have taken the mean of the $S_A{\to}I$

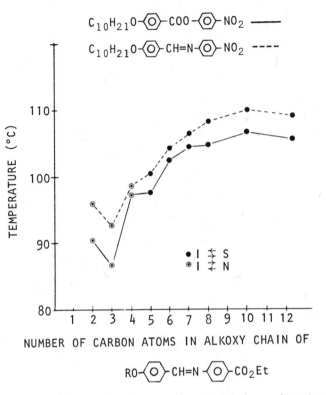

Figure 7.   Comparison of ester vs. anil linkage in elevating $S_A \rightarrow I$ temperature in binary mesophase systems.

temperatures for each of the two pure components and subtracted this value from the observed $S_A \rightarrow I$ maximum as indicated in the previous figure. We have found that the stoichiometry of maximum $S_A \rightarrow I$ temperature is 50:50 mole percent for these systems. Thus, we are justified in using the mean as described above.

This is similar in principle to the Job plot (20) used in molecular complex chemistry. The mean value described above yields a number which would be expected if one encountered ideal solution behavior, i.e. a linear variation of thermal behavior with composition. The numbers plotted in the figure above (ΔT) represent the degree of non-linearity of the mixture. The curves are strikingly similar in shape with the ester curve lying above the Schiff's base curve; both curves maximizing at an alkoxy chain length of about ten carbon atoms. One possible explanation for the fact that the ester curve is above the Schiff's base curve is that the molecules comprising the ester mesophase are more available for intermolecular charge transfer interactions than Schiff's base molecules due to the less efficient intramolecular charge transfer interaction in

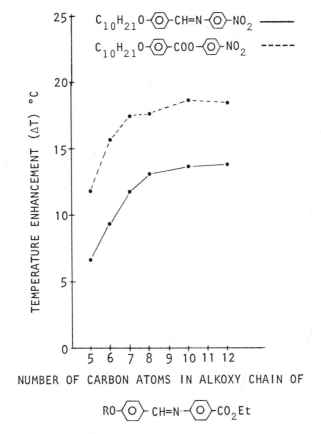

Figure 8. Comparison of ester vs. anil linkage in terms of degree
of enhancement of $S_A \rightarrow I$ temperature.

the ester molecules. Intramolecular charge transfer interactions
should reduce the inherent Lewis acidity and basicity of aromatic
rings within the same molecule due to conjugative interactions
between the rings. As the Schiff's base molecules are more effi-
ciently π-conjugated than the esters, they will not participate to
as great an extent in an intermolecular complex; thus, the lower
curve for the Schiff's bases. Although an increase in dipolar
character should results from an intramolecular charge transfer
interaction, it is apparently of less consequence than the stabilization due to molecular complex formation.

It should be noted that arguments concerning structure- property relationships in the mesophase based on temperatures alone
are to be used with the utmost caution. Calorimetric studies, in
which one obtains the quantities ΔH and ΔS for these transitions,
are in progress.

REFERENCES

1. H. Sackmann and D. Demus, Z. Physik. Chem. (Leipzig), 224, 177 (1963) and H. Sackmann and D. Demus, ibid, 230, 285 (1965).

2. J. S. Dave, P. R. Patel and K. L. Vasanth, Indian J. Chem., 4, 505 (1966); Mol. Cryst. Liquid Cryst., 8, 93 (1969).

3. J. P. Schroeder and D. C. Schroeder, J. Org. Chem., 33, 591 (1968).

4. We use the term "enhanced" to described maxima in transition lines, following the useage of J. S. Dave and R. A. Vora in Liquid Crystals and Plastic Crystals, Vol. 1, G. W. Gray and P. A. Winsor, eds., Ellis Horwood Ltd., Chichester, England, 1974, p. 167.

5. J. W. Park, C. S. Bak and M. M. Labes, J. Amer. Chem. Soc., 97, 4398 (1975).

6. J. W. Park and M. M. Labes, Mol. Cryst. Liquid Cryst., Letters, 34, 147 (1977).

7. C. S. Oh, Mol. Cryst. Liquid Cryst., in press.

8. A. C. Griffin and J. F. Johnson, J. Amer. Chem. Soc., 99, 4859 (1977).

9. The term "bimolecular smectic" is used to describe a molecular arrangement in which d>ℓ. It is our feeling that "bimolecular" used to describe one smectic layer in which two molecules must participate, is preferable to "bilayered" smectic which might be interpreted as two layers.

10. D. L. Fishel and P. R. Patel, Mol. Cryst. Liquid Cryst., 17, 139 (1972).

11. L. Kofler and A. Kofler, Thermomikromethoden, Verlag Chemie, Weinheim, 1954.

12. D. Coates and G. W. Gray, J. Chem. Soc., Perkin II, 300 (1976).

13. A. C. Griffin and N. W. Buckley, to be published.

14. R. Foster, Organic Charge-Transfer Complexes, Academic Press, London, 1969.

A. C. G. wishes to thank Research Corporation for generous support of this work.

15. W. B. Person and R. S. Mulliken, Molecular Complexes: A Lecture and Reprint Volume, Wiley, New York, 1969.

16. M. E. Neubert, L. T. Carlino, R. D'Sidocky and D. L. Fishel, Liquid Crystals and Ordered Fluids, Vol. 2, J. F. Johnson and R. S. Porter, eds., Plenum Press, New York, 1974, p. 293.

17. P. E. Cladis, R. K. Bogardus, W. B. Daniels, and G. N. Taylor, manuscript submitted for publication.

18. A. C. Griffin, unpublished work.

19. Reference 14, p. 199.

20. P. Job, Compt. Rend., 180, 925 (1925); Ann. Chim. [10], 9, 113 (1928); Ann. Chim. [11], 6, 97 (1936).

21. M. J. S. Dewar and A. C. Griffin, J. Amer. Chem. Soc., 97, 6662 (1975).

LIQUID CRYSTALLINE INTERACTIONS

BETWEEN CHOLESTERYL ESTERS AND PHOSPHOLIPIDS

Martin J. Janiak and Carson R. Loomis

Biophysics Section, Department of Medicine
Boston University School of Medicine
Boston, Massachusetts 02118

Among the many organic compounds which form liquid crystalline states, phospholipids and cholesteryl esters are naturally occurring, biologically important molecules. Phospholipids are the major lipid components of cell membranes (1) and together with cholesteryl esters comprise the bulk of lipids found in the circulating lipoproteins of the serum (2). The liquid crystalline nature of these lipids has been shown to be intrinsically important in the structure of both membranes and low density serum lipoproteins (3-5). Certain diseases, in particular atherosclerosis, are associated with the accumulation of phospholipid and cholesteryl ester (6,7). In atherosclerosis, these lipids have been shown to be primarily in their liquid crystalline states at physiological temperature (8).

Essential to understanding the functional role of phospholipid and cholesteryl ester in normal and diseased processes is a knowledge of their interaction in the liquid crystalline state. In this paper we describe results on the bulk phase behavior of mixtures of phosphatidylcholine (lecithin) and three separate cholesteryl esters as a function of lipid concentration, hydration and temperature. Two polyunsaturated cholesteryl esters, cholesteryl linolenate and cholesteryl linoleate, have been studied with the naturally occurring phospholipid, egg yolk lecithin. To correlate chain length and unsaturation dependence of the fatty acid moiety, the interaction of dimyristoyl lecithin and cholesteryl myristate was also examined. Comparison of the phase behavior of these three systems leads to several general conclusions regarding lecithin-cholesteryl ester liquid crystalline interactions.

METHODS

Cholesteryl ester (CE) and lecithin (LEC) were weighed into a
flask and dissolved in choloroform-methanol to give bulk solutions
containing a specific weight ratio of CE:LEC (X% CE : 100 - X% LEC)
where CE is either cholesteryl linolenate (CLn), cholesteryl linoleate (CL) or cholesteryl myristate (CM) and LEC is either egg
lecithin (EL) or dimyristoyl lecithin (DML). Each specific weight
ratio of CE:LEC was prepared as a series of mixtures of increasing
water content by methods previously described (9). The composition
of each mixture is referred to as a% water : (100 -a)% total lipid,
where the ratio of CE:LEC in the total lipid is given by (X% CE :
100 - X% LEC). Hydrated mixtures containing CLn:EL and CL:EL were
equilibrated for 4 - 6 hours at 45 to 50°C. For the CM:DML system,
the samples were briefly heated to 90°C to melt the ester to a
liquid and then equilibrated at 50°C. Each mixture was examined
by polarizing light microscopy and x-ray diffraction in a manner
similar to that previously described (10).

RESULTS

Hydrated lecithin (greater than 5 weight percent water) forms a
"neat" textured liquid crystalline phase at temperatures above
the order-disorder transition (9), the temperature where the hydrocarbon chains of lecithin melt from an ordered paracrystalline state
to a disordered liquid-like state (11-13). This transition occurs
at -7°C for hydrated egg lecithin and 23°C for hydrated dimyristoyl
lecithin. A single "neat" textured phase is exhibited by both egg
lecithin and dimyristoyl lecithin at water contents up to 40 weight
percent (see Figure 1b). At water contents exceeding 40 percent,
an additional phase of excess water is present. This "neat"
textured phase consists of a two-dimensional array of lecithin
molecules packed as a bimolecular leaflet with intercalated layers
of water (9).

On adding cholesteryl ester to egg lecithin or dimyristoyl
lecithin, a single "neat" textured lamellar phase is observed over
a narrow concentration range of cholesteryl ester and water (see
Figures 1a,b and 2). For example, mixtures containing a CE:LEC
ratio less than 5.0:95.0 exhibit a homogeneous single phase between
15 and 35 weight percent water. Water contents less than 15 or
greater than 35 percent contained two phases, the additional phase
being cholesteryl ester. Mixtures containing greater than 40
weight percent water contained three phases: a "neat" textured
lamellar phase, a pure cholesteryl ester phase and an excess water
phase.

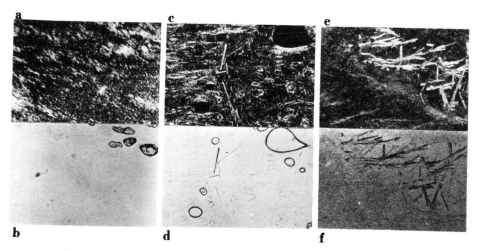

FIGURE 1 Liquid crystalline textures of cholesteryl ester-phospholipid mixtures observed by polarizing light microscopy (shown here cholesteryl linoleate-egg lecithin-water at 23°C). A single phase displaying a homogeneous unbroken field by straight light in (a) exhibits a neat textured appearance under crossed polarizers (b). Two phases are present in (c) straight light and (d) crossed polarizers, a neat textured lamellar liquid crystalline phase and cholesteryl linoleate. Most of the ester has crystallized as needles interspersed in the neat textured phase but a few droplets contain uncrystallized ester in metastable state. Three phases are present in (e) straight light and (f) crossed polarizers, the third phase being excess water.

In all three CE:LEC systems, no mixture with a weight ratio greater than 5.0:95.0 contained less than two phases. All mixtures studied contained either a "neat" textured phase and a cholesteryl ester phase (Figures 1c, d) or three phases, the additional phase being excess water (Figures 1e, f).

Hydrated mixtures of CLn:EL (Figure 2a) and CL:EL (Figure 2b) were studied over the temperature range of 0 to 60°C. Mixtures containing two or three phases demonstrated reversible smectic to cholesteric and cholesteric to isotropic phase transitions of cholesteryl ester. These transitions were observed at 29 and 30°C or 34 and 36.5°C for mixtures containing cholesteryl linolenate and cholesteryl linoleate, respectively. Mixtures containing large amounts of cholesteryl ester often exhibited needle shaped crystals which melted at the appropriate temperature for a given cholesteryl ester, 36°C for cholesteryl linolenate and 42°C for cholesteryl

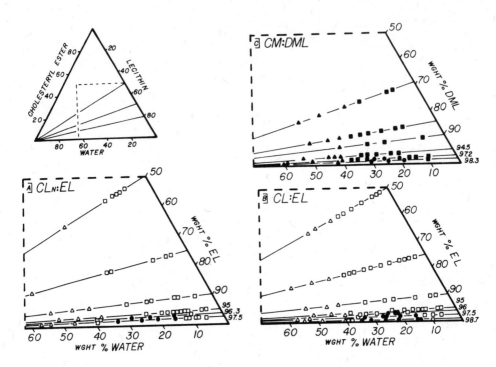

FIGURE 2   The number of phases and microscopically observed textures in mixtures of cholesteryl ester-lecithin-water.   (a-c) an
expanded scale of the lower right hand corner of the diagram.   (a)
cholesteryl linolenate-egg lecithin-water at $23°C$; (b) cholesteryl
linoleate-egg lecithin-water at $23°C$; (c) cholesteryl myristate-
dimyristoyl lecithin-water at $37°C$.   Each mixture is represented
by a symbol denoting the number of phases present and their textures: (●) one phase, neat textured liquid crystal; (□) two phases,
neat textured liquid crystal and the smectic phase of cholesteryl
ester; (■) two phases, neat textured liquid crystal and crystals
of cholesteryl ester; (△) three phases, neat textured liquid
crystal, the smectic phase of cholesteryl ester and excess water;
(▲) three phases, neat textured phase, crystals of cholesteryl
ester and excess water.

linoleate (14).

Hydrated mixtures containing CM:DML (Figure 2c) were examined between 25°C and 90°C. When a cholesteryl myristate phase was present, all mesophase transitions were observed reversibly; crystal to smectic at 70.5°C, smectic to cholesteric at 79.5°C and cholesteric to isotropic at 85.5°C. At 37°C, a cholesteryl myristate phase is present as a crystalline phase.

Thus, only limited solubility of cholesteryl ester is observed in the "neat" textured lamellar liquid crystalline phase of hydrated lecithin. In addition, the temperatures of the phase transitions associated with the excess phase of the cholesteryl ester in these ternary mixtures are identical to those of pure cholesteryl ester indicating little or no incorporation of lecithin into the cholesteryl ester phase occurs.

X-ray diffraction was performed on all hydrated mixtures at 23°C (for CLn:EL and CL:EL) or 37°C (for CM:DML). X-ray diffraction patterns from hydrated lecithin (both EL and DML) at these temperatures are characterized by several low angle diffraction maxima (Figure 3a) in the ratio 1 : 1/2 : 1/3 : 1/4.... which defines the interlamellar repeating unit (i.e., the thickness of the lipid bilayer and adjacent water layer). In the wide angle region, only a broad diffuse band centered at 4.6 Å is present over the entire range of water contents studied.

All mixtures containing cholesteryl ester exhibited similar x-ray diffraction patterns as observed for lecithin alone, as shown in Figure 3b. In addition, mixtures containing greater than 5 weight percent ester produced several additional reflections arising from the crystal phase of excess cholesteryl ester. The reflections associated with the crystalline phase of ester were absent when the sample was examined at temperatures corresponding to the smectic liquid crystalline state of cholesteryl ester. At these temperatures, only the single characteristic intensity maximum of the smectic phase was observed together with reflections arising from the lamellar liquid crystalline phase (Figure 3c). These occur at 36Å for CLn and CL and 34Å for CM.

For each ternary system, x-ray diffraction studies were performed as a function of water content for each constant lipid ratio (X% CE : 100 - X% LEC). In this fashion, the effect of cholesteryl ester on the swelling behavior of lecithin could be determined. From the swelling curves, the interlamellar repeating unit, d, was obtained as a function of water content and lines corresponding to equal values of the interlamellar repeat were defined on a triangular co-ordinate diagram for a given composition of the three components. The lines defining constant values of the interlamellar

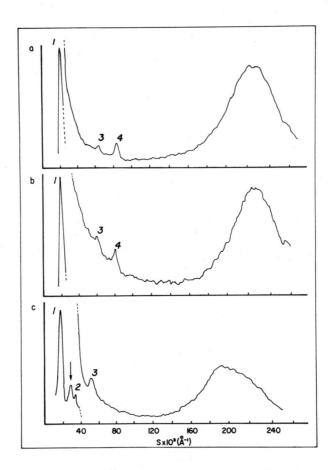

FIGURE 3 Representative microdensitometer traces of x-ray
diffraction patterns from various mixtures of similar water content. (a) hydrated egg lecithin containing 10 percent water; first
order reflection at 48.5 Å. (b) mixture containing 12.9% water
(5% CLn:95% EL); first order reflection at 49.7 Å. (c) mixture
containing 12.0% water (75% CLn: 25% EL); first order reflection
56.9 Å. The intensity maximum at 35 Å (arrow) arises from the
smectic phase of cholesteryl ester. A diffuse band is observed
at 4.6 Å in (a) and (b). In (c), a more diffuse band observed at
4.85 Å appears asymmetric probably due to an intensity contribution
from the smectic phase of cholesteryl ester at high angle.

repeat begin at the apex corresponding to 100 weight percent cholesteryl ester (Figure 4) and define equivalent values as a function of composition (9, 15).

For all three ternary systems, the lines in the upper part of the triangle are linear. For mixtures containing CLn:EL (Figure 4a), a discontinuity occurs at the constant lipid ratio of 3.75:96.25. The constant d lines show a significant bulge towards the water apex of the triangle, the magnitude diminishing with increasing water content. At the maxima value of d, this bulge is no longer present but the discontinuity occurs at 2.5 CLn:97.5 EL. These results indicate that the liquid crystalline structure of lecithin is affected on incorporation of CLn and small amounts of this ester remain incorporated in the presence of excess water.

In contrast, mixtures containing CL:EL do not exhibit this bulge toward the water apex (Figure 4b). In this case, the discontinuity approaches the lecithin-water axis as the water content increases until no discontinuity is observed at the swelling limit of the lecithin. Thus, mixtures containing CL appear to have only a small effect on the lamellar liquid crystalline phase of lecithin and no ester remains incorporated above the lecithin swelling limit.

Mixtures containing CM appear to produce no effect on the interlamellar repeat of the liquid crystalline phase of DML (Figure 4c). However, microscopy data (Figure 2c) indicate that up to 4.0% CM is incorporated into the lamellar liquid crystal. The absence of any marked structural effect is apparently due to the strong similarity in the molecular lengths of the two molecules. The length of the CM steroid nucleus is 8.86Å (16) and is very similar to the length of the hydrocarbon chain of DML and CM in the liquid crystalline state (9.9Å). Thus, the ester molecule can be easily accommodated into the lamellar liquid crystal without significant perturbation of the structure.

DISCUSSION

From the microscopy and x-ray diffraction data presented, the three component phase diagrams of cholesteryl ester-lecithin-water can be constructed (Figure 5). For mixtures which contained both the metastable liquid crystalline phases and crystalline phase of cholesteryl ester, the crystalline phase is taken to represent the true equilibrium state. A number of different zones can be delineated for each ternary system studied:

Zone I, a single phase of lamellar liquid crystal. For CLn:EL (Figure 5a), the incorporation of ester increases with increasing water content to a maximum value of∿ 4 weight percent

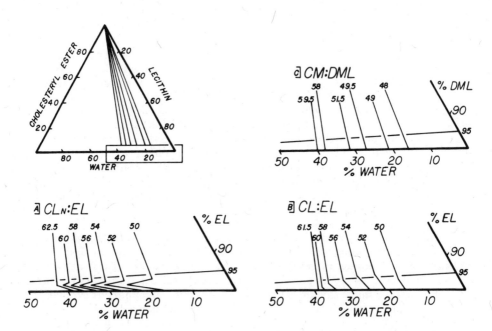

FIGURE 4   Lines defining equal values of interlamellar repeat (d) as a function of composition (the lower right portion of the triangle has been expanded for each ternary system). The zone to the left of the line representing the largest value of d is invariant and corresponds to the largest value observed for all mixtures studied in this zone. The difference in the maximum value of d for mixtures of CLn:EL and CL:EL is due to a small difference in the fatty acid composition of lecithin used in each study.

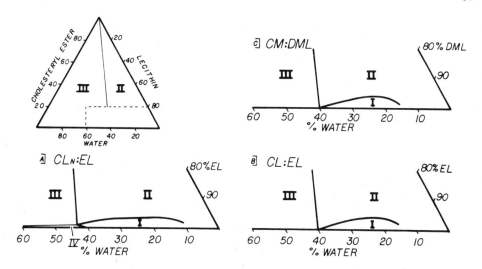

FIGURE 5   Ternary phase diagram of cholesteryl ester-lecithin-water (the lower right portion of the triangle has been expanded for each ternary system).   (a) CLn-EL-water at 23°C;   (b) CL-EL-water at 23°C;   (c) CM-DML-water at 37°C.

between 15 and 30 percent water. Ester incorporation decreases thereafter to a minimum of 1.5% at 40% water. For CL:EL (Figure 5b), incorporation of ester likewise increases with increasing water content, to a maximum of 4.5% between 20 and 25% water before decreasing to zero at 40% water. The ternary system of CM:DML (Figure 5c), behaves in identical fashion as CL:EL except that only ~4% ester is maximally incorporated.

Zone II, two phases, lamellar liquid crystal and crystalline cholesteryl ester.

Zone III, three phases, lamellar liquid crystal, crystalline cholesteryl ester and excess water.

Zone IV, two phases, lamellar liquid crystal and excess water. This zone exists for CLn:EL only where some ester remains incorporated above maximum swelling of lecithin.

For comparison, Figure 6 shows the maximum incorporation of

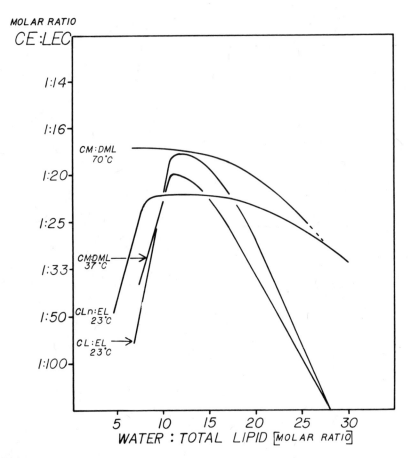

FIGURE 6   Maximum incorporation of cholesteryl ester (molar ratio of CE:LEC) as a function of water content (molar ratio of water to total lipid, CE + LEC).

cholesteryl ester into the lecithin lamellar liquid crystalline phase as a function of water concentration. In order to compare the three ternary systems directly, the data is plotted as the molar ratio of CE:LEC on the abscissa and of water to total lipid (CE + LEC) on the ordinate.

The ternary systems containing CM:DML (37°C) and CL:EL (23°C) exhibit very similar behavior over all hydration values, both systems incorporating no cholesteryl ester at a water to lipid ratio of ∿28:1. The somewhat increased ratio of incorporation for CL:EL (1:18) as compared to CM:DML (1:20) may be due to the effect of temperature. The results for CL:EL were obtained at a

temperature $30^{\circ}C$ above the order-disorder transition temperature of
EL $(-7^{\circ}C)$, while for CM:DML results were obtained $14^{\circ}C$ above the
transition of DML $(23^{\circ}C)$. This would indicate that temperature
induced disorder relative to the order-disorder transition
temperature allows for increased incorporation of ester in mixtures
containing CM:DML at higher temperature and incorporation of large
amounts of CLn in the high temperature phases of anhydrous lecithin
(10).

Incorporation of cholesteryl ester in mixtures of CLn:EL
exhibits a different hydration dependence. Although the molar
ratios of CE:LEC are quantitatively different, the CLn:EL system
displays a hydration dependence similar to CM:DML at higher
temperature, above a water to lipid ratio of 8:1. At a water to
lipid ratio of ∿28:1, both systems incorporate comparable amounts
of ester (∿1:30). This observation qualitatively suggests that
the degree of unsaturation may be equivalent to thermal disordering
of the system.

These observations suggest the following regarding the interaction of cholesteryl ester with phospholipid: 1.) the incorporation of cholesteryl ester is a temperature dependent phenomenon
requiring the existence of a thermally disordered liquid crystalline
structure of lecithin; 2.) the extent of incorporation is temperature dependent; increasing amounts of cholesteryl ester can be
incorporated at higher temperatures relative to the order-disorder
transition; 3.) the structural effect of thermal disordering of
phospholipid bilayers (bilayer thickness, surface area, etc.) may
also be induced by the presence of highly unsaturated cholesteryl
esters. This effect, while not apparent for the cis di-unsaturated
ester, cholesteryl linoleate, is observed for cholesteryl linolenate (all cis $\Delta 9$, 12, 15 octadecatrienoate) suggesting potentially
similar behavior may also occur for other highly unsaturated
cholesteryl esters.

In biological systems, we would predict a low solubility of
cholesteryl esters in phospholipid rich structures such as cell
membranes. In addition, cholesteryl esters when present should
exist primarily as a separate phase. The appearance of cholesteryl
esters in certain biological systems is consistent with these
conclusions. Studies on the lipid composition of many natural
membrane systems have indicated little or no detectable cholesteryl
ester is present (1, 18, 19). In low density serum lipoprotein,
which contains a large amount of cholesteryl ester, the proposed
structure contains a cholesteryl ester rich core in which the
molecules exist in a smectic liquid crystalline state at
physiological temperature (4, 5, 20). Finally the separation of
cholesteryl esters into an individual phase in the intimal layer
of large arterial vessels may represent one of the earliest stages

in the development of atherosclerosis (21). Thus, the absence or
abundance of cholesteryl esters in various biological systems and
in association with certain pathological states may be, in large
part, determined by their liquid crystalline interactions with
neighboring lipid components.

ACKNOWLEDGEMENT

We are indebted to Donald M. Small and G. Graham Shipley,
in whose laboratories this work was carried out, for making
possible this contribution. This work supported by U.S.Public
Health Service grants AM 11453 and HL – 18623 (D.M. Small, principal
investigator).

The authors' present addresses are: M.J. Janiak, Department
of Molecular Biophysics and Biochemistry, Yale University, New
Haven, Connecticut; C.R. Loomis, Department of Biochemistry, Duke
University Medical Center, Durham, North Carolina.

REFERENCES

1. Rouser, G., Nelson, G.J., Fleischer, S. and Simon, G.
    Biological Membranes, Vol. I, Academic Press (1968).

2. Blood Lipids and Lipoproteins: Quantitation, Composition and
    Metabolism (Nelson, G.J., ed.) Wiley-Interscience, N.Y. (1972).

3. Shipley, G.G., Biological Membranes, Vol. II, Academic Press,
    N.Y. (1973).

4. Deckelbaum, R.J., Shipley, G.G.; Small, D.M., Lees, R.S. and
    George, A.K. Science. (1975) 190,392.

5. Atkinson, D., Deckelbaum, R.J., Small, D.M. and Shipley, G.G.
    Proc. Nat. Acad. Sci. (1977) 74, 1042.

6. Lipid Storage Diseases (J. Bernsohn and H.J. Grossman, eds.)
    Academic Press, N.Y.(1971).

7. Atherosclerosis (R.J. Jones, ed.) Springer-Verlag, N.Y. (1970).

8. Katz, S.S., Shipley,G.G., and Small, D.M. J. Clin. Invest.
    (1976) 58, 200.

9. Janiak, M.J., Loomis, C.R., Shipley, G.G. and Small, D.M.
    J. Mol. Biol. (1974) 86, 325.

10. Loomis, C.R., Janiak, M.J., Small, D.M. and Shipley, G.G. J. Mol. Biol. (1976) 86, 309.

11. Reiss-Husson, F.J.Mol. Biol. (1967) 25, 363.

12. Small, D.M. J. Lip. Res. (1967) 8, 551.

13. Tardieu, A., Luzzati, V. and Reman, F.C. J. Mol. Biol. (1973) 75, 711.

14. Small, D.M. Surface Chemistry of Biological Systems, Plenum Press, (1970).

15. Lecuyer, H. and Dervichian, D.G. J. Mol. Biol. (1969) 45, 39.

16. Craven, B. and DeTitta, G.T. J. Chem. Soc. (1976) Perkins II, 814.

17. Luzzati, V. Biological Membranes; Vol. II, Academic Press, N.Y. (1968).

18. Ashworth, L.A.E. and Green, C. Science (1966) 152, 210.

19. Zambrano, F., Fleischer, S. and Fleischer, B. Biochim. Biophys. Acta (1970) 380, 357.

20. Tardieu, A., Mateu, L., Sardet, C., Weiss, B., Luzzati, V., Aggerbeck, L. and Scanu, A.M. J. Mol. Biol. (1976) 101, 129.

21. Small, D.M. and Shipley, G.G. Science (1974) 185, 222.

POLYMERS OF β-LACTAM ANTIBIOTICS: STRUCTURAL, CHEMICAL AND

BIOLOGICAL PROPERTIES

M.K. Stanfield, B. L. Warren, F.H. Wilson

Department of Biochemistry, Tulane Medical School

1430 Tulane Avenue, New Orleans, LA. 70112

ABSTRACT

Polymers of benzylpenicillin, ampicillin and cephaloridine have been isolated, physically examined and tested for allergenicity. Five of eleven penicillin sensitive subjects were found to give a positive response in the lymphocyte transformation test (LTT) while two tested individuals showed release of histamine from basophils. Two penicillin sensitive subjects also gave positive LTT with cephaloridine polymer.

INTRODUCTION

The β-lactam antibiotics are among the most valuable weapons in a physicians' arsenal. The penicillins are essentially non-toxic with the very important exception of allergic reactions which can vary from a rash to anaphylatic shock and death.

Penicillin sensitivity can be exhibited as all the clinical classes of allergic reactions (types I, II, III and IV) (Stewart 1967, Knudsen et al. 1970, Redmond and Levine 1968, Freed 1975, Levine 1971). However, most of these reactions seem to be either type I or type III reactions or perhaps mixtures of these (Redmond and Levine 1968, Basoma et al. 1976). Therefore an assay which can demonstrate IgE antibody responsible for type I reactions (immediate reactions such as histamine assay (May et al. 1970) or type III reactions e.g., passive cutaneous anaphylaxis (PCA (Ovary 1958), are often employed in penicillin allergy studies. To detect type IV reactions (cellular hypersensitivity reactions), assay of

lymphocyte transformation (Halpern et al. 1967) or the migration
inhibition test (Oritz-Oritz et al. 1974) are used.

This study has made use of the lymphocyte transformation test
to measure the allergenicity (type IV) of polymers of three β-
lactam antibiotics, benzylpenicillin, ampicillin and cephaloridine.
The histamine release assay was used to measure type I allergini-
city to benzylpenicillin.

MATERIALS AND METHODS

Subjects

20 subjects were chosen on the basis of personal history of
allergic reactions with penicillin and negative history of allergy
reactions with penicillin and negative history of allergy to the
tetracyclines. In all cases a physician had observed the reaction
e.g., skin rash, edema, or anaphylaxis, and specifically warned
the patient against receiving further treatment with penicillin.
Descriptions of the allergic reaction were recorded but were often
lacking in detail as to the type of penicillin administered, the
time of onset of the allergic response after administration, etc.
The subjects were 90% Caucasian and 10% non-Caucasian, 40% male
and 60% female. The average age was 25 with a range of 19-52
years. Persons receiving medication were excluded from the study.
All were in good health at the time of testing. The time since
the subjects had last received penicillin ranged from one year to
20 years.

Preparation and Isolation of Polymers for Bioassay

Benzylpenicillin in non-buffered powder form was a gift from
Squibb, Inc. Cephaloridine (Loridine) in non-buffered powder form
was purchased from Eli Lilly and Co. Alpha-amino benzylpenicillin
(ampicillin) was purchased from Ayerst Labs, Inc. as a non-buffered
powder. Polymers were prepared from the penicillins by adding one
gram of parent material to one ml of sterile, glass distilled water.
Loridine polymer was similarly prepared but a greater dilution
(1 g/10 ml of water) was required for total solubility. Solutions
were placed in sterile, capped glass tubes in the dark at room
temperature for 14 days then frozen at -20°C until fractionation.

To fractionate penicillin G or cephaloridine polymers, one
to two ml of penicillin G or Loridine polymer solution were placed
on a Pharmacia SR 25/100 column (2.5 cm X 100 cm) containing 40
grams of G-25 medium Sephadex run at a rate of 5 ml/min. Glass
distilled water was used as eluent. Eluent fractions of 3 ml
volume were collected then pooled to yield three kinds of polymer
as determined by time on the column, color, and spectroscopy.

Both cephaloridine and penicillin G pool I consisted of fractions
60 ml-90 ml, MW > 5000; pool II was made up of volumes from 120-
150 ml, MW 1000-5000; and pool III was made of volumes from 165 ml-
195 ml, MW < 1000. The three pools were then lyophilized and
stored at -20°C until needed. Ampicillin polymers were formed in
the same manner as the other polymers but were not fractionated.
Then the total solution was lyophilized and stored at -20°C until
needed.

Polymeric material used in characterization studies (see
below and results section) were obtained in a slightly different
manner. Glass distilled water was added to the antibiotics benzylpenicillin or ampicillin, and the solution immediately chromatographed.

Polymer Characterization

Solid disk IR spectroscopy was performed using a fast scan
from 400 cm^{-1} to 700/cm^{-1} on the Beckman Acculab I Infra-Red
Spectrophotometer. Dlear disks were obtained by mixing 9 mg
spectroscopic grade potassium bromide powder (Matheson, Coleman
and Bell) with 2 mg of sample then subjecting the mixture under
vacuum to 14,000 lb/sq. in. pressure using a Carver Lab Press.

Aqueous solutions of the various eluted fractions from the
Sephadex G-25 column were scanned in the near ultraviolet (320 nm-
220 nm) using the Beckman DB. 3 ml of eluted material were scanned
as a representative sample of every 20 ml of material eluted from
the column.

Nuclear Magnetic Resonance spectra were obtained by dissolving 30 mg of sample in 0.4 ml of D_2O or DMSO (dimethyl sulfoxide)
at room temperature using Varian Associates A60 spectrometer. An
internal standard of either tetramethylsilane or for D_2O samples,
3-(trimethylsilyl)-1-propanesulfonic acid, sodium salt was used.

Histamine Release Assay

Histamine release was measured by a modification of the method
of May et al. (1970). All glassware was washed in a non-fluores-
cent detergent-Sparkleen[R] (Calgon). White cell separation from
the blood was performed in plastic tubes. For cell separation 6 ml
of whole heparinized blood (100 USP units heparin/10 ml of blood)
was layered onto 3 ml of methylcellulose-Hypaque[R] (130 ml of a 1%
methylcellulose solution mixed with 50 ml of a 50% Hypaque[R]
solution). 50% sodium Hypaque[R] was purchased from Winthrop Labs
and Methocel A-25 was a gift from Dow Chemical Co.

The white cells in the plasma were centrifuged for 8 minutes
at 170 g's. The supernatant was decanted and the pellet was

resuspended in 37°C tris-CM buffer (3.75 g Trizma, 6.95 g NaCl, .37 g KCl, .09 g CaCl$_2$·2H$_2$O) and .235 g MgCl$_2$·6H$_2$O in 1 liter of distilled water) to give one ml more than needed for the experiment. One ml of cells was added to 0.1 ml of polymeric antibotic or starting material in the various dilutions to be run, using tris-CM buffer as diluent. 0.1 ml of AB+ serum (taken from a normal, non-penicillin sensitive person and stored at -20°C until needed) was also added. At this time, blanks, which account for fluorescence due to the antigen itself, the tris-CM buffer, the serum and fluorescent portions of the cells which were non-specifically washed free into the medium, were run. Also, dilutions of the internal standard (300 µg/1) were set up using histamine dihydrochloride. Further, one tube designated "total histamine" was set up without any antigen but with one ml of cells and 0.1 ml of 5% serum. All tubes were then incubated in a water bath at 37°C for 40 min. 0.9 ml of the incubated cells in the total histamine tube were transferred to a salt (300 mg NaCl) N-butanol (1.25 ml) mixture. All other tubes were centrifuged at 600 g to pellet the cells and 0.9 ml of the supernatant was transferred to correspondingly numbered tubes containing the salt-butanol mixture. An aqueous-organic extraction of the histamine was performed by adding 100 µl of 3 N NaOH to each tube and mixing vigorously for exactly 3 min. followed by centrifugation at 900 g for 3 min. Tubes were mixed briefly and centrifuged again at 900 g for 8 min. Into tubes with corresponding numbers containing 0.12 N Hcl and 1.9 ml of N-heptane, 1 ml of the butanol layer was transferred. These tubes were vigorously mixed for exactly 1 min then centrifuged at 600 g for 5 min. The heptane layer was removed by suction and 0.5 ml of the HCl layer was placed in a glass culture tube and allowed to cool in an ice bath for 20 min. At this time the dilutions of histamine external standard (200 µg/1) were diluted with HCl and placed in the ice bath. Then 200 µl of 0.75 N NaOH and 60 µl of 0.05% OPT (o-phthaldialdehyde, obtained from Calbiochem) were added to the mixture which was chilled in an ice bath for 40 min. Finally, 100 µl of 1.25 N H$_3$PO$_4$ was added to the mixture followed by incubation at room temperature for exactly 20 min. and the fluorescence of this resulting solution measured in an Aminco-Bowman spectrophotofluorometer at 360 nm excitation and 450 nm emission. The per cent histamine release was calculated by comparing the fluorescence of any tube incubated with polymer to the fluorescence of the tube containing 1 ml of disrupted cells (total histamine). Both tubes were also corrected for any fluorescence not due to histamine by subtracting the value of the corresponding blanks. Cell viability tests were also performed to insure that no non-specific release of histamine due to cell damage was recorded as a positive.

Lymphocyte Transformation Test

The technique utilized was that of Valentine and Waithe (1971)

as modified by Wilson (personal communication). The entire procedure was carried out using sterile technique. Antigenic dilutions were set up in triplicate.

Lymphocytes from 40 ml of blood were separated from other
blood cells by layering 6 ml of heparinized, (100 USP units
heparin/10 ml of blood) saline diluted, (1 part blood: 2 parts
physiological saline) blood onto 2 ml of Ficoll-Hypaque[R] (Winthrop
Labs) and 3.2 ml of distilled water were mixed to give a solution
with specific gravity of 1.078-1.080. The tubes were centrifuged
for 10 min at 450 g. Lymphocytes, which appeared as a band in
the plasma, were drawn off using a pasteur pipette. They were
placed in a screw-top volumetric test tube and centrifuged for
8 min at 500 g. The supernatant was decanted, the pellet resuspended in 2 ml of 37° C Roswell Park Memorial Institute (RPMI-
1640 from Grant Island Biological Co., with glutamine but without
antibiotics) and counted in an hemocytometer. Cells were diluted
in a solution composed of 20% AB+ serum, 1% 1:200 chlortetra-
cycline HCl (GIBCO) and 79% RPMI to yield a final concentration
of 5×10^5 cells per ml. One ml of cell suspension was added to
0.1 ml of polymeric penicillin G, ampicillin or Loridine, or
parent antibiotic (starting material) and diluted with sterile
physiological saline in culture tubes. After incubation in a 10%
carbon dioxide, 90% air atmosphere at 36° C for 6 days, 1 μC_i in
10 μl of ^{3}H-thymidine was added to each tube except those used to
determine cell death. On the 7th day of incubation, the cells
were harvested by centrifuging at 450 g at 4° C for 15 min. The
supernatant was decanted and 1 ml cold physiological saline was
added then mixed. Tubes were centrifuged at 450 g for 15 min at
4° C, the resulting supernatant decanted, and 5 ml 5% trichloro-
acetic acid (TCA) added followed by mixing. Tubes were centrifuged for 30 min at 450 g at 4° C, the supernatant decanted, and the
remaining material washed twice with 2 ml cold absolute methanol
by centrifugation at 450 g for 15 min at 4° C. The supernatant
was decanted and 0.5 ml Soluene 350 (Packard Instrument Co.) was
added. Tubes were incubated at 56° C for 2 hours and the contents
of each tube added directly into the scintillation vial. The
culture tubes were washed twice with 5 ml of scintillation cocktail
(0.2 g of POPOP and 10 g of PPO in 2 liters of scintillation grade
toluene) into scintillation vials yielding a total volume of 10.5 ml
in the vials. Radioactivity was counted in a Packard Tri-Carb
Liquid scintillation spectrometer model # 3320. Per cent efficiency
was determined and DPM calculated using the formula (CPM X Efficiency) 100 = DPM. Values obtained for triplicate tests were examined and if one of the three varied from the other two by five
times or more, that value was excluded. Stimulation of polymer
challenged cells beyond the subject's own control cells was calculated and statistically significant increases were determined using
"Student's t test" with a confidence limit of 0.2%

RESULTS

Polymerization of Benzylpenicillin and Ampicillin

The polymers formed on the spontaneous degradation of penicillin G (PGO) and ampicillin (AMP) were separated into fractions
of varying molecular weights by column chromatography on Sephadex
G25 for spectroscopy. The range of molecular weights, from greater
than 5000 (14 - 15 residues) to less than 1000 (1 - 3 residues),
fractionated as follows: Fraction I > 5000; Fraction II = 3500 -
4000; Fraction III - 2500 - 3000; Fraction IV = 1000 - 1500; and
Fraction V < 1000 which were used for spectral studies.

Some of the degradation products which are thought to take
part in the polymerization process (Clarke et al. 1949) are shown
below:

Degradation of Penicillins

Ultraviolet Spectroscopy

Ultraviolet spectroscopy was used to identify some of these
degradation products and to monitor the column effluent samples.
The presence of benzylpenicillenic acid (λ_{max} 322 mμ) in Fract. III
was seen. The higher molecular weight molecules were obtained
only in very small amounts, (milligrams from a 100 g sample).

Infrared Spectroscopy

Infrared studies of the polymers of benzylpenicillin and
ampicillin show a definite loss of clarity of the spectra with
increasing molecular weight, together with a substantial change in

the band at 1785-1790 cm^{-1} where the highly strained β-lactam
carbonyl absorbs. This suggests that penicilloic acid is of
primary importance in the polymeric structure. The two absorption
bands due to the amide of the side chain, the amide I band at 1665
cm^{-1} and amide II band near 1490 cm^{-1} were greatly diminished in
the polymers. Their intensity in Fraction II when compared to the
still moderately strong carboxylate anion absorption near 1610
cm^{-1} was greatly reduced. Although the same technique was used
to prepare discs of ampicillin and its polymers for IR spectroscopy,
it was not possible to obtain spectra from these with the same
degree of resolution as were obtained from penicillin G and its
polymers. The spectra of ampicillin Fractions IV and V (molecular
weight, < 1500) closely approximated, but were not as sharply defined as those of starting material prior to chromatography. The
β-lactam absorption at 1765 cm^{-1} was relatively strong and there
was a fairly clear fingerprint region. Fraction I (M W > 5000)
was poorly resolved and showed no absorption near 1760 cm^{-1}, indicating loss of β-lactam ring structure. With decreasing molecular weight, the spectra more closely resembled unfractionated
material. The spectrum of Fraction VI, which followed the monomeric material, showed a return to polymeric character with
moderate retention of the β-lactam absorption peak. A final
fraction obtained only in trace amounts, showed major changes in
the overall spectrum and no β-lactam absorption.

Nuclear Magnetic Spectroscopy

II VI   I   III   IV

Benzylpenicillin Assignments

On exchange of the amide hydrogen with deuterium in the D$_2$O, the
coupling with the β-lactam ring hydrogens decreased collapsing
the multiplet near 4.6 τ. Figure 1 illustrates the 60 MHz spectra
of benzylpenicillin (PGO) and various polymeric fractions together
with a sample of benzylpenicilloic acid (BPO). The dimethyl-
sulfoxide absorption near 7.6 τ was clearly seen in all spectra
where this solvent is employed. Figure 2 illustrates the ampicillin
polymers and its penicilloic acid derivative (AAPO). The changing
nature of the spectra with increasing molecular weight is clearly
seen. Some of these effects were due to increasing viscosity and

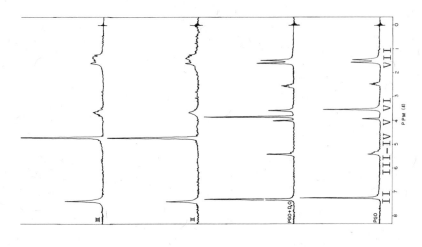

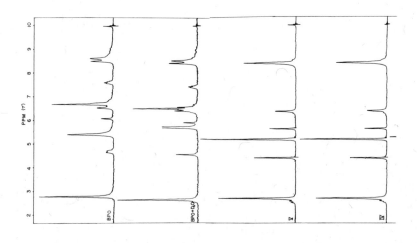

FIGURE 1 – NMR Spectra of Benzylpenicillin (PGO), its Polymers and Benzylpenicilloic Acid (BPO)

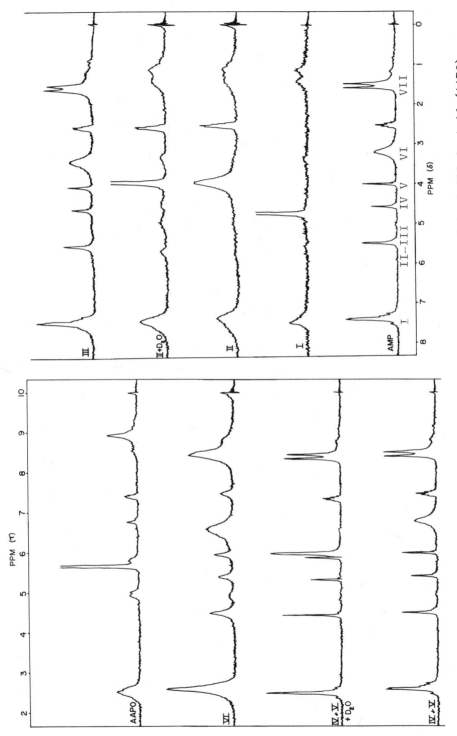

FIGURE 2 — NMR Spectra of Ampicillin (AMP), its Polymers and Ampicilloic Acid (AAPO)

D_2O was added in an attempt to minimize this contribution.   The
ampicillin assignments are as follows:

Figure 2 illustrates the NMR results for ampicillin and its
polymers. These assignments are the same as those made by Green
et al[7]. The amide hydrogen absorption near 1.3 τ is not shown in
either figure.

The higher molecular weight polymers exhibited the broad, ill-
defined bands characteristic of polymers of mixed structure.   In
the lower molecular weight materials the distinct signals for the
non-equivalent methyl groups, the thiazolidine ring hydrogen and
the benzylic hydrogens appeared.   The hydrogens attached to
nitrogen were also clearly shown in DMSO solution.   In most
penicillins studied, the C5 and C6 hydrogens of the β-lactam ring
have very similar chemical shifts[8].   In the lower molecular weight
samples examined the typical ABX pattern near 4.5 τ appeared.

For purposes of these spectral studies, both benzylpenicillin
and ampicillin were separated into a larger number of fractions
than were those samples used for bioassay purposes.   The chromatographic method of separation was, however, the same in all cases.

Treatment with heavy metal ions greatly enhanced the rate of
degradation of benzylpenicillin.   However, the NMR showed that the
β-lactam ring is maintained intact for as long as an hour at ambient
temperature.   At the same time, the down field methyl peak slowly
disappeared with the appearance of a new peak up field from the
original methyl signals.   It is proposed that the metal ion chelates
with the sulfur and the exocyclic N6 of the side chain amide.   This
would momentarily stabilize the β-lactam ring while leading to new
methyl signal and aiding the rupture of the thiazolidine ring.

                        Histamine Release Assay

The basis for testing the allergic type I response of an individual to a given antigen by the histamine release assay is as
follows: when challenged by an antigen to which a subject has been
sensitized, the IgE antibody bound to the surface of the basophil
"recognizes" the antigen and causes the release of histamine.   About

70% of penicillin sensitive people would be expected to have IgE
bound to their basophils (Basoma 1976). If this IgE is specific
for penicillin polymer, or if polymer can cross react with this
IgE (perhaps specific for penicillin-protein conjugate) then
histamine should be released, and the amount of histamine released
can be measured. Due to their high degree of fluorescence,
ampicillin and cepharloidine could not be used in this assay.

Table I shows the result of the histamine release assay with
a control subject while Table II shows a typical result with penicillin sensitive individual.

TABLE I

HISTAMINE RELEASE ASSAY

CONTROL SUBJECT

Benzylpenicillin Polymer Fraction	Dilution (w/v)	% Histamine Release
I	1:10	22
I	1:20	7
I	1:40	18
I	1:80	18
II	1:1	21
II	1:2	10
II	1:4	10
II	1:8	8
III	1:1	7
III	1:2	11
III	1:4	18
III	1:8	20

The lack of a dose response relationship between the amount of
histamine release and the antigenic dilutions (Lichtenstein et al.
1964) in the penicillin sensitive subject's data can be explained
by the use of antigenic dilutions well above the dose response
range. The high concentrations were used to insure the release of
histamine, as this study was concerned with whether polymer could
elicit histamine release at all rather than the reproducibility
of a certain level of histamine release with a given polymeric
dilution. Thus, most histamine release in these studies should
have been maximal.

35 other histamine assays were run but an insufficient total
histamine level was obtained, therefore, it was not possible to
meaningfully measure percent histamine release. Frequencies of

TABLE II

HISTAMINE RELEASE ASSAY

PENICILLIN SENSITIVE

Benzylpenicillin Polymer Fraction	Dilution (w/v)	% Histamine Release
I	1:10	76
I	1:20	140
I	1:40	100
I	1:80	31
II	1:1	128
II	1:2	97
II	1:4	100
II	1:8	80
III	1:1	96
III	1:2	82
III	1:4	70
III	1:8	80

Lymphocyte Transformation Test

This assay is based on the fact that T lymphocytes causing allergic responses to penicillin have receptors which could specifically recognize penicillin polymer or might cross react with penicillin polymer. Thus, upon exposure to specific antigens, these T lymphocytes would transform (which would include replication of their DNA) and produce lymphokines causing "stand-by" T cells to also transform. The degree of transformation can be monitored by measuring ^{3}H-thymidine utilization. The more transformation occurring, the greater the radioactivity measured. A known stimulant of transformation, phytohemagglutinin A (PHA) is included in these studies as a measure of the ability of the subject's cells to transform.

The results of the LTT are shown in Table III (control subject) and Table IV (penicillin sensitive).

Five of eleven sensitive individuals gave positive LTT's with one or more dilutions of one or more antibiotic polymers. All other polymeric dilutions in all other subjects gave no lymphoblastic transformation. No frequencies of delayed hypersensitivity were calculated as too few subjects were tested in this study.

TABLE III

LYMPHOCYTE TRANSFORMATION TEST

CONTROL SUBJECT

Benzylpenicillin Polymer	Dilution (w/v)	Ave. DPM	Stimulation Index
Penicillin G I	1:10	2330	1.5
I	1:100	1573	1.0
I	1:1000	1698	1.0
Penicillin G II	1:1	1020	0.6
II	1:10	1349	0.8
II	1:100	1762	1.1
II	1:1000	1793	1.1
Penicillin G III	1:1	877	0.5
III	1:10	2258	1.5
III	1:100	2442	1.5
III	1:1000	2255	1.5
PHA	1:2000	127899	76.6
Control	-	1670	-

DISCUSSION

Some suggestions for the structures of the polymers can be made on the basis of the spectral data. Two major degradation products are probably involved in the benzylpenicillin polymers. The presence of penicillenic acid which can form a disulfide bond is suggested by the disappearance of the amide absorption in the IR spectra of the higher molecular weight polymers and the UV absorption at 322 mμ. The penicilloic acid residues as the other major component is suggested by the NMR data and earlier chemical tests (Stewart et al. 1970). That other components are present is shown by the appearance of absorptions in the NMR spectra of the polymers not shown by penicilloic acid. The polymers appear to be heterogeneous but are mostly certainly ordered structures with great water solubility even when the MW is over 5000.

TABLE IV

LYMPHOCYTE TRANSFORMATION TEST
PENICILLIN SENSITIVE SUBJECT

Compound		Dilution (w/v)	Ave. DPM	Stimulation Index
Penicillin G	I	1:10	9266	15.0
	I	1:100	829	1.2
	I	1:1000	1120	1.7
Penicillin G	II	1:1	441	0.6
	II	1:10	736	1.1
	II	1:100	487	0.7
	II	1:1000	555	0.8
Penicillin G	III	1:1	671	1.0
	III	1:10	684	1.1
	III	1:100	409	0.6
	III	1:1000	555	0.8
Monomer		1:10	335	0.5
		1:100	529	0.8
		1:1000	443	0.7
Cephaloridine	I	1:20	675	1.0
	I	1:200	519	0.8
	I	1:2000	308	0.5
Cephaloridine	II	1:20	693	1.0
	II	1:200	480	0.8
	II	1:2000	382	0.6
Cephaloridine	III	1:20	998	1.6
	III	1:200	7993	12.0
	III	1:2000	646	1.0
Monomer		1:20	926	1.5
		1:200	480	0.7
		1:2000	293	0.5
Ampicillin Polymer		1:40	7523	9.1
		1:400	754	1.3
		1:800	958	1.5
Ampicillin Monomer		1:10	613	0.9
		1:100	424	0.7
		1:1000	588	0.9
PHA		1:2000	194251	311.0
Control		–	625	–

The ampicillin polymers all appear to involve the reactive
benzylamine group. Ampicillin polymerizes very readily in solution with almost total loss of the β-lactam structure in the
heavier polymers as confirmed by the IR spectra. In DMSO the
polymeric solutions are so viscous as to make the NMR bands extremely broad. Addition of D_2O sharpened the bands by decreasing
the viscosity but caused some material to precipitate. This tends
to confirm the non-homogeneous nature of the polymers and suggests
that the NMR reflects the lower molecular weight materials in this
pool of chromatographic fractions. The majority of the polymer is
probably formed through amide links between the carboxyl group
formed on opening of the β-lactam ring and the benzylamine. Cyclic
dimers are also possible but there is very little evidence for the
presence of penicillenic acid residues in the ampicillin polymers.

The structures of the cephaloridine polymers are still under
study. Preliminary spectral data suggests major loss of β-lactam
structure in the higher MW polymer.

For use in the biological assays, polymeric materials were
deliberately formed by permitting the antibiotics to stand for 14
days in solution prior to chromatography in the case of benzyl-
pencillin and cephaloridine. Ampicillin was not fractionated
prior to use in the tests. This work has been directed toward
confirming the polymers' abilities to elicit an allergic response
when cells from penicillin sensitive individuals were challenged
with these materials. These studies have shown that histamine
release can be stimulated by the polymeric materials derived from
benzylpenicillin. (Ampicillin and cephaloridine can not be tested
using this assay as they give excessive fluoresence.) This confirms
the idea that polymers formed from the penicillin group of drugs
play a role in immediate IgE mediated allergic reactions in
penicillin sensitive individuals.

The use of the lymphocyte transformation test (LTT) to
better characterize the role of benzylpenicillin, ampicillin, and
cephaloridine polymers in delayed hypersensitivity was suggested
by the ability of ampicillin polymer to induce delayed hypersensitivity reactions in baboons (Munro et al. 1976). Also, Halpern
et al. (1967) and Assem and Vickers (1975) had shown in vitro
blastogenic transformation of lymphocytes from penicillin sensitive subjects mediated by crude penicillin or penicillin-protein
conjugates. The demonstration of lymphocyte transformation
mediated by penicillin polymers in the work reported here confirms
the role of polymers in delayed hypersensitivity reactions to
penicillins.

Earlier work had shown that there was a possibility that a
hypersensitivity reaction could be caused by a form of penicillin

polymer bound to protein (ampicillin polymer-bovine gamma globulin stimulated formation of antibodies in experimental animals Shaltiel et al. 1971). Similar polymer-protein complexes could be formed in the LTT using foreign protein from the AB$^+$ serum or autologous protein derived from cell death and disintegration. While there is no information about the role these complexes play in the stimulation of delayed hypersensitivity, they could be the materials acting as antigen or as elicitor of T-lymphocyte transformation. The role of these polymer-protein complexes in delayed hypersensitivity to penicillin remains to be elucidated.

The nature of these assays do not allow differentiation between cross reactions of penicillin polymers with immunological receptors specific for penicillin polymer-protein conjugates, for penicillin-protein conjugates, or reaction with receptors specific for the polymer. Some of the subjects tested were sensitized to penicillin prior to 1967 after which protein-penicillin conjugates were routinely removed by industry from penicillin preparations. These individuals may have been sensitized by these conjugates or penicillin itself and yet their cells still respond to polymer. This suggests the strong possibility of cross reactivity among various possible antigens.

The ability of cephaloridine polymer to cross react with anti-penicillin antibodies has been previously demonstrated (Dewdney et al. 1971, Munro et al. 1976). The current results suggest that cephaloridine might be cross-reacting with the same T-lymphocyte receptors as the penicillin polymer. However, this remains speculative until confirmed by tests with additional subjects.

The results of these in vitro studies suggest that the allergic responses of some penicillin sensitive individuals may be due to polymers present in these antibiotics. Thus, the importance of polymers in human penicillin allergy has been demonstrated, but much work remains to be done to fully elucidate their structure and the nature of their role in penicillin allergy.

REFERENCES

Assem, E.S.K. and Vickers, M.R., (1975) Investigation of the Response to Some Hapenic Determinants in Penicillin Allergy by Skin and in vitro Allergy Tests. Clin. All. 1 43.

Basoma, A., Villalmonzo, I.G., Campos, A., Pelaez, A., Berglund, A., (1976) IgE Antibodies Against Penicillin as Determined by RAST. J. All. Cl. 57 (3) 214.

Clark, H.T., Johnson, J.R., and Robinson, Sir R., Ed., (1949) "The Chemistry Penicillin", Princeton University Press, Princeton, N. J.

Davis, B.D., Dulbecco, R., Eisen, H.N., Ginsberg, H.S., Wood, W.B., McCarty, M. (1974) Microbiology. Harper and Row; Hagerstown, Md. p. 582.

Freed, J. (1975) Penicillin Allergy. J. Okla. St. Med. Ass. 68 (4) 108.

Halpern, B., Ky, N.T., Amache, N. (1967) Diagnosis of Drug Allergy in vitro with the Lymphocyte Transformation Test. J. Allergy 40 168.

Knudsen, E.T., Dewdney, J.M., Trafford, J.A.P. (1970) Reduction in Incidence of Ampicilin Rash by Purification of Ampicillin. Br. Med. J. 1 469.

Levine, B.B. (1971) Atopy and Mouse Models. Int. Arch. All. 41 88.

May, C.D., Lyman, M., Alberto, R., Cheng, J. (1970) Procedure for Immunochemical Study of Histamine Release from Leukocytes with Small Volumes of Blood. J. Allergy 46 #1 12.

Munro, A.C., Dewdney, J.M., Smith, H., Wheeler, A.C., (1976) Antigenic Properties of Polymers Formed by Beta-Lactam Antibiotics. Int. Archs. All. 50 192.

Oritz-Oritz, L., Zamacona, G., Garmilla, C., Arellano, M.T. (1974) Migration Inhibition Test in Leukocytes from Patients Allergic to Penicillin. Journal of Allergy 113 993.

Ovary, Z. (1958) Immediate Reactions in the Skin of Experimental Animals Provoked by Antigen Antibody Interaction. Prog. All. 5 459. (Karger, Basel 1958)

Redman, A.P., Levine, B.B. (1968) Delayed Skin Reactions to Benzyl-penicillin In Man. Int. Arch. Allerg. 33 193.

Shaltiel, S., Mizrahi, R., Sela, M., (1971) On the Immunological Properties of Penicillin. Proc. Roy. Soc. Lond. B 179 411

Stewart, G.T. (1967) Allergenic Residues in Penicillins. Lancet i 1177.

Stewart, G.T., Butcher, B.T., Wagle, S.S., Stanfield, M.K. (1970) Biopolymerization of Peptide Antibiotics. Liquid Crystals and Ordered Fluids. Plenum Press 33.

Valentine, T., Waithe, B. (1971) <u>In vitro</u> Methods in Cell Mediated
    Immunity. Bloom and Glade (Eds.) Acad. Press, N.Y., N.Y.

Wilson, L. Personal Communication.

MAGNETIC RELAXATION OF SMALL MOLECULES IN ORDERED FLUIDS

AND LIQUID CRYSTALS

B. M. Fung

Department of Chemistry
University of Oklahoma
Norman, Oklahoma 73019

John H. Johnson and E. A. Grula

School of Biological Sciences
Oklahoma State University
Stillwater, Oklahoma 74074

Abstract. The anisotropic motion and magnetic relaxation of small
molecules in ordered fluids and liquid crystals are discussed.
The molecules concerned are those with non-zero orientation factor,
considerable local rotational diffusion and rapid internal rotation.
Fourier intensities and relaxation times are calculated and
expressed in terms of the orientation factor, the rotational
correlation times, and the angle between the major axis and the
vector of a dipolar pair or a quadrupolar axis. Special cases
of molecules with and without internal rotation are considered.
Examples of deuteron relaxation of $CDCl_3$ in a thermotropic liquid
crystal and in lipid vesicles, and C_6D_6 in lipid vesicles and
biological membranes are discussed.

INTRODUCTION

The study of magnetic relaxation offers valuable information
on molecular motion. The theory of magnetic relaxation in thermotropic liquid crystals has been fairly well developed.[1-6] In
particular, Doane and co-workers[5] and Freed[6] have recently
presented comprehensive theories on magnetic relaxation in liquid
crystals.

In addition to studying large molecules that form the
liquid crystalline ordering matrix, the nuclear relaxation of

small molecules in ordered fluids and liquid crystals has also
attracted the attention of many investigators.[7-13] Studies of the
latter type are important for solvent molecules in lyotropic liquid
crystals, solutions of small molecules in thermotropic liquid
crystals, and small organic molecules dissolved in lipid vesicles
and cell membranes. In order to interpret the results, theories
of both Doane[5] and Freed[6] can be used. Although Freed's treatment
is more rigorous, the result depends upon the choice of a reorienting potential. Since we are going to use the approximation
of neglecting long-range fluctuations, application of the theory
of Doane and co-workers[5] would be more straightforward. We will
develop equations applicable to the magnetic relaxation of mobile
small molecules with non-zero orientation factors. Effects of
internal rotation[14] will be included in the treatment. Since
small solvent molecules in lyotropic liquid crystals may be
rapidly exchanging between different sites,[9,15,16] the following
analysis would not be directly applicable to these systems.

THEORY

For a quadrupolar nucleus of spin $= 1$ and for a pair of
dipolar nuclei with the same spin I, the spin-lattice relaxation
time (T_1), spin-spin relaxation time (T_2) and the spin-lattice
relaxation time in the rotating framework ($T_{1\rho}$) are related to
the Fourier intensities ($J(\omega)$) by:[17,18]

$$\frac{1}{T_1} = K[J_1(\omega_o) + J_2(2\omega_o)] \tag{1a}$$

$$\frac{1}{T_2} = K[\frac{1}{4}J_0(0) + \frac{5}{2}J_1(\omega_o) + \frac{1}{4}J_2(2\omega_o)] \tag{1b}$$

$$\frac{1}{T_{1\rho}} = K[\frac{3}{8}J_2(2\omega_1) + \frac{5}{2}J_1(\omega_o) + \frac{1}{4}J_2(2\omega_o)]. \tag{1c}$$

where $K = \frac{3}{2}I(I+1)\gamma^4\hbar^2 r^{-6}$ $\qquad\qquad$ (2a)

for dipolar interaction and

$$K = \frac{9\pi^2}{8}(1 + \eta^2/3)\ (e^2qQ/h)^2 \tag{2b}$$

for quadrupolar interaction. In (1), ω_o is the Larmor frequency
and ω_1 is the angular frequency of the rotating field. In (2),
γ is the gyromagnetic ratio, r is the distance between the two
nuclei, η is the asymmetry parameter, and e^2qQ/h is the quadrupole
coupling constant. The Fourier intensities are given by[17]

$$J_i(\omega) = \int_{-\infty}^{\infty} <F_i^*(t+\tau)F_i(t)>\exp(-i\omega\tau)d\tau, \tag{3}$$

where $F_i(\tau)$'s are the time-dependent spherical harmonic functions of order two which describe the major axis of the molecule in the laboratory frame. If the major axis forms an angle θ with the z axis (the direction of the magnetic field) at time t, and an angle Δ with the quadrupolar axis or dipolar vector, and the latter undergoes a stoichastic reorientation about the major axis with a correlation time τ', it can be shown that[5,14]

$$<F_i^*(t+\tau)F_i(t)> = \sum_{j=0}^{2} <f_{ij}(\theta_o)><F_i'^*(t+\tau)F_i'(t)>, \qquad (4)$$

where

$$<f> = \frac{1}{8}\begin{pmatrix} <8-24\sin^2\theta+18\sin^4\theta> & 144<\sin^2\theta-\sin^4\theta> & 9<\sin^4\theta> \\ 2<\sin^2\theta-\sin^4\theta> & <8-20\sin^2\theta+16\sin^4\theta> & <2\sin^2\theta-\sin^4\theta> \\ 2<\sin^4\theta> & 16<2\sin^2\theta-\sin^4\theta> & <8-8\sin^2\theta+\sin^4\theta> \end{pmatrix}$$

$$\approx \frac{1}{5}\begin{pmatrix} 1+4S & 12(1-S) & 3(1-S) \\ (1-S)/6 & 3S+2 & (1-S)/2 \\ 2(1-S)/3 & 8(1-S) & 2+3S \end{pmatrix} \qquad (5)$$

$$<F_0'^*(t+\tau)\cdot F_0'(t)> = (3\cos^2\Delta-1), \qquad (6a)$$

$$<F_1'^*(t+\tau)\cdot F_1'(t)> = \frac{1}{4}\sin^2(2\Delta)\exp(-|\tau'|/\tau_c), \qquad (6b)$$

$$<F_2'^*(t+\tau)\cdot F_2'(t)> = \sin^4\Delta\cdot\exp(-|4\tau'|/\tau_c), \qquad (6c)$$

and $S = <3\cos^2\theta-1>/2$, $\qquad (7)$

S being the orientation factor of the major axis of the small molecule. In equations (4), (5), and (6), two approximations have been used. The first is that local rotational diffusional motion of the small molecule is large, and long range or collective fluctuations[1,2,5] can be neglected. The second approximation is to set $<\cos^4\theta>\approx(1+4S)/5$.[5]

Simplifying equations (3)-(7), the Fourier intensities can now be expressed in terms of the orientation factor, the correlation times and ω:

$$J_0(0) = \frac{4}{5}[A(1+4S)(2\tau_1)+B(1-S)(2\tau_2)+C(1-S)(2\tau_3)], \qquad (8a)$$

$$J_1(\omega) = \frac{2}{15}[A(1-S)\frac{2\tau_1}{1+\omega^2\tau_1^2}+B(2+3S)\frac{\tau_2}{1+\omega^2\tau_2^2}$$
$$+C(1-S)\frac{2\tau_3}{1+\omega^2\tau_3^2}], \qquad (8b)$$

and $\quad J_2(2\omega) = \dfrac{8}{15}[A(1-S)\dfrac{2\tau_1}{1+4\omega^2\tau_1^2} +B(1-S)\dfrac{2\tau_2}{1+4\omega^2\tau_2^2}$

$$+C(2+3S)\dfrac{\tau_3}{1+4\omega^2\tau_3^2}], \qquad\qquad (8c)$$

where $\tau_1 = \tau$ = rotational correlation time of the major axis,

τ' = internal rotational correlation time about the major axis,

$$\tau_2 = (1/\tau_1 + 1/\tau')^{-1}, \qquad\qquad (9a)$$

$$\tau_3 = (1/\tau_1 + 4/\tau')^{-1}, \qquad\qquad (9b)$$

$$A = (3 \cos^2\Delta-1)^2/4, \qquad\qquad (10a)$$

$$B = 3 \sin^2(2\Delta)/4, \qquad\qquad (10b)$$

and $\quad C = 3 \sin^4\Delta/4. \qquad\qquad (10c)$

Thus, the relaxation times of an oriented small molecule can be readily calculated by substituting equation (8) into (1). In an isotropic liquid, S=0 and equation (8) is reduced to equation (21) in reference 14.

Strictly speaking, equation (8) is applicable only when the director of the ordering matrix is parallel to the magnetic field. However, we have found that deuteron T_1's of small molecules in a lyotropic phase and in a smectic A phase did not depend upon the angle between the ordering matrix and the magnetic field (Table I). The reason for this is not clear to us. Because of this experimental finding, we will not include the angular dependence of T_1 and T_2 in the following discussion.

THE ABSENCE OF INTERNAL ROTATION AND EXAMPLES

A special case of the present problem is a molecule in which the quadrupolar axis or the dipolar vector coincides with the major axis and internal rotation does not affect the relaxation, for example $CDCl_3$ and $HC\equiv CH$. In this case, $\Delta=0$ and $\tau'=\infty$. The Fourier intensities become

$$J_0(0) = \dfrac{8}{5}(1+4S)\tau, \qquad\qquad (11a)$$

$$J_1(\omega) = \dfrac{4}{15}(1-S)\dfrac{\tau}{1+\omega^2\tau^2} , \qquad\qquad (11b)$$

Table I. Deuteron spin-lattice relaxation time at 9.21 MHz (in
          sec) as a function of the orientation of the sample with
          respect to the magnetic field

	0°	30°	60°	90°	120°	150°	180°	random[a]
$CDCl_3$-PBLG[b]	0.400	0.398	0.400	0.410	0.414	0.415	0.430	0.426
C_6D_6 in BBAA[c]	0.30	0.30	0.29	0.29	0.28	0.33	0.28	0.28

a. The samples had no macroscopic ordering.

b. PBLG = poly-γ-benzyl-L-glutamate (MW 200,000). The data were
    obtained for a sample with a "mole ratio" of 0.215 (monomeric
    BLG to $CDCl_3$) at 9.21 MHz and 25°C. The viscosity of the
    sample was so large that the re-orientation of the molecules
    after the rotation of the sample (ref. 15) was very slow.
    The sample was first aligned in the magnetic field overnight,
    then it was rotated to a proper angle to determine T_1. After
    each experiment, the sample was rotated back to 0° for at
    least 10 min. to prevent the re-orientation of the PBLG mole-
    cules. The experimental uncertainty of T_1 was about 5%.

c. BBAA = 4-n-butyloxybenzilidine-4'-acetoaniline. The data
    were obtained for a sample containing ca. 6 mole % C_6D_6
    supercooled to 65°C in the magnetic field at 9.21 MHz. The
    modulation of the free induction decay by quadrupolar splitting
    obeyed the $(3 \cos^2\theta - 1)$ relation, indicating that the sample
    was in its smectic A phase and aligned. The intensity of a
    peak at 0.4-0.5 ms after the 90° pulse (11 μs) was measured
    for the T_1 determination. The experimental uncertainty of
    T_1 was about 10%.

and $J_2(2\omega) = \dfrac{16}{15}(1-S) \dfrac{\tau}{1+4\omega^2\tau^2}.$ (11c)

In non-spinning liquid crystal solutions, the proton and
deuteron spectra of solute molecules show multiple transitions.
Simple relaxation times such as those expressed in equation (1)
cannot be directly applied, and relaxation matrix elements have
to be used to define parameters obtained in selective relaxation
experiments.[12] For example, a quadrupolar nucleus with spin I=1
the following quantities are defined:[12]

$$R_1 = K[J_1(\omega_o) + J_2(2\omega_o)],$$ (12a)

$$R_1' = 3K J_1(\omega_o),$$ (12b)

$$R_2 = K[\frac{1}{4}J_0(0) + \frac{3}{2}J_1(\omega_0) + \frac{1}{4}J_2(2\omega_0)], \tag{12c}$$

and $\quad R_2' = K J_1(\omega_0). \tag{12d}$

In the case of $\omega_0\tau<<1$, substitution of equations (11) into (12) yields

$$R_1 : R_1' : R_2 : R_2' = 10(1-S) : 6(1-S) : (8+7S) : 2(1-S). \tag{13}$$

For a solution of 4.5 mole% of $CDCl_3$ in N-(p-methoxybenzyl-idene)-p-butylaniline (MBBA) at 13°C, S was found to be 0.117 ± 0.001.[12] The calculated ratios are $R_1 : R_1' : R_2 : R_2' = 1.00 :$ 0.60 : 1.00 : 0.20. The experimental ratios are $R_1 : R_1' : R_2 =$ 1.00 : 0.57 : 0.83 (R_2' could not be determined).[12] Considering the approximations used in obtaining equation (13) and the fact that the experimental uncertainties were 10% for R_1, 5% for R_1' and 7% for R_2,[12] the two sets of data are in reasonable agreement. It is interesting to note that equation (13) yields $R_1 \geq R_2'$ for $S \geq 2/17$, and $R_1<R_2'$ for $S>2/17$.

If the proton or deuteron signal shows no splitting, T_1 and T_2 can be defined by equation (1). Then,

$$\frac{1}{T_1} = R_1, \tag{14a}$$

and $\quad \dfrac{1}{T_2} = R_2 + R_2'. \tag{14b}$

For $\omega_0\tau<<1$,

$$\frac{T_2}{T_1} = \frac{2(1-S)}{2+S}, \tag{15}$$

which indicates that $T_2 \geq T_1$ for positive values of S if the line does not split. Equation (15) is most useful for obtaining the orientation factor S for small organic molecules in micellar systems, in which there is no dipolar or quadrupolar splitting in the NMR spectrum. For example, for a dispersion of 3.5% $CDCl_3$ and 5% (by weight) dipalmitoyl lecithin in water at 26°C, the deutero-chloroform molecules dissolve in the lipid bilayers in the vesicles, which undergo random motion. We found that T_1 = 0.215 ± 0.010 sec and T_2 = 0.042 ± 0.002 sec for deuteron in $CDCl_3$. According to equation (15), S = 0.73, a value that cannot be obtained from con-tinuous wave measurements. Using the value of e^2qQ/h = 168 kHz,[19] one finds τ = 1.25 x 10^{-11} sec. This satisfies the condition $\omega_0\tau<<1$. For $CDCl_3$ in hexadecane, an isotropic medium with a hydrocarbon chain length comparable to that of the lipid, T_1 =

$T_2 = 0.81 \pm 0.02$ sec and $\tau = 4.64 \times 10^{-12}$ sec at 26°C.

THE PRESENCE OF INTERNAL ROTATION AND EXAMPLES

A good example of a molecule with internal rotation is deutero-benzene. Here, $\Delta = \pi/2$. Under the condition of $\omega_o \tau \ll 1$, the substitution of (8) into (1) yields

$$\frac{1}{T_1} = \frac{K}{3}(1-S)\tau_1 + \frac{K}{4}(1+S)\tau', \qquad (16a)$$

and

$$\frac{1}{T_2} = \frac{K}{6}(2+S)\tau_1 + \frac{K}{8}(2-S)\tau'. \qquad (16b)$$

The internal rotation of benzene about the six-fold axis is fast and rather independent of the viscosity of the solvent,[21] and the correlation time of the reorientation $\tau' = 6.4 \times 10^{-13}$ sec. Deuteron relaxation of C_6D_6 in 4,4'-n-hexyloxyazoxybenzene has been studied by the non-selective pulse technique.[7] Substituting the experimental values of $e^2qQ/h = 187$ kHz,[20] $S = 0.082$ and $T_1 = 2.8 \times 10^{-3}$ sec into (16), it was found that $\tau_1 = 3.0 \times 10^{-9}$ sec and $T_2 = 2.5 \times 10^{-3}$ sec. A value of $\tau_c = 0.50 \times 10^{-9}$ sec was obtained in ref. 7 by treating benzene as a spherical molecule, which is clearly incorrect. A value of $T_2^* = 0.75 \times 10^{-3}$ sec was obtained from linewidth measurement.[7] This is probably much smaller than the true value of T_2 because the effects of field inhomogeneity and dipolar splitting would cause appreciable line broadening.

Equation (16) can be used to calculate the values of S and τ_1 for C_6D_6 partitioned in lipid vesicles, intact bacterial cells (predominantly in the cell membranes) and isolated cell membrane preparations. The results of experimental T_1 and T_2 and calculated values of S and τ_1 are listed in Table II.

Although the values of T_1, T_2, S and τ_1 are slightly dependent upon the benzene/lipid mole ratio,[22] it can be observed from Table II that the motional freedom of benzene displays the trend: intact bacterial cells > cell membranes > lipid vesicles. The orientation of benzene with respect to the lipid chains has the trend: intact bacterial cells $\sim$ cell membranes > lipid vesicles. These data suggest that the hydrocarbon regions of bacterial cells and isolated cell membranes, due to the presence of a variety of hydrocarbon chain lengths, isomeric forms, and chain configurations, are probably less compact than those in dipalmitoyl lecithin vesicles. The use of deuterobenzene to study membrane structure may be complementary to spin probes. The concentration needed for the measurement of deuteron NMR is considerably larger than that for ESR. However, because of the small size and flat structure of the benzene molecule, the perturbation to the local environment may be much less

Table II. Deuteron relaxation times (9.21 MHz and 25°C), calculated
           orientation factor and rotational correlation time of
           the C_6 axis for C_6D_6

	Hexadecane	Dipalmitoyl lecithin vesicles[a]	Bacteria[b]	Cell membranes[b]
T_1, sec	0.72±0.02	0.25±0.01	0.69±0.04	0.17±0.02
T_2, sec	0.73±0.02	0.0024±0.0002	0.073±0.005	0.016±0.002
S	0	0.99	0.85	0.87
τ_1, sec	1.03×10^{-11}	2.2×10^{-9}	7.5×10^{-11}	3.5×10^{-10}

a. Lecithin/C_6D_6 mole ratio = 0.26.

b. Micrococcus lysodeikticus

than that of bulky spin probes.

A puzzling point in the results obtained from the study of
$CDCl_3$ and C_6D_6 in vesicles is the rather large values of S derived.
Intuitively one would expect the benzene ring to orient more or
less parallel to the chains, with $|S| \le 0.5$. In the calculation,
the effect of collective fluctuations was neglected and $\omega_0 \tau_1 << 1$
was assumed. It is possible that these approximations may cause
errors in the analysis, and further study must be made to assess
these effects.

Acknowledgement. BMF is a recipient of a Research Career
Development Award from the National Institutes of Health.

REFERENCES

1. P. Pincus, Solid State Commun. 7, 415(1969).

2. J. W. Doane and D. L. Johnson, Chem. Phys. Lett. 6, 291(1970).

3. W. Wölfel, F. Noack and M. Stohrer, Z. Naturforsch. 30a,
    437(1975).

4. R. Blinc, N. Luzar and M. Burger, J. Chem. Phys. 63, 3445(1975).

5. P. Ukleja, J. Pirs and J. W. Doane, Phys. Rev. A14, 414(1976).

6. J. H. Freed, J. Chem. Phys. 66, 4183(1977).

7. Y. Egozy, A. Loewenstein and B. L. Silver, Mol. Phys. 19,

177(1970).

8. B. M. Fung and T. H. Martin, J. Chem. Phys. $\underline{61}$, 1698(1974).

9. B. M. Fung and T. H. Martin, Liquid Crystals and Ordered Fluids,
    Vol. 2 (Eds. J. F. Johnson and R. S. Porter), Plenum Press,
    New York, 1974, p. 267.

10. W. A. Hines and E. T. Samulski, ibid., p. 257.

11. J. Courieu, C. L. Mayne and D. M. Grant, J. Chem. Phys. $\underline{66}$,
    2669(1977).

12. R. R. Vold and R. L. Vold, J. Chem. Phys. $\underline{66}$, 4018(1977).

13. J. H. Johnson, E. A. Grula, K. D. Berlin and B. M. Fung,
    "Proceedings of Symposium on Cellular Function and Molecular
    Biology," in press.

14. D. E. Woessner, J. Chem. Phys. $\underline{36}$, 1(1962).

15. R. D. Orwoll and R. L. Vold, J. Amer. Chem. Soc. $\underline{93}$, 5335(1971).

16. J. P. Jacobsen and K. Schaumburg, J. Mag. Res. $\underline{24}$, 173(1976).

17. A. Abragam, The Principles of Nuclear Magnetism, Chapter 9,
    Oxford University Press, 1961.

18. G. P. Jones, Phys. Rev. $\underline{148}$, 332(1966).

19. J. L. Ragle and K. L. Sherk, J. Chem. Phys. $\underline{50}$, 3553(1969).

20. P. Pyykkö and U. Lähteenmäki, Ann Univ. Turkuensis AI, No.
    93 (1966).

21. D. R. Bauer, G. R. Alms, J. I. Brauman, and R. Pecora, J.
    Chem. Phys. $\underline{61}$, 2255(1974).

22. B. M. Fung and T. W. McGaughy, to be published.

ABSORBANCE AND PITCH RELATIONSHIPS IN DICHROIC

GUEST-HOST LIQUID CRYSTAL SYSTEMS

H.S. Cole, Jr. and S. Aftergut

General Electric Corporate Research and Development
Schenectady, New York   12301

ABSTRACT

The absorbance of a dichroic dye dissolved in chiral nematic hosts in the Grandjean texture increased with decreasing pitch of the host. Absorbance was higher for hosts of low birefringence and high order parameter. Boundary condition, homeotropic or parallel (homogeneous), had no appreciable effect on absorbance, but threshold voltage was higher for the parallel than homeotropic boundary. The results agreed with theoretical plots relating absorbance to pitch.

INTRODUCTION

Dichroic (or pleochroic) dyes are constituted of anisotropic molecules whose absorption is a function of their orientation with respect to incident light. In the dye studied in the present work, the vector for the n-π^* transition resulting from the absorption of light in the visible region of the spectrum lies parallel to the long molecular axis. Maximum and minimum absorption is obtained when the dye molecules are oriented so that their long axes are, respectively, perpendicular and parallel to the direction of incident light. The orientation can be brought about by dissolving the dye as a guest in a liquid crystal host where the dye molecules tend to align with their long axes parallel to those of the liquid crystal molecules. Since the orientation of liquid crystals can be switched by electric fields, such guest-host systems have been proposed as a means of light-to-dark transitions for information display applications.[1]

117

White and Taylor[2] have pointed out that a short-pitch cholesteric liquid crystal in the Grandjean texture is a particularly
useful host for enhancing absorption of unpolarized light. This
is due to the fact that the light propagates in two elliptically
polarized modes which are both absorbed by the dye. In contrast,
only one of the polarized modes is absorbed in a long-pitch host.
The parameters of the host controlling absorption are the pitch
(period) and birefringence. Other factors include the order parameter of the dye and the usual parameters of dye concentration,
absorption coefficient and layer thickness governed by Beer's and
Lambert's laws.

The present study reports on the effect of pitch, birefringence and boundary conditions of liquid crystal hosts on the absorption of guest-host systems, and compares the results with
theoretical expressions derived by Saupe.[3]

EXPERIMENTAL

Absorption spectra of the guest-host materials were obtained
in a Cary 14 spectrophotometer. The data were corrected for absorption by the cell and liquid crystal host and reflection losses
by subtracting the absorbance of a control cell devoid of dye from
the absorbance of the cell containing the guest-host material.

The dichroic dye, 4,4'-bis(4-N,N-dimethylaminonaphthylazo)-
azobenzene*, was used at a concentration of 1% in a host comprised
of a mixture of an optically active liquid crystal (CB-15) and
one of the following nematic liquid crystals: E-7, ROTN-101, and
Nematic Phase 1132 TNC Licristal®. CB-15 and E-7 were obtained
from BDH Ltd. and are believed to be, respectively, p-act-amyl-
p'-cyanobiphenyl and a mixture of p-pentyl-, p-heptyl-, and p-
octyloxy-p'-cyanobiphenyl and p-cyano-p"-terphenyl. ROTN-101,
obtained from Hoffman LaRoche, is a mixture of esters. Licristal
1132, obtained through the courtesy of E. Merck, is a mixture of
three phenylcyclohexanes and one biphenylcyclohexane. All the
hosts had a positive dielectric anisotropy.

The pitch of the host was adjusted by varying the concentration of CB-15, and the pitch was measured with the aid of a wedge
cell as described by Kassubek and Meier.[4]   The pitch was measured at a single concentration of CB-15 in each nematic host,
and the pitch of the other compositions was calculated on the assumption that the pitch varies inversely with the weight fraction
of CB-15. A concentration of 5% CB-15 gave a pitch of 3.2 µm in
E-7 and 2.2 µm in ROTN-101. A concentration of 4% CB-15 gave a
pitch of 3.3 µm in Licristal 1132.

---

*Patent applied for

Principal refractive indices $n_{\shortparallel}$ and $n_\perp$ were as follows: For E-7, 1.774 and 1.528 at 550 nm and $25^\circ C$, interpolated from data obtained by H.A. Tarry.[5] For ROTN-101, 1.671 and 1.509 at 500 nm and $25^\circ C$. For Licristal 1132, the only available value was $\Delta n = 0.14$ at 589 nm and $20^\circ C$, as reported by the manufacturer.

The cells were constructed from glass coated with indium tin oxide and sealed with epoxy. Cell spacing was $12.5 \pm 0.5 \ \mu m$. In cells with homeotropic boundary, the boundary was established by treating the cell surfaces with silane coupling agent (Dow XZ-2-2300). Parallel (homogeneous) alignment was provided either by evaporation of SiO at an incidence angle of 60° or by unidirectional rubbing of a thin polymer layer coated on the substrate.

RESULTS

To determine the effect of pitch on absorbance, a constant amount (1%) of dichroic dye was dissolved in liquid crystal hosts whose pitch was controlled by varying the concentration of the optically active biphenyl derivative CB-15. Three hosts of different birefringence $\Delta n = n_{\shortparallel}-n_\perp$ were employed: a) a mixture of cyanobiphenyls, E-7, $\Delta n = 0.246$, b) a mixture of cyanophenyl benzoate esters, ROTN-101, $\Delta n = 0.16$ and c) a mixture of cyanophenylcyclohexanes, Licristal 1132, $\Delta n = 0.14$. Absorbance was measured in 12.5 μm thick cells whose inner surfaces had been treated to promote parallel (homogeneous) and homeotropic alignment. In the quiescent state prior to application of an electric field, samples with both types of boundaries appeared to be in the Grandjean state, and absorbance in this state was always measured before a field was applied.

Figure 1 shows the absorbance of the dye dissolved in a host prepared from E-7 and CB-15 for the case of homeotropic boundary. Curve A is the spectrum of the quiescent state at $25^\circ C$, curve B is that of the activated state, at $25^\circ C$, produced by application of about three times the threshold voltage, and curve C is that of the field-free isotropic state at $72^\circ C$.

Figure 2 shows the transmission at the absorption peak of 542 nm of the dye in E-7 and CB-15 as a function of pitch for parallel and homeotropic boundaries. The transmission in the activated state is approximately constant at 50%. The slight spread around 50% may be attributed to a combination of factors such as small variations in liquid crystal thickness, dye concentration and order parameter. The transmission in the field-free state increases markedly with an increase in pitch.

Figure 3 shows the transmission at the absorption peak of 535 nm of the dye in the host ROTN-101 plus CB-15. As in the other

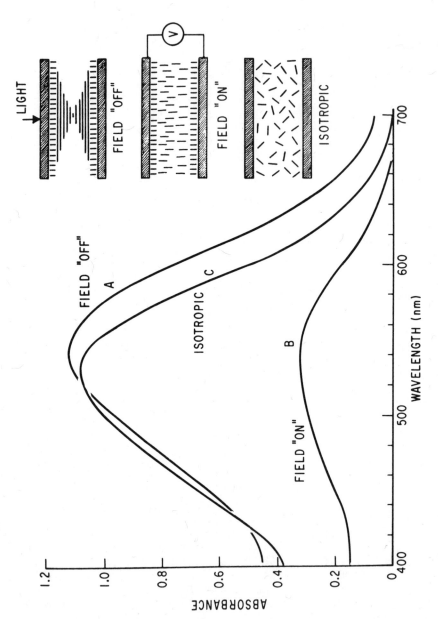

Figure 1.   Absorption spectra with unpolarized light of dichroic dye in host E-7 plus CB-15 of 3.2 μm pitch.   Curve A, V = 0, 25°C; curve B, V = 20V, 25°C; curve C, V = 0, 72°C.

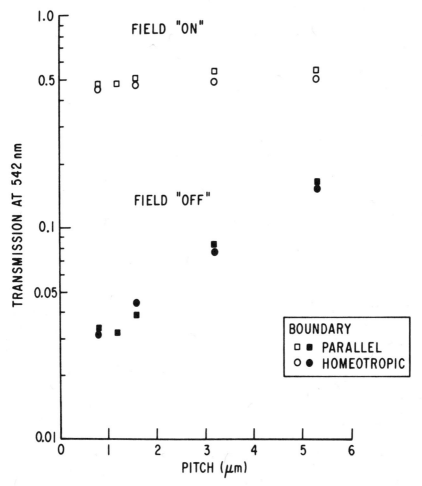

Figure 2.   Transmission of dichroic dye in host E-7 plus CB-15
            (Δn = 0.24) as function of pitch and boundary condition
            in the presence and absence of electric field.

host, the transmission in the activated state is approximately
constant at 50% and that of the Grandjean state increases with
the pitch.

The transmission in the Licristal plus CB-15 host was measured at only one pitch, 3.3 μm, at the absorption peak of 540 nm.
With homeotropic boundary, the transmission was 0.57 in the activated state and 0.066 in the quiescent state.

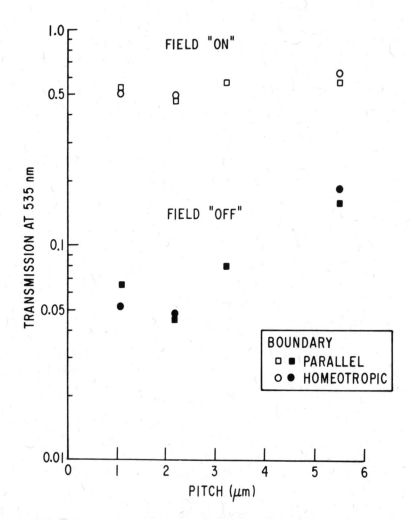

Figure 3.   Transmission of dichroic dye in host ROTN-101 plus CB-
            15 (Δn = 0.16) as function of pitch and boundary con-
            dition in the presence and absence of electric field.

    The order parameter S was obtained from the dichroic ratio
$R = D_\parallel/D_\perp$ where $D_\parallel$ and $D_\perp$ are the polarized absorbances parallel
and perpendicular to the nematic director. The dichroic ratio
was determined by two experimental methods. In the first method,
the dye was dissolved in the nonchiral hosts (without CB-15), and
the absorption was measured in cells with parallel boundary using

polarizers oriented parallel and perpendicular to the nematic director. The results shown in Figure 4 indicate that the order
parameter is 0.75 in E-7 and 0.71 in ROTN-101 as computed at the
respective absorption peaks of 542 nm and 535 nm. The order parameter in Licristal 1132 was 0.75 at the peak of 540 nm.

A second method was employed to determine the dichroic ratio
and order parameter for hosts containing CB-15 for the following
reasons. First, the first method is inapplicable because it is
not possible to achieve the required unidirectional alignment.
Second, order parameter is a function of reduced temperature.
Addition of substantial amounts of CB-15 lowers the clearing point
and raises the reduced temperature at the ambient temperature at
which absorbance was studied. Consequently, the order parameter
tends to decrease with decreasing pitch. The second method, which
did not employ polarized light, utilized the activated and iso-
tropic absorbance and assumed that $D_{iso} = (D_\| + 2D_\perp)/3$. $D_\perp$ is
equivalent to the absorbance of the activated homeotropic state
and $D_\|$ is computed from D_{iso}. Since the experiments yielded D_{iso}

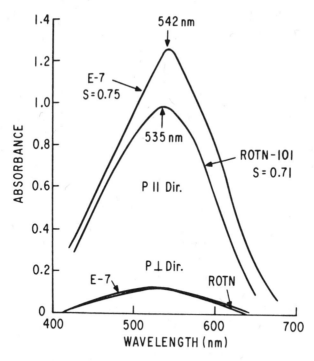

Figure 4.   Polarized absorption spectra of dichroic dye in hosts
            E-7 and ROTN-101 measured with polarizer parallel and
            perpendicular to nematic director.

at elevated temperatures above the clearing point, D_{iso} had to
be normalized to the same ambient temperature at which $D_\perp$ was measured. The temperature correction factor was derived from a measurement of the temperature dependence of the absorbance of a solution of the dye in host CB-15 which is isotropic above 0°C and
has a structure similar to the components of E-7. In the temperature range of interest, the absorbance at 542 nm decreased at an
average rate of 0.215% per $^\circ$C with increasing temperature. The
same correction factor was used for ROTN-101.

The threshold voltage as a function of liquid crystal thickness and boundary condition is shown in Figure 5 for a mixture
of 5% CB-15 in E-7 containing 1% of dye. The threshold is higher
for the parallel than homeotropic boundary.

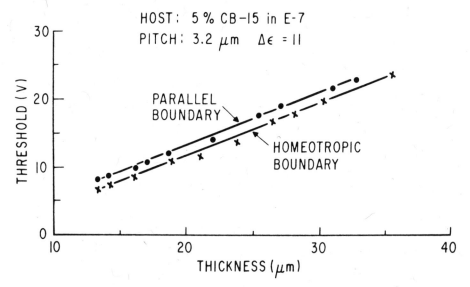

Figure 5.   Threshold voltage as function of thickness and boundary
            condition for host E-7 plus CB-15 (pitch = 3.2 μm) con-
            taining dichroic dye.

DISCUSSION OF RESULTS

Ordinary (unpolarized) light entering a liquid crystal layer
is propagated in polarized modes. For a discussion of the propagation of these modes in cholesteric liquid crystals of the Grandjean texture, it is convenient to consider three cases differing
in the relative magnitude of pitch p with respect to the wavelength
of light λ.

Case 1: $p \gg \lambda$. This case has been treated by Mauguin[6] in connectin with nematic layers[7] subjected to a mechanical twist and by others (e.g. Berreman)[7] for $\pi/2$ twist cells. The polarized modes follow the twist, and the nematic simply behaves as a waveguide. Consequently, dichroic dyes dissolved in a long-pitch host absorb only a single mode; absorption of unpolarized light is a maximum of 50% and contrast ratio is limited to 2:1.

Case 2: $p \simeq \lambda/n$. This case, previously analyzed by Ferga-son[8] among others, gives rise to the well-known iridescent colors due to Bragg reflection.

Case 3: $p > \lambda/n$. In this case, treated earlier by White and Taylor[2] and pertinent to the present work, both normal modes have a component parallel to the nematic director and the major axis of the dichroic dye molecule. Therefore, absorption of un-polarized light can be greater than 50%. The absorption depends on the pitch of the cholesteric which, along with the principal refractive indices $n_{\shortparallel}$ and $n_{\perp}$, determines the eccentricity of the polarized modes. The other important factor is the order parameter $S = \frac{1}{2}(3 \cos^2 \theta - 1)$ where θ is the average angle between the absorption axis (transition moment vector) of the dichroic dye and the nematic director.

The transmission of unpolarized light incident normal to a Grandjean layer (parallel to the cholesteric helix) is given by

$$T_{off} = \tfrac{1}{2}(e^{-\alpha_1 d} + e^{-\alpha_2 d}) \tag{1}$$

where $\alpha_{1,2}$ is the attenuation constant for the two normal modes and d is the thickness.

According to the treatment by Saupe[3], for weakly absorbed radiation outside the reflection band

$$\alpha_{1,2} = \frac{\bar{\alpha}}{\mu_{\pm}\lambda'} \left\{ 1 \pm \frac{\gamma\delta + 2\lambda'^2}{(\delta^2 + 4\lambda'^2)^{\frac{1}{2}}} \right\} \tag{2}$$

where the symbols in (2) are defined as follows:

$$\bar{\alpha} = \frac{1}{2(\bar{\epsilon})^{\frac{1}{2}}} (n_{\shortparallel}\alpha_{\shortparallel} + n_{\perp}\alpha_{\perp}) \tag{3}$$

$$\bar{\epsilon} = \frac{n_{\shortparallel}^2 + n_{\perp}^2}{2} \tag{4}$$

$$\alpha_{\shortparallel} = (2S + 1)\alpha_o \tag{5}$$

$$\alpha_{\perp} = (1-S)\alpha_o \tag{6}$$

$$\gamma = \frac{n_{\shortparallel}\alpha_{\shortparallel} - n_{\perp}\alpha_{\perp}}{n_{\shortparallel}\alpha_{\shortparallel} + n_{\perp}\alpha_{\perp}} \tag{7}$$

$$\delta = \frac{n_{\shortparallel}^2 - n_{\perp}^2}{n_{\shortparallel}^2 + n_{\perp}^2} \tag{8}$$

$$\lambda' = \frac{\lambda}{p(\bar{\epsilon})^{\frac{1}{2}}} \tag{9}$$

$$\mu_{\pm}\lambda' = \left\{ 1 + \lambda'^2 \pm (\delta^2 + 4\lambda'^2)^{\frac{1}{2}} \right\}^{\frac{1}{2}} \tag{10}$$

α_o is the absorption coefficient of the dye in the isotropic sol-
vent, $\alpha_{\shortparallel}$ and $\alpha_{\perp}$ are the principal absorption coefficients, and
λ and λ' are the wavelength of incident light and reduced wavelength.

The transmission of the activated state is given by

$$T_{on} = e^{-\alpha_{\perp}d} = e^{-(1-S)\alpha_o d} \tag{11}$$

The effect of S on T_{off} as a function of pitch is shown in
Figure 6. The calculation was made for T_{on} = 0.5, λ = 550 nm,
$n_{\shortparallel}$ = 1.77 and $n_{\perp}$ = 1.53. The transmission increases with increasing pitch and tends toward a constant value in the long-pitch re-
gime. The role of birefringence for constant S of 0.75 at λ = 550
nm and T_{on} = 0.5 is shown in Figure 7 where Δn = 0.14, 0.20 and
0.24 corresponding to $n_{\shortparallel}$ = 1.69, 1.75 and 1.77, respectively.
An increase in birefringence results in increased transmission
of the Grandjean texture.

Experimental data were checked against theory by computing
T_{off} in equation (1) for each experimental value of T_{on}. The
product $\alpha_o d$ was first evaluated for each T_{on} value using equation
(11), and this product was employed in the calculation of T_{off}.
A comparison of experimental and theoretical T_{off} data is shown
in Table 1 for the case of parallel boundary. The order parameters

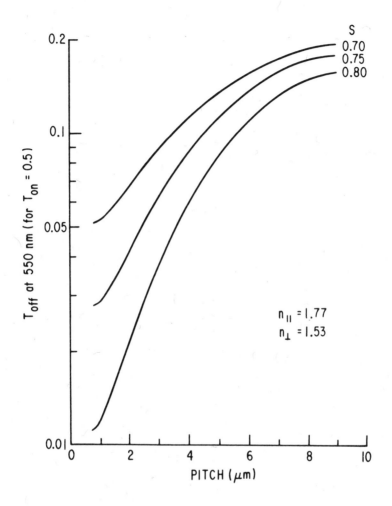

Figure 6.   Semilog plots of transmission T_{off} in the Grandjean
texture at 550 nm vs. pitch as function of order param-
eter for field-on transmission of 50% and $n_{\shortparallel}$, $n_{\perp}$ of
1.77, 1.53, computed from equation (1).

shown in the Table were determined from D_{iso} with the following
exceptions: for the compositions of longest pitch, the order parameters measured in the nonchiral hosts were assumed, and for ROTN

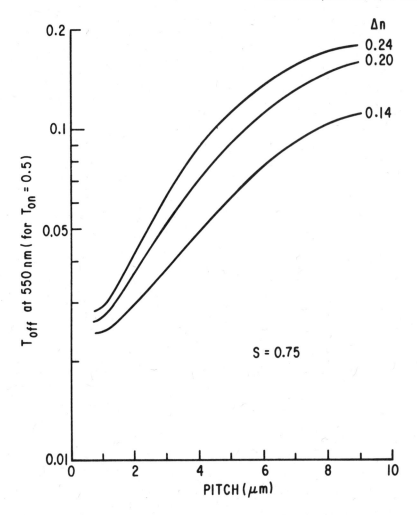

Figure 7.   Semilog plots of transmission T_{off} in the Grandjean
            texture at 550 nm vs. pitch as function of birefrin-
            gence Δn = 0.14, 0.20, 0.24 corresponding to $n_{\parallel}$ = 1.69,
            1.75, 1.77, at field-on transmission of 50%, computed
            from equation (1).

with pitch of 3.7 and 3.2 µm, the order parameters were assumed
to be the same as for the other pitch values. The birefringence of
the materials was assumed to be the same as that for the host without CB-15. For both hosts, transmission T_{off} decreases with decreasing pitch. The relatively high T_{off} values for the shortest
pitch in E-7 and ROTN-101 are due, respectively, to a lower order

Table 1

Comparison of Experimental and Theoretical Transmission

Host	Pitch (μm)	Δn	S	Experimental T_{on}	T_{off}	Theory T_{off}	Deviation from Theory
E-7	5.3	0.24	0.75[a]	0.54	0.156	0.142	11%
E-7	3.2	0.24	0.74	0.54	0.084	0.093	-10%
E-7	1.6	0.24	0.74	0.50	0.039	0.041	- 5%
E-7	1.1	0.24	0.74	0.47	0.032	0.026	23%
E-7	0.8	0.24	0.72	0.47	0.034	0.031	10%
ROTN-101	5.5	0.16	0.71[a]	0.47	0.156	0.085	83%
ROTN-101	3.7	0.16	0.71[b]	0.45	0.049	0.052	- 6%
ROTN-101	3.2	0.16	0.71[b]	0.56	0.079	0.095	-17%
ROTN-101	2.2	0.16	0.71	0.47	0.048	0.041	17%
ROTN-101	1.1	0.16	0.71	0.54	0.066	0.061	8%

a) Measured in nonchiral host.   b) Assumed value.

parameter (0.72) and higher T_{on} transmission. The high T_{on} value for ROTN with 3.2 μm pitch also accounts for the higher T_{off} value. In general, higher order parameter results in lower T_{off} at constant T_{on}. Except for two cases in Table 1, deviation from theory is less than 20%. Some of the data in this paper were compared earlier [9] with the equations of White and Taylor. [2] The pitch reported [9] earlier for mixtures of E-7 and CB-15 was in error and is 20% shorter. The present theory is in better agreement with the data.

The role played by birefringence is illustrated in Table 2 where experimental data and theoretical T_{off} values at constant pitch are shown for parallel boundary conditions. The order parameter of the guest-dye systems was about the same for E-7 and Licristal 1132, and T_{off} is lower for the latter at comparable activated state transmission as expected from its lower Δn. The birefringence of ROTN is nearly the same as that of Licristal 1132 but T_{off} for ROTN is higher than for Licristal because the former had a lower order parameter.

Figures 2 and 3 indicate the effect of boundary condition on transmission. Samples with homeotropic boundary are expected

Table 2

Effect of Birefringence on Transmission

Host	Δn	S	P (μm)	Experimental T_{on}	T_{off}	Theory T_{off}
E-7	0.24	0.74	3.2	0.54	0.084	0.093
Licristal 1132	0.14	0.75	3.3	0.57	0.066	0.068
ROTN-101	0.16	0.71	3.2	0.56	0.079	0.095

to have higher transmission than those with parallel boundary as is the trend for the ROTN host (Figure 3). The coherence length of the boundary layer is estimated at approximately one-half of the pitch. In the short-pitch regime, the theoretical increase of transmission amounts to only a few percent. The data for the E-7 host (Figure 2) show the opposite trend, and no explanation for this can be offered.

The influence of host on order parameter and spectral absorption is evident in Figure 4 which indicates higher order parameter and bathochromic shift in the biphenyl liquid crystal.

In practical display applications, contrast ratio, defined as T_{on}/T_{off} is an important performance characteristic. It is evident that contrast ratio is a function of transmission and can therefore be traded off against transmission. An arbitrarily high contrast ratio can be obtained by reducing T_{on}, e.g. by increasing dye concentration or liquid crystal thickness.

ACKNOWLEDGMENT

We thank Prof. A. Saupe for a theoretical treatment of the optics of Grandjean textures, Dr. D.E. Castleberry and C.R. Stein for technical discussion and Integrated Display Systems Inc. for partial financial support of this work.

REFERENCES

1.   G.H. Heilmeier and L.A. Zanoni, Appl. Phys. Lett. 13, 91 (1968); G.H. Heilmeier, J.A. Castellano and L.A. Zanoni, Mol. Cryst. Liq. Cryst. 8, 293 (1969).

2.   D.L. White and G.N. Taylor, J. Appl. Phys. 45, 4718 (1974).

3.   A. Saupe (private communication, to be published).

4.   P. Kassubek and G. Meier, Mol. Cryst. Liq. Cryst. 8, 305 (1969).

5.   A.H. Tarry (Report by Royal Signals and Radar Establishment).

6.   C. Maugin, Bull. Soc. Franc. Miner. 34, 71 (1911).

7.   D.W. Berreman, J. Opt. Soc. Am. 63, 1374 (1973).

8.   J.L. Fergason, Appl. Opt. 7, 1729 (1968).

9.   H.S. Cole, Jr. and S. Aftergut, Appl. Phys. Lett. 31, 58 (1977).

THE SELF CONTROL OF LIPID BILAYERS

Donald L. Melchior and Joseph M. Steim

Department of Chemistry
Brown University
Providence, R. I.   02912

INTRODUCTION

A characteristic of nearly all organisms is their ability to accomodate the fatty acid composition of their lipids to growth temperature (7,8,10,23). The principal biochemical adjustment is an increase in the fraction of low-melting-point fatty acids (usually unsaturates) in lipids synthesized at lower growth temperatures. The obvious physical result of this adjustment is a depression of lipid melting points. Physiologically, the alteration of fatty acid composition tends to maintain constant bilayer fluidity at various growth temperatures (27). Maintainence of membrane fluidity becomes especially important in organisms such as *Acholeplasma laidlawii* (15,29) and *Escherichia coli* (16,21,30), whose membranes can undergo a transition at or just below growth temperature. In such cases, the shift in the fatty acid spectrum at lower growth temperatures is absolutely essential to maintain a functional fluid state. As membranes pass through a transition from high to low temperature they become progressively more crystalline. The increased order can give rise to aberrant behavior, including changes in enzyme kinetics, cell leakage, cessation of cell division, and even cell lysis (16,19,21). If these and other calamitous effects are to be avoided, the increase in unsaturation or the inclusion of other fatty acids of low-melting-point becomes a physiological necessity. A control mechanism is implied, one that senses temperature and the physical state of the membrane and directs the incorporation of proportionally more unsaturated or other low-melting fatty acids into membrane lipids as the temperature decreases. This control of membrane transition temperatures has been directly demonstrated by calorimetry in several microorganisms, including A. *laidlawii* (15),

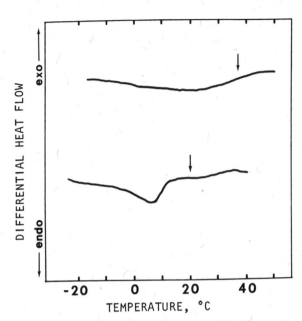

Figure 1. The effect of growth temperature upon the position of the cytoplasmic membrane transition in E. coli W945 grown in minimal medium. Calorimeter scans were run in 50% ethylene glycol-water. In (a), from cells grown at 37°, the transition extends from about -10 to 40°. In (b), from cells grown at 20°, it extends from about -5 to 15°. In both cases the membranes are fully fluid at or just below growth temperature.

which incorporates exogenous fatty acids from the growth medium, and *E. coli* (30), which biosynthesizes fatty acids endogenously. In *A. laidlawii* B, whether grown at 37° or 25°, the membranes were almost entirely fluid at growth temperature but underwent a transition and became progressively more crystalline as the temperature was lowered below that of growth. That is, the order-disorder membrane transition was maintained in the neighborhood of the growth temperature, but was shifted down at lower growth temperature to maintain membrane fluidity. A similar downshift of the membrane transition in *E. coli* is shown in Figure 1, where a thermogram of cells grown in minimal medium at 20° is compared to a thermogram of the same organism grown at 37°.

Although the detailed mechanism or mechanisms for temperature modulation of membrane fatty acid composition have not been worked out in any organism (5), it has become clear that control can take

place at several, possibly interrelated, levels. One level is the
biosynthesis of unsaturated fatty acids. In some cases, desaturase
activity appears to be governed by the solubility of oxygen, which
serves as an eventual electron acceptor (1). In others, such as
Bacillus megaterium (4), enzyme synthesis is affected by temperature. Fatty acid desaturase is not synthesized in this organism
at 35° but is strongly induced at 20°. Temperature also has a direct effect upon the desaturase protein itself, which, once synthesized at low temperature, undergoes rapid irreversible inactivation
at higher temperatures. Direct temperature effects upon enzyme activity have also been found in E. *coli*, which produces monoenoic
fatty acids via dehydration of the growing acyl chain within the
fatty acid synthetase system itself. Deprived of glycerol in order
to uncouple phosphatidic acid synthesis from fatty acid synthesis,
E. *coli* accumulates large quantities of free fatty acids which become proportionally more unsaturated as temperature is decreased
(2). A reversible dependence of β-hydroxydecanoyl thioester de-
hydrase upon temperature may be implicated.

In addition to effects of temperature upon the biosynthesis
of unsaturated fatty acids, another level of control is apparently
at the site of phospholipid synthesis in the membrane. Temperature-
dependent selection of saturated and unsaturated fatty acid CoA
by membrane-bound acyl transferase, which catalyzes the esterification of glycerophosphate, has been demonstrated by Sinensky (26).
Presented with a mixture of oleoyl and palmitoyl CoA, cell-free
E. *coli* acyl CoA:glycerophosphate acyltransferase produces increasingly greater proportions of unsaturated lysophosphatidic acid at
lower temperature. The acyltransferase apparently possesses a
"pre-programmed" selective temperature response. It is the molecular nature of this temperature "program", and its possible inter-
relationships with fatty acid biosynthesis, that we shall consider
in more detail.

Crucial to the understanding of the temperature-dependent selection process at the membrane level is the realization that fatty acids seem to be selected on the basis of melting point, a thermodynamic property that only indirectly reflects molecular structure. Although unsaturation is the usual route to low melting point,
the same goal is attained in some organisms by employing structural
alternatives, such as branched chains in many gram-positive bacteria or cyclopropane-containing chains in many gram-negative bacteria (35). A more convincing argument that the principal consideration is thermodynamic rather than structural is based on the fact
that a given organism, if forced to do so, will chose any low-melting exogenously supplied fatty acid to accomplish its goal of lowering transition temperatures and controlling fluidity. The best
illustration is again A. *laidlawii* (12,14,15,19), which lacks desaturase activity and, as pointed out earlier, incorporates large

amounts of exogenous fatty acids into its membrane lipids. Fatty
acids of progressively lower melting points are required as the
growth temperature is decreased. *Cis*-unsaturates serve the purpose
admirably even at the lowest temperatures, but growth is also normal if *cis*-unsaturates are replaced by branched-chain or cyclopro-
pane fatty acids or by elaidate, an unnatural *trans*-unsaturated compound. Unsaturated fatty acid auxotrophs of E. *coli* show similar
behavior, and will accept elaidate or even bromostearate (3,24).
If in fact the temperature sensing selection mechanism within the
membrane is thermodynamically determined, and depends upon the properties of an ensemble of molecules, it is difficult to imagine it
to be based upon enzyme specificity. The binding of substrates to
enzymes reflects the molecular structure of the ligand, and interaction occurs on a one-to-one basis, so that strictly thermodynamic
properties have no meaning in such interactions.

     In accord with this thermodynamic point of view, we shall pre-
sent evidence that the temperature "program" of acyltransferase in
A. *laidlawii*, and possibly in other organisms as well, is an innate
property of the bilayer in which the protein is imbedded rather than
a property of the protein itself. Our argument is based upon correlations of the pattern of incorporation of palmitate and oleate,
from the incubation medium into membrane polar lipids, with the
physical state of the membrane bilayer determined calorimetrically
(16,18). Furthermore, the physical binding of free fatty acids to
lipid bilayers formed from extracted membrane lipids shows the same
temperature dependence shown by the enzymatic process in live cells.
Similar temperature-dependent binding is also characteristic of model
bilayers containing synthetic lecithin.

EXPERIMENTAL

     The temperature "program" for the selection of exogenous fatty
acids by A. *laidlawii* was revealed by plotting the ratios of [^{14}C]-
palmitate to [^{3}H]oleate (palmitate/oleate ratio) simultaneously incorporated into the membrane lipids (26), while the physical state
of the membrane bilayer was followed by differential scanning calorimetry (17,31). Comparison of the palmitate/oleate ratios with
the differential scanning calorimeter scans shows the temperature
program to correlate with the physical state of the bilayer, which
in turn depends upon the fatty acid composition of the membrane
lipids. As a consequence, the program and resultant palmitate/
oleate plots can be changed at will by changing the fatty acid
composition. The greatest alteration in the physical state of a
membrane of fixed fatty acid composition occurs during a transition, where the bilayer passes from an ordered state at low temperature to a disordered state at higher temperatures. Below the
transition lateral diffusion is inhibited and the lipids are crys-

tallized within the plane of the membrane, while above it free di-
fussion and mixing takes place (16,19). The progression from one
state to the other can be sudden or quite gradual, depending upon
the fatty acid heterogeneity of the lipids. The greatest changes
in the palmitate/oleate ratio of incorporated fatty acids occur at
the beginning of the transition, where fatty acid selection becomes
strongly temperature dependent. At temperatures below the transition, the ratio of palmitate to oleate incorporated is nearly independent of temperature, while during the transition it is strongly
temperature dependent. The palmitate/oleate ratio thus shows a pronounced change in slope at the low-temperature end of the transition.

These phenomena are illustrated in the following two figures.
Figure 2 is a plot of the palmitate/oleate ratio of incorporated
fatty acids and the extent of membrane melt determined calorimetrically in A. *laidlawii* grown in unsupplemented tryptose. There is
a clear correlation between the two. Below the transition, which
begins gradually at about 15°, fatty acid selection is unaffected
by temperature. During the transition, however, selectivity becomes
strongly dependent upon temperature, and the palmitate/oleate plot
roughly parallels the fraction melted. The increase of palmitate
with respect to oleate is approximately eight-times greater at 40°,
toward the end of the transition, than at its beginning. New protein synthesis cannot account for the results, since chloramphenicol-
treated cells give identical patterns.

The effects of shifting the membrane transition temperature are
shown in Figure 3, plots of the palmitate/oleate ratio and extent
of transition in A. *laidlawii* grown in oleate-enriched and palmitate-
enriched media. The transition in oleate cells is lowered compared
to tryptose-grown cells, beginning at about 0° and ending at about
20°. From 20° to 40° the membrane bilayer is completely fluid.
Over the entire range, from 0° to 40°, the temperature program has
a positive slope and selectivity is strongly temperature dependent.
Thus the increasing preference for palmitate over oleate with increasing temperature appears to be characteristic of fluid membranes.
It vanishes when the membrane solidifies. The temperature-indepen-
dence of selectivity by membranes below their transition temperature
is clearly emphasized in Figure 3d, a plot of the palmitate/oleate
ratio for palmitate-enriched cells with a high transition temperature.

Figures 2 and 3, taken together, define the relation between
the thermal response of fatty acid incorporation and the state of
the lipids in the membrane. When the membrane is crystalline, temperature dependence is weak or absent. As the membrane enters a
transition, and passes from the ordered state to the fluid state,
selectivity becomes temperature dependent. When the membrane of
oleate-enriched cells is fully fluid, above its transition, select-

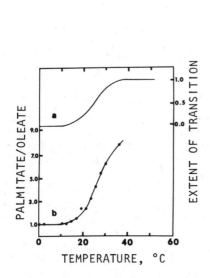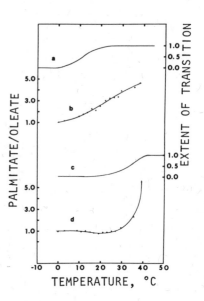

Figure 2. *Correlation of the temperature dependence of fatty acid incorporation into membrane lipids with the state of the membrane bilayer in* A. laidlawii *cells grown in unsupplemented tryptose medium. Extent of transition (curve a) is obtained from the integrated calorimeter scan of the transition in membranes, and progresses from 0.0 for a fully ordered bilayer to 1.0 for a fully fluid bilayer. Palmitate/oleate (curve b) is the ratio of [*^{14}*C]palmitate [*^{3}*H]oleate activities incorporated into the lipids of the same cells, and is normalized to 1.0 at the starting temperature.*

Figure 3. *The effect of altering the state of the membrane bilayer upon the incorporation of fatty acids into membrane lipids. For cells grown in an oleate-enriched medium, the extent of transition (curve a) shows the membrane to be undergoing a transition below 20° and fully fluid thereafter. The accompanying ratio of [*^{14}*C]-palmitate to [*^{3}*H]oleate incorporation (cuve b) is strongly temperature dependent both during and after the transition. In palmitate-enriched cells, the extent of transition (curve c) shows the entire physiological temperature range. The accompanying palmitate/oleate incorporation ratio (curve d) is nearly independent of temperature.*

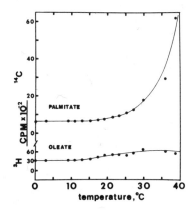 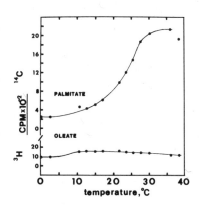

Figure 4.   The simultaneous incorporation of [^{14}C]palmitate and
[^{3}H]oleate into the lipids of palmitate-enriched A. laidlawii cells.
The ratios plotted in Figures 2 and 3 were obtained from the data
presented in this and Figures 5 and 6.   The ordinate scales are chos-
en so that relative increases in cpm of palmitate and oleate are
comparable (i.e. doubling of both palmitate and oleate produces
identical increments in the ordinate).

Figure 5.   The simultaneous incorporation of [^{14}C]palmitate and
[^{3}H]oleate into the lipids of cells grown in unsupplemented tryptose
medium.

ivity continues to be temperature dependent.

In Figures 2 and 3, the palmitate/oleate ratios have been normalized to 1.0 at 0° for convenience in presentation. The origin
of the temperature dependence of the ratios is revealed by the absolute counts of palmitate and oleate. Incorporation by cells grown
in palmitate-enriched medium is shown in Figure 4. At temperatures
below the transition few counts of palmitate are incorporated, and
the incorporation is independent of temperature. Upon the onset of
the transition a large increase in the rate of incorporation of palmitate begins, accompanied by a relatively much smaller proportional
increase in oleate incorporation. By 37°, palmitate incorporation
has increased by a factor of ten, while oleate incorporation has
little more than doubled. The shapes of the two curves differ, with
the oleate curve flattening at higher temperatures. In cells grown
in unsupplemented tryptose medium (Figure 5), the proportional increase of palmitate incorporation is again more strongly temperature
dependent than oleate, except that the entire curve is shifted to
lower temperatures to again coincide with the transition. Oleate

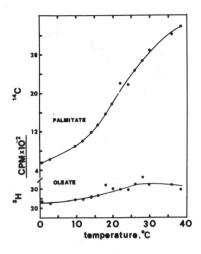
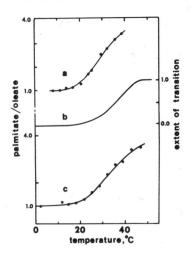

Figure 6. The simultaneous incorporation of [^{14}C]palmitate and [^{3}H]oleate into the lipids of oleate-enriched cells.

Figure 7. Correlations between the palmitate/oleate ratio incorporated into cellular membrane lipids (curve a), the extent of transition in membranes (curve b), and the palmitate/oleate ratio of fatty acids physically bound to bilayers of extracted membrane lipids (curve c). Both incorporation and binding curves reflect the state of the bilayer, and are identical within experimental error. The cells are grown in unsupplemented tryptose medium.

incorporation again increases only slightly with temperature, and even appears to decrease again at higher temperatures to a rate near that observed at 0°. The rate of palmitate incorporation becomes constant at the highest temperature and may even decrease. This effect is not seen in oleate-grown cells (Figure 6), and might be explained by endogenous synthesis of saturated fatty acids. Nevertheless, it does not obscure the correlation between the calorimetrically observed transition and the shift in uptake curves. The same pattern of a strong temperature response in palmitate uptake and a much weaker relative response in oleate uptake is repeated by oleate-grown cells. The rate of palmitate uptake is no longer temperature-independent at lower temperatures as it is in palmitate-enriched cells and cells grown in unsupplemented tryptose medium, since even at 0° the membrane transition has already begun. At temperatures above the transition, where the membrane is completely fluid, the rate of palmitate uptake continues to be strongly temperature dependent. The rate of oleate uptake again seems to plateau at higher temperatures.

The possibility of artefacts must be considered, since the
state of the labeled fatty acids in the incubation medium is ill-
defined. Some may be in solution, some bound to other components
in the medium, and others could exist as droplets or micelles. All
could experience physical changes with temperature which might affect their availability to the cells. However, the plots correlate
well with calorimeter scans of the membranes and can be changed dras-
tically without changing the composition of the incubation medium
used during incorporation. Thus, the plots apparently reflect events
associated with the cells and not with unexpected changes in the
medium. Rate-limiting steps occurring in the medium or exhaustion
of the supply of labeled fatty acids are also not implicated, since
the palmitate/oleate ratio of incorporated fatty acids did not change
with increased incubation time (24 min, 4 times the usual procedure)
at constant cell concentration or with cell concentration (increased
by a factor of 4) at constant incubation time of 6 min. Ordinarily,
approximately 5% of the total labeled fatty acids added to the medium were incorporated during incubation. Endogenous synthesis of
saturated fatty acids in A. *laidlawii* is low compared to uptake
from exogenous sources. It can be discounted as a perturbing factor, since it would be expected to flatten the curves to some extent but could not account for the observed differences between
curves.

The correlation between the state of the membrane bilayer and
the pattern of incorporation of palmitate and oleate into membrane
lipids appears to be mimicked by the physical binding of fatty acids
by lipid bilayers. For binding experiments, small pieces of filter
paper were impregnated with lipids dissolved in cholorform, dried,
incubated in buffer containing oleate and palmitate, then finally
blotted to remove excess liquid and counted. Calorimetry verified
that bilayers were formed within the paper. The palmitate/oleate
ratio of labeled fatty acids incorporated into the membrane lipids
of cells grown in unsupplemented tryptose medium is shown in Figure
7, together with the extent of the membrane transition seen by calorimetry and the palmitate/oleate ratio of fatty acids physically
bound by lipids extracted from the membrane. All data come from
the same culture of A. *laidlawii* grown on unsupplemented tryptose
medium. As seen in Figure 2, incorporation again parallels the
transition. However, selective binding of fatty acids by bilayers
of extracted lipids, which undergo the same transition as bilayers
within membranes, behaves in the same way. Within the limits of
experimental error, incorporation and binding curves are identical.
It is not known whether selective binding is kinetically or thermodynamically determined, but preliminary data suggests a thermodynamic mechanism since the same behavior is observed for experiments
of 1-4 h binding. Absolute counts of fatty acids bound by extracted
lipids are plotted in Figure 8. The pattern is quite similar to that
seen in incorporation by live cells. Palmitate binding is not

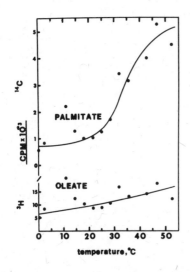

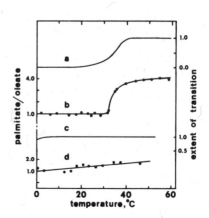

Figure 8.   The simultaneous binding of [^{14}C]palmitate and [^{3}H]oleate by bilayers of lipids extracted from the membranes of A. laidlawii grown in unsupplemented tryptose medium.   Scatter is greater than in studies of incorporation by cellular lipids because of variations in the amount of lipid deposited in filter paper strips, but the scatter is eliminated in ratio plots of binding (Figure 7b).

Figure 9.   Selective binding of fatty acids by phosphatidylcholine bilayers.   A mixture of 25% egg phosphatidylcholine and 75% dipal-mitoyl phosphatidylcholine gives the transition shown in curve a and the [^{14}C]palmitate/[^{3}H]oleate binding ratio in curve b.   Pure egg phosphatidylcholine has the transition seen in curve c and the binding ratio in curve d.   Since the transition in egg phosphatidyl-choline occurs almost entirely below 0°, the bilayers are fluid over nearly the entire temperature range shown.

strongly dependent upon temperature below the transition, but at the onset of the transition becomes strongly temperature dependent. Oleate binding, on the other hand, is proportionally less responsive to temperature.

Transition-dependent selective binding of fatty acids is not confined to bilayers of A. laidlawii lipids.   The same phenomenon occurs with mixed phosphatidylcholine.   Figure 9a,b displays calori-metric and binding data for bilayers containing 75% dipalmitoyl phosphatidylcholine and 25% egg phosphatidylcholine, which, because of its saturation, broadens the dipalmitoyl phosphatidylcholine transition and lowers it below its customary 41°.   Below the transi-

tion selective binding is again independent of temperature, as it
is in binding by A. *laidlawii* lipids and in incorporation into lipids by cells. The binding curve again changes with the onset of
the transition, and becomes strongly temperature dependent during
it. Above the transition, the completely fluid bilayers bind proportionally far more palmitate than below, and the palmitate/oleate
plot continues to show a decreased but still positive slope. Although the shape of the palmitate/oleate plot does not appear to
correspond as closely to the extent of the transition as it does
in the case of incorporation by cells or binding by extracted membrane lipids, the phenomenon of transition-dependent selectivity
is displayed by all three systems.

The temperature-dependent selective binding shown by fluid
lamellar phosphatidylcholine in Figure 9b (i.e. a positive slope
from 40-55°) is also characteristic of pure egg phosphatidylcholine,
which is far more unsaturated than the egg phosphatidylcholine/dipal-
mitoyl phosphatidylcholine mixture. The egg phosphatidylcholine
transition, shown in Figure 9c, is 95% complete at 0°, so that the
bilayers are essentially fluid over the entire 0-50° range. The
binding curve is featureless and linear, with a positive slope. Although their slopes differ, the behavior of palmitate/oleate plots
of fatty acid binding by egg phosphatidylcholine and fluid bilayers
in oleate-enriched A. *laidlawii* (Figure 3a,b) are analogous even
though the lipid of A. *laidlawii* contains no phosphatidylcholine
(22).

DISCUSSION

The pattern of incorporation of palmitate and oleate into
A. *laidlawii* lipids is consistent with the thermal response of most
other organisms, whose lipids become progressively enriched in unsaturated fatty acids as growth temperature decreases. However, although the palmitate/oleate ratio of incorporated fatty acids increases rapidly with increasing temperature, such behavior does not
occur until the onset of the membrane transition.

The palmitate/oleate ratios of incorporated fatty acids are
nearly independent of temperature at temperatures below the membrane
transition, but at the beginning of the transition begin to increase,
and during the transition they continue increasing in a roughly exponential manner. In the fully fluid membrane of oleate-enriched
cells (Figure 3), the palmitate/oleate ratio continues to remain
strongly temperature dependent at temperatures well above the completion of the transition. Absolute incorporations reveal that the
large increases in palmitate/oleate ratios result mostly from large
relative increases in palmitate rather than in oleate incorporation.
Compared to palmitate, the relative changes in oleate are only weakly

temperature dependent. In at least one case, that of cells grown
in ordinary tryptose medium (Figure 5), the rate of incorporation
of oleate at higher temperatures decreases with increasing temperature. Similar decreases may occur at higher temperatures in both
palmitate-enriched (Figure 4) and oleate-enriched cells (Figure 6).

These results can be explained in terms of temperature-dependent enzyme activities. One possibility is a transacylase whose
temperature dependence is far more pronounced for straight-chain
saturated fatty acids such as palmitate. Alternatively, two enzymes
might be active in A. *laidlawii*, one of which, specific for straight-
chain saturates, is more temperature dependent than the other. Similarly, temperature-dependent selectivity might be explained by the
thermal response of acyl CoA synthetase. If such speculations are
correct, the absence of temperature dependence in the palmitate/
oleate ratios and in the absolute incorporations at temperatures below the membrane transition may simply reflect the fact that growth
does not occur below the transition (14). Cells which are not growing would not be expected to be synthesizing new lipids at an appreciable rate. Alternatively, one could argue that the transacylase
enzyme or enzymes become inactivated in an ordered bilayer matrix.

Although these or other mechanisms which explain temperature-
dependent selectivity in terms of the thermal response of fatty acid
specific enzymes are consistent with the incorporation studies, the
physical binding of palmitate and oleate by extracted lipids and
model phosphatidylcholine bilayers suggest a viable alternative.
The remarkable similarities shared by fatty acid incorporation into
membrane lipids in live cells and fatty acid binding by extracted
membrane lipids suggest that the pattern of uptake from the growth
medium may be determined by the physical properties of the membrane
lipid bilayer. The palmitate/oleate plots of both incorporation
and binding are both strongly temperature dependent, while oleate
incorporation and binding are not (Figure 8). Furthermore, both incorporation and binding are less temperature-dependent at temperatures below the transition than above.

Temperature-dependent selective binding of fatty acids by bilayers apparently is not limited to the lipids of A. *laidlawii*. It
is also a property of phosphatidylcholine, which does not occur in
A. *laidlawii* or bacteria but is a major phospholipid of animal and
plant membranes. A. *laidlawii* contains only glycolipids and acidic
phospholipids (22) which bear a negative charge, while phosphatidylcholine is zwitterionic. However, in spite of very different surface charges, the effect of temperature upon fatty acid binding is
essentially the same in both cases. Evidently selectivity depends
largely upon the changing apolar hydrocarbon core of the bilayer
rather than the polar interface. Since transition-mediated selective binding seems to be a general property of bilayers, the control

of membrane fluidity and transition temperatures by this mechanism
may be common in Nature. The temperature program of the acylase
enzymes in membranes of E. coli K12 (26) correlates very well with
calorimetric determinations of the inner membrane transition in that
organism (16).

The mechanism for selective binding is unknown. It could be
thermodynamically or kinetically controlled. If it is an equilibrium phenomenon, as preliminary evidence suggests, it could be explained in terms of the temperature dependence of the partition of
fatty acids between the apolar interior of the bilayer and the ex-
terior aqueous medium. The relative affinity for long-chain saturated acids compared to unsaturated acids is considerably greater
for a fluid bilayer than for the same bilayer in an ordered state.
Thus this affinity continuously increases as the bilayer progresses
through its order-disorder transition and the fraction of fluidity
increases. Once the bilayer is above its transition, at least in
oleate-enriched cells, selectivity continues to be temperature dependent. The thermal response in this fully fluid region may depend upon both the lipid classes in the membrane and their fatty
acid composition. Lateral phase separation (19,25) of membrane lipids may affect the changes in incorporation and binding which occur
during the transition by changing the lipid composition of the fluid
portions of the bilayer, since, as the membrane solidifies, the
shrinking areas of fluidity become progressively enriched in unsaturated lipids. If the esterifying enzymes are free to diffuse within the membrane they may be excluded from solidified portions of
the bilayer and remain with the changing fluid regions (19,34) where
they would have access to fatty acids or fatty acid derivatives.
Although incorporation and binding studies have been carried out
only with palmitate and oleate, we regard oleate as a model compound for other fatty acids of low melting point. Any structural
modification of a long-chain saturated fatty acid would affect both
solubility and melting point.

It is interesting that the temperature dependence of selective
binding of fatty acids by bilayers resembles that of the solubility
of fatty acids in organic solvents (28). Figure 10A shows the
temperature dependence of the solubility of an homologous series
of saturated fatty acids in benzene. Lower-melting fatty acids are
more soluble than those of higher melting points. As in binding
by lipid bilayers, the curves for high-melters are more strongly
temperature dependent than those of the low-melters. This effect
is illustrated in Figure 10B, where the ratios of the solubilities
of a high-melter, stearate, to a low-melter, myristate, is plotted
as a function of temperature. Although the mechanism of fatty acid
binding to lipid bilayers is unknown and two fatty acids were presented simultaneously to the bilayers in contrast to the solubility
data taken with separate fatty acids at equilibrium, the solubility

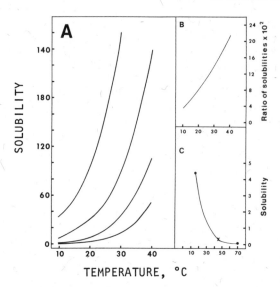

Figure 10. Solubilities of fatty acids in organic solvents.
A: Solubility, in grams per 100 g of benzene as a function of
temperature for lauric acid (12:0), myristic acid (14:0), palmitic
acid (16:0), and stearic acid (18:0). Absolute solubilities vary
inversely with melting point, but the solubilities of the higher-
melting fatty acids are more temperature dependent.
B: Ratio of the solubilities in benzene of a higher-melting-point
fatty acid, stearate, to a lower-melting-point fatty acid, myristate. The ratio decreases with decreasing temperature, as it
does with fatty acid binding to bilayers.
C: Solubilities, in weight percent, of 18-carbon fatty acids in
diethyl ether at -30°: o oleic acid (cis-Δ9, 18:1; mp 16.3°),
x elaidic acid (trans-Δ9, 18:1; mp 45.5°), ● stearic acid (18:0;
mp 69.6°). Again, solubilities vary inversely with melting point.

and binding curves behave similarly. Figure 10C shows the solubilities of three 18-carbon fatty acids in diethyl ether: oleic
acid (cis-Δ9, 18:1), elaidic acid (trans-Δ9, 18:1), and stearic
acid (18:0). Again, solubility is a function of melting point.

If the pattern of fatty acid incorporation can be explained
in terms of the physical properties of the bilayer, it is unnec-
essary to invoke an enzymatic mechanism for the selection of fatty
acids to control transition temperatures and maintain fluidity in
membranes of A. laidlawii. Suppose the acylase protein or pro-
teins are imbedded in the membrane bilayer, and are responsible

solely for catalyzing the esterification of membrane lipids. That is, they have little or no innate ability to distinguish between various fatty acids or to change their selectivity for fatty acids with changing temperature. They accept and use fatty acids or fatty acid derivatives supplied to them by the bilayer. The selective temperature program is a property of the bilayer itself. In a sense, the acylase enzyme is a lipoprotein, comprising a protein imbedded in a bilayer. Catalytic activity resides in the protein, but the bilayer provides temperature-programmed selectivity.

Such selectivity in fatty acid incorporation would tend to maintain bilayer fluidity constant as growth temperature changes. Furthermore, since the slopes of palmitate/oleate plots change abrutply toward the low temperature end of the transition, selective binding could provide a homeostatic mechanism, operating at the membrane level, to sense the thermodynamic state of the membrane and make appropriate adjustments to maintain the transition below or largely below growth temperature. In this sense, lipid bilayers would be capable of controlling their own physical state. In cells growing largely upon exogenous fatty acids, selective binding could be the major means of controlling membrane transitions. Similar considerations might also account for positional specificity, since the melting point of the acid esterified to the β carbon is usually lower than that of the α carbon of the phospholipid (6,13,20).

Although the model just suggested is based upon the uptake of exogenous fatty acids by A. *laidlawii*, the same model might be generalized to explain at least some temperature effects upon the composition of membrane lipids synthesized from endogenously biosynthesized fatty acids. That an interrelationship exists between selectivity at the membrane level and the synthetase system is demonstrated in E. *coli* cells, which produce longer chain fatty acids when uncoupled from phospholipid synthesis than when lipid synthesis is allowed to take place normally (2). The intermediary for linking synthesis and desaturation of fatty acids with lipid synthesis in the membrane may be the bilayer as a temperature dependent selective sink. Sumper and Traüble (32) have in fact verified that long-chain acyl CoA molecules will bind to and dissolve in dimyristoyl lecithin bilayers and E. *coli* membranes, where they are free to diffuse about and encounter appropriate membrane-bound enzymes. Furthermore, the efficacy of bilayers as an acceptor for fatty acid CoA is known to depend upon environmental conditions. Phosphatidic acid becomes an increasingly effective acceptor as ionic strength is increased, presumably because of charge neutralization by counter ions (33). Since environmental temperature also modifies the efficacy of bilayers as acceptors (16,18), such modification may provide the basis for temperature-dependent control.

Consider the regulation of desaturase activity, and suppose that a desaturase exists that competes with the bilayer for the

saturated end product of fatty acid synthetase. If at low temperatures the bilayer is in a relatively crystalline state that does
not readily bind saturated fatty acids, the substrate will be operated upon instead by the desaturase. The resulting unsaturated
fatty acid can now easily enter the bilayer and be incorporated into the membrane lipids. As the bilayer becomes more fluid, it more
successfully competes with the desaturase for saturated fatty acid
CoA and the spectrum of fatty acids entering the membrane resevoir
shifts toward increasing saturation. A case of control of fatty
acid unsaturation by the thermodynamic state of the membrane has
been reported in *Tetrahymena pyriformia* (9, 11).

   A second thermal effect which might be explained by the concept of the bilayer as a temperature-programmed sink is the shortening of biosynthesized fatty acids at lower growth temperature.
Shortening occurs in many organisms, including A. *laidlawii* deprived of exogenous sources of fatty acids (M. Tourtellotte, personal communication). The progressive tendency already demonstrated in A. *laidlawii* for the bilayer to accept relatively more unsaturated than saturated fatty acids as temperature is lowered is
again central to the argument. If in fact thermodynamic properties
rather than specific molecular structure is the predominant factor
affecting acceptability of a fatty acid derivative by the bilayer,
one might expect that shorter-chain molecules, like unsaturated
long-chain ones, would at low temperature be more acceptable than
saturated long-chain ones. Long-chain saturated molecules that are
not accepted could be desaturated before being accepted. An alternative possibility is suggested by the work of Sumper (33), who
demonstrated that dimyristoyl lecithin, acting as a fatty acid CoA
sink, reverses the inhibition of fatty acid synthesis from acetyl
CoA in a system containing fatty acid synthetase and acetyl-CoA
carboxylase from yeast. The promotion of synthesis by added lecithin apparently arises from competitive reversible binding of palmitoyl or stearoyl CoA by the lipid bilayers and acetyl CoA carboxylase. Furthermore, fatty acid chain length depended upon inhibition of carboxylase by palmitoyl CoA. Increased inhibition led
to an increased rate of synthesis of fatty acids of shorter chain
lengths. Although Sumper directed his attention toward an explanation for the chain-shortening effect of anerobiosis, the same
point of view can be exterded to thermal effects if the temperature-
dependent properties of the bilayer sink are kept in mind. At
lower temperatures, as relatively more longer-chain saturated CoA
molecules are excluded from the bilayer and accumulate, acetyl-CoA
synthetase inhibition by the accumulated longer-chain compounds
would cause a shift toward the biosynthesis of shorter chains.

REFERENCES

1. Brown, C. M. and Rose, A. H., *J. Bacteriol.* 99, 371 (1969).

2. Cronan, J. E., *J. Biol. Chem.* 250, 7054 (1975).

3. Fox, C. F., Law, J. H., Tsulsogashi, N., and Wilson, G., *Proc. Natl. Acad. Sci. U. S. A.* 67, 598 (1970).

4. Fulco, A. J., *Biochim. Biophys. Acta* 218, 558 (1970).

5. Fulco, A. J., *Ann. Rev. Biochem.* 43, 215 (1973).

6. Hildebrand, J. G. and Law, J. H., *Biochemistry* 3, 1304 (1969).

7. Irving, L., Schmidt-Nielson, K. and Abrahamsen, N. S. B., *Physiol. Zool.* 30, 93 (1956).

8. Johnston, P. V. and Roots, B. I., *Comp. Biochem. Physiol.* 11, 303 (1964).

9. Kasai, R., Kitazima, Y., Martin, C. E., Nozawa, Y., Skviver, L. and Thompson, G. A., Jr., *Biochemistry* 15, 5228 (1976).

10. Marr, A. G. and Ingraham, J. L., *J. Bacteriol.* 84, 1260 (1962).

11. Martin, C. E., Hiramitsu, K., Kitazima, Y., Nozawa, Y., Skviver, L., and Thompson, G. A., Jr., *Biochemistry* 15, 5218 (1976).

12. McElhaney, R. and Tourtellotte, M. E., *Science* 164, 433 (1969).

13. McElhaney, R. N. and Tourtellotte, M. E., *Biochim. Biophys. Acta* 202, 120 (1970).

14. McElhaney, R. N., *J. Mol. Biol.* 84, 145 (1974).

15. Melchior D. L., Morowitz, H. J., Sturtevant, J. M., and Tsong, T-Y., *Biochim. Biophys. Acta* 210, 114 (1970).

16. Melchior, D. L., and Steim, J. M., *Ann. Rev. Biophys. Bioeng.* 5, 205 (1976).

17. Melchior, D. L., Scavitto, F. J., Walsh, M. T., and Steim, J. M., *Thermochim. Acta* 18, 43 (1977).

18. Melchior, D. L. and Steim, J. M., *Biochim. Biophys. Acta* 466, 148 (1977).

19. Melchior, D. L. and Steim, J. M. In "Prog. Surf. Memb. Sci." (D. A. Cadenhead and J. F. Danielli eds.) Vol. 13, Academic

Press, New York   (1977).

20. Okuyama, H., Yamada, K., Ikezawa and Wakil, S. J., *J. Biol. Chem. 251*, 2487   (1976).

21 Overath, P. and Thilo, L., *MPT International Review of Science: Biochemistry Series 2*, in press.

22. Razin, S. In "Prog. Surf. Memb. Sci." (D. A. Cadenhead, J. F. Danielli, and M. D. Rosenberg eds.) Vol. 9, Academic Press New York, 257   (1975).

23. Rose, A. H., ed. "Thermobiology", Academic Press, New York (1967).

24. Schairer, H. V. and Overath, P., *J. Mol. Biol. 44*, 209   (1969).

25. Shimshick, E. J. and McConnell, H. M., *Biochemistry 12*, 2351 (1973).

26. Sinensky, M., *J. Bacteriol. 106*, 449   (1971).

27. Sinensky, M., *Proc. Natl. Acad. Sci. U. S. A. 71*, 522   (1974).

28. Singleton, W. S. In "Fatty Acids, Part I" (K. S. Markley, ed.) Second Edition, 609, Interscience Publishers, Inc. New York (1960).

29. Steim, J. M., Tourtellotte, M. E., Reinert, J. C., McElhaney, R. N., and Rader, R. L., *Proc. Natl. Acad. Sci. U. S. A. 63*, 103, (1969).

30. Steim, J. M. In "Mitochondria/Biomembranes: (S. A. Van Den Berg, P. Borst, L. L. M. Van Deenen, J. C. Riemersma, E. C. Slater, and J. M. Taeger, eds.) 185, North-Holland, Amsterdam (1972).

31. Steim, J. M. In "Methods in Enzymology" (S. Fleischer and L. Packer, eds.) Vol. XXIIB, 262, Academic Press, New York   (1974).

32. Sumper, M. and Traüble, H. *FEBS letters 30*, 469   (1973).

33. Sumper, M., *Eur. J. Biochem. 49*, 469   (1974).

34. Tourtellotte, M. E. In "Membrane Molecular Biology" (C. F. Fox and A. O. Keith, eds.) 439, Sinaven Associates, Stamford, CT (1972).

35. Wakil, S. J., ed. "Lipid Metabolism", Academic Press, New York (1970).

TRANSITIONS WITHIN THE CHOLESTERIC MESOPHASE OF CHOLESTERYL STEARATE

Lidia Roche Farmer[+] and Paul F. Waters

Gillette Research Institute and The American University

Rockville, Maryland 20850   Washington, D. C. 20016

INTRODUCTION

Implicit in the measurement of the thermodynamic properties of the substances which undergo phase changes is the assumption that the transitions are equilibrium phenomena, i.e., that time is not a variable. In the measurement of the freezing points of small molecules by cooling at a steady rate, supercooling invariably occurs if the crystal structure is complex. The formation of the mesophases of the monotropic esters of cholesterol, whose structures are indeed complex, requires sensible time for optimizing molecular orientations.

In this work the approach to equilibrium was monitored by means of the capacitance of an air-dielectric capacitor incorporated into a cell in which the thermodynamic properties and the intensities of reflected light can be measured. Within the cholesteric mesophase small isobaric temperature changes and small isothermal pressure changes imposed on the sample require times of the order of hours for equilibration. It is this deliberate measurement technique which revealed the transitions reported here.

RESULTS

The material, apparatus, cell-filling and cell-calibration procedures are described elsewhere[1].

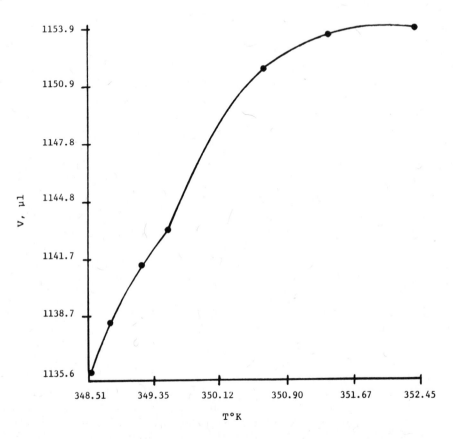

Fig. 1. Volume of cholesteryl stearate vs. temperature in the cholesteric mesophase at one atmosphere.

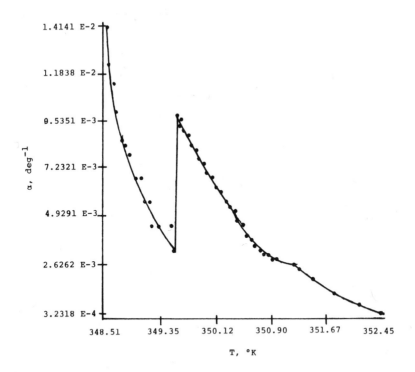

Fig. 2. Volume coefficient of expansion
of cholesteryl stearate vs. temperature in the
cholesteric mesophase at one atmosphere.

Volume Coefficient of Expansion

    The volume, V, of a sample of cholesteryl stearate was measured
in the cell as a function of temperature, T, at 1 atm by means of
the calibrated microsyringe incorporated into the cell. The volume
was measurable to a precision of 0.1 µl and the maximum temperature
fluctuation recorded was 0.05°K. Each volume point was recorded
along with the capacitance until the latter reached a constant
value. A plot of V vs. T is shown in Fig. 1.

The volume coefficient of expansion, α, given by:

$$\alpha = \frac{1}{V}\left(\frac{\delta V}{\delta T}\right)_P$$

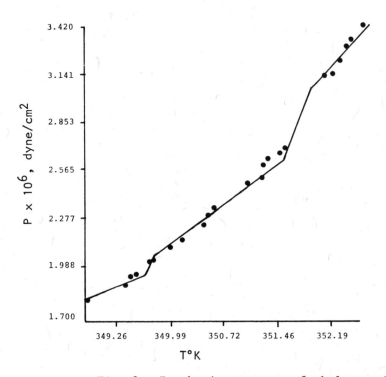

Fig. 3. Isochoric pressure of cholesteryl stearate vs. temperature in the cholesteric mesophase.

was determined at selected volumes, using values of the thermal coefficient of the volume derived from the graphical differentiation of Fig. 1. Values of α vs. T are plotted in Fig. 2.

Cohesive Energy

The pressure, P, of a sample of cholesteryl stearate was determined in the closed cell at each temperature from the capacitance vs. pressure calibration curve, after achieving a constant capacitance reading following a change in temperature. In Fig. 3, P is plotted vs. T.

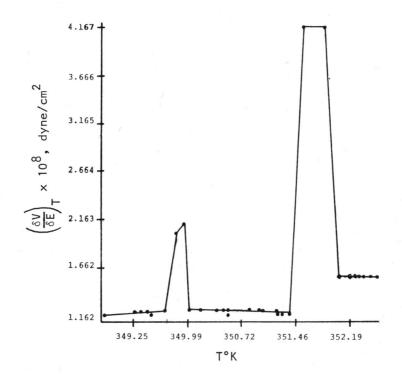

Fig. 4.   Cohesive energy of cholesteryl
stearate vs. temperature in the cholesteric
mesophase.

The cohesive energy, $\left(\dfrac{\delta E}{\delta V}\right)_T$, is given by:

$$\left(\frac{\delta E}{\delta V}\right)_T = T\left(\frac{\delta P}{\delta T}\right)_V - P$$

where E is the internal energy and P is the environmental pressure.
The cohesive energy was calculated at each temperature using values
of the thermal pressure determined by graphical differentiation of
Fig. 3.   The calculated cohesive energy terms are shown plotted vs.
T in Fig. 4.

Compressibility

Consideration of the preceeding figures suggested that the compressibility should be measured at several temperatures; one of these should be between 349.50° and 350.00°K and another between 351.00 and 352.00°K. The pressure of a sample of cholesteryl stearate was determined in the closed cell by means of the pressure vs. capacitance calibration curve, after the capacitance stabilized and the volume was measured with the calibrated microsyringe at 349.75° and 351.45°K. Plots of V vs. P at these temperatures are shown in Figs. 5 and 7, respectively.

The compressibility, β, defined by:

$$\beta = -\frac{1}{V}\left(\frac{\delta V}{\delta P}\right)_T$$

was determined at selected volumes at each of the two temperatures

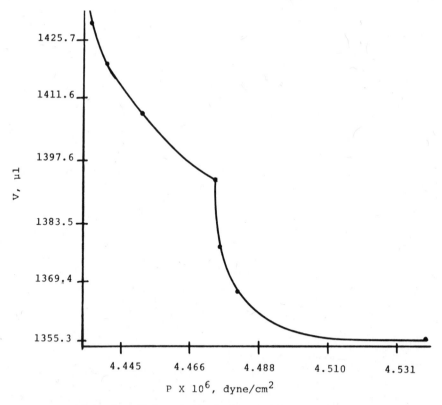

Fig. 5. Volume of cholesteryl stearate vs. pressure in the cholesteric mesophase at 349.75°.

using values of the pressure coefficient of the volume derived from the graphical differentiation of Figs. 5 and 7. The calculated β values are plotted in Figs. 6 and 8.

Reflectance

After equilibrium was achieved at each point in the measurement of volume as a function of temperature (Fig. 1) an optical scanning of the relative intensity of the reflected unpolarized incident light was carried out in a modified Brice-Phoenix light scattering photometer. The angle of incidence, i, and the angle of reflection, r, were each 22.5°. Typical data are shown in Fig. 9, where the recorder reading from the photometer, in millivolts, is plotted vs. the wavelength of incident unpolarized light, λ. The wavelength of maximum intensity was determined at each temperature and a graph of these values vs. T was constructed. The thermal coefficient of the wavelength of maximum intensity of reflected light was derived from this plot by graphical differentiation. The derivative is plotted vs. T in Fig. 10.

DISCUSSION

While it would be desirable to have more data points in the critical regions in Fig. 1, an analysis of the graph reveals that transitions do occur at temperatures in the vicinity of 349.6°K and 351.1°K. The data of Fig. 2, derived from Fig. 1, show a transition at 349.6°K and the probability of another at 351.2°K.

Examination of Fig. 4, derived from Fig. 3, reveals discontinuities at 349.9°K and 351.8°K. These appear to corroborate the transitions found in Fig. 1 and the temperature shifts are in the correct direction for this system. Bagley and Scigliano examined the cohesive energy of polymethylmethacrylate and they found a sharp rise in that parameter immediately below the glass transition temperature and a decrease above the transition temperature.[2] This behavior through a second-order transition point is similar to that shown in Fig. 4. Further, the breaks in Figs. 5 and 7 occur at 349.8°K and 4.475 atm and 351.5°K and 6.175 atm, respectively. The temperature shifts to higher values at higher pressures are in the expected directions.

The thermal coefficient of the wavelength of maximum intensity of reflected light exhibits a maximum at 349.6°K. The second transition is not confirmed in this measurement.

Nevertheless, the slope changes in the graphs of the thermodynamic state functions and in α and β indicate that cholesteryl stearate exhibits two second-order transitions within the cholesteric

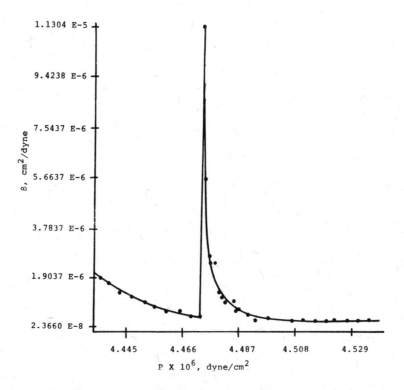

Fig. 6.   Compressibility of cholesteryl stearate
vs. pressure   in the cholesteric mesophase at 349.75°K.

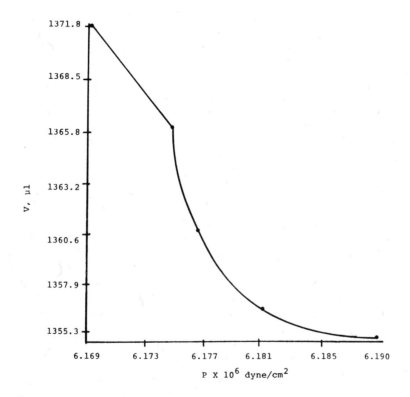

Fig. 7. Volume of cholesteryl stearate vs. pressure in the cholesteric mesophase at 351.45°K.

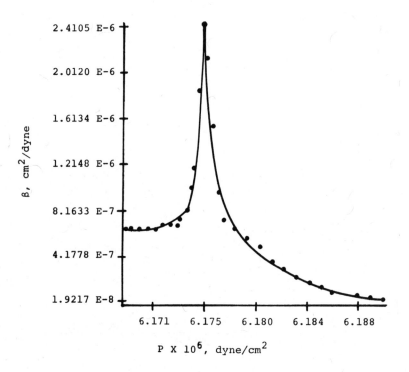

Fig. 8.   Compressibility of cholesteryl stearate
vs. pressure in the cholesteric mesophase at 351.45°K.

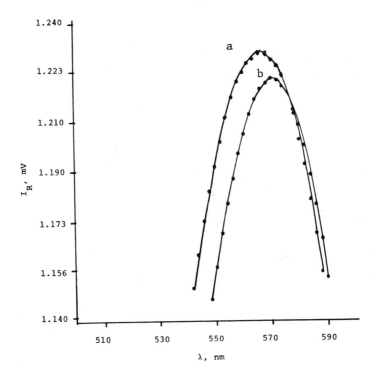

Fig. 9. Intensity of light reflected (i = r = 22.5°)
by cholesteryl stearate vs. wavelength of incident light in
the cholesteric mesophase. a: T = 348.72°K; b: T = 349.35°K.

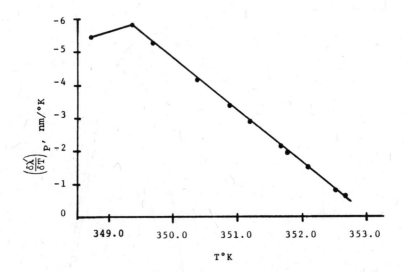

Fig. 10. The temperature derivative of the wavelength of maximum intensity of light reflected (i = r = 22.5°) vs. temperature of cholesteryl stearate in the cholesteric mesophase.

mesophase. Such a transition was noted in cholesteryl myristate.[3]

The variation of α in Fig. 2, near the temperature limits of the cholesteric mesophase, is similar to its behavior in the isotropic phase, where it was shown to exhibit about a four-fold decrease upon cooling through the transition temperature.[1] Immediately above the isotropic-cholesteric transition temperature (352.65°K), $\alpha = 1.44 \times 10^{-3}$ deg^{-1} at 353.15°K, while below the transition temperature in the cholesteric mesophase it is 3.23×10^{-4} deg^{-1} at 352.60°K. In the low temperature region of the cholesteric mesophase, $\alpha = 1.41 \times 10^{-2}$ deg^{-1} at 348.72°K. Below the cholesteric-smectic transition temperature (348.65°K), $\alpha = 2.93 \times 10^{-3}$ deg^{-1} at 347.18°K. Thus, the lower value of α, below and close to a transition temperature, can be associated with greater molecular order. Then the four-fold increase in α just above the cholesteric-smectic transition implies a disordering in this region of the cholesteric mesophase. We interpret this to mean that there is a pretransitional effect immediately above the cholesteric-smectic transition temperature, in which the cholesteric structure is perturbed, in part, while incipient smectic regions form. This would be analogous to the observation of cholesteric-like structures in the isotropic region of some cholesteryl esters.[1,4,5,6] Keating predicted that a pretransitional effect should occur above the cholesteric-smectic transition temperature and he ascribed it to the Lorentz field.[7] The reflectance measurements support this view. At 352.60°K λ' is 567 nm and the intensity of light reflected (i = r = 22.5°) is 8.7 mV. At 348.72°K λ' is 571 nm and the intensity is 1.2 mV which indicates that there is a significant disruption of the cholesteric organization just prior to the onset of the cholesteric-smectic transition.

REFERENCES

1. Waters, P. F. and Farmer, L. R., Colloid and Surface Science, Vol V, 97 (1976).
2. Bagley, E. B. and Scigliano, J. M., Polym. Eng. Sci., 11 (4), 320 (1971).
3. Barrall, E. M., Porter, R. S. and Johnson, J. F., Mol. Cryst., 3, 103 (1967).
4. Elser, W., Pohlmann, J. L. W. and Boyd, P. F., Mol. Cryst. and Liq. Cryst., 20, 77 (1973).
5. Coates, D. and Gray, G. W., Phys. Lett., 45A, 115 (1973).
6. Coates, D. and Gray, G. W., Phys. Lett., 51A, 335 (1975).
7. Keating, P. N., Mol. Cryst. and Liq. Cryst., 8, 315 (1969).

†Gillette Research Fellow, The American University.

DOMAINS DUE TO ELECTRIC AND MAGNETIC FIELDS IN BULK SAMPLES OF LIQUID CRYSTALS

E. F. Carr

Physics Department, University of Maine

Orono, Maine 04473

ABSTRACT

Domains are created by electric or magnetic fields in bulk
samples initially well aligned. The fields are applied perpendicular to the nematic director, and domains form which appear to
be separated by walls. The possibility of inversion walls are
considered. When applying electric fields (conduction regime) the
space-charge density should be a maximum at the walls because of
the conductivity anisotropy. The forces, due to the interaction
of the electric field with the space-charge at the walls, tend to
shear the sample. Because of shear flow, the director associated
with the sample between the walls is turned toward the electric
field. This mechanism is also likely to be involved in the
dynamic scattering mode.

INTRODUCTION

A mechanism for Williams domains[1] has been proposed[2] by considering an interaction[3] between an applied electric field and the
space-charge associated with the conductivity anisotropy, and this
model has been verified experimentally [4,5]. Although there is
fluid motion, the alignment of the molecules is stable for a small
range of low voltages applied to the electrodes. The work discussed here deals with molecular alignment and fluid motion at
higher voltages. The objective is to show that when high voltages
are applied, the initial behavior of the nematic material appears
to be similar to that giving rise to Williams domains, but walls
(defects) quickly form leading to shear flow alignment. This
applies to the dynamic scattering mode in thin samples and what

has often been referred to as anomalous alignment in bulk samples.
The results presented here involve bulk samples of nematics
exhibiting negative dielectric and positive conductivity anisotropies, but they will be related to work on thin samples to
appear elsewhere.[6]

EXPERIMENTAL

A schematic diagram of the apparatus is shown in Figure 1.
An unpolarized laser beam (approximately 1 mw) was directed
parallel to the free surface. The light which was scattered by
the liquid crystal in a direction perpendicular to the beam was
sufficient for observation of the phenomena at the free surface,
and the transmitted light was adequate for observations at the
rear electrode-to-liquid crystal interface. Figures 2, 3 and 5
involved observations of the free surface and the electrode-to-
liquid crystal interface simultaneously. The electrode separation
for the work discussed here was 0.5cm except for the results in
Figure 7 where it was 0.15cm. The camera and microscope could
be rotated for independent observations of the free surface or
electrode-to-liquid crystal interface.

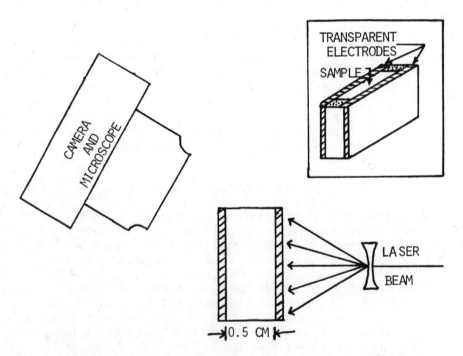

Figure 1. Schematic diagram for optical observations

The nematic mixtures used in this investigation were purchased from Eastman Organic Chemicals (Cat. No. 11643) and E. Merck (Cat. No. 5A), and all measurements were made at room temperature. The dielectric anistropy ($\varepsilon_{\shortparallel}' - \varepsilon_{\perp}'$) was -1.6, and the ratio of the conductivities ($\sigma_{\shortparallel}/\sigma_{\perp}$) parallel and perpendicular to the director was 2.0 for the Eastman material. The resistivity was approximately 10^9 ohm-cm. The dielectric anisotropy ($\varepsilon_{\shortparallel}' - \varepsilon_{\perp}'$) was -0.2 for the Merck material. This material was only used for the results in Figure 7.

DOMAINS DUE TO A 50 HZ ELECTRIC FIELD

A photograph of the sample holder (filled with the sample), with the director aligned parallel to both the free surface and the conductive coated (gold) glass electrode, is shown in Figure 2a. A magnetic field was used to align the director as described, but was removed before applying an electric field. The electrode separation was 0.5 cm, and the experimental setup is shown in Figure 1. The photograph in Figure 2b was taken after a 50 volt (50 Hz) source had been applied for 7 min. Domains with a separation of approximately 800 microns can be seen at the free surface and the electrode-to-liquid crystal interface. The time lapse between turning off the magnetic field and turning on the electric field could normally be several minutes without appreciably affecting the results.

In some respects the domains in Figure 2b resemble Williams domains[1], but many of the conditions normally satisfied when observing Williams domains were not satisfied here. The applied voltage was higher and the sample was much thicker in this case. The creation of Williams domains depends primarily on the voltage applied to the electrodes; while in our case the results depend primarily on the electric field intensity, and the alignment is changing while the field is applied. Conductivity measurements have indicated that the average angle of rotation of the director might be as much as 30° when the photograph in 2b was taken, which is larger than one would expect for a normal Williams pattern. When the electric field was left on much longer than 7 min it became increasingly evident that walls (defects) were being created. This was determined from observations of the free surface. After 7 min the pattern at the electrode-to-liquid crystal interface became less clear. Earlier work[8] using an electrode separation of 0.15 cm had indicated that domains created by electric fields were similar to those created by magnetic fields, but it was necessary to wait several minutes before making observations at the electrode-to-liquid crystal interface when using an electric field. This will be discussed later.

(a)

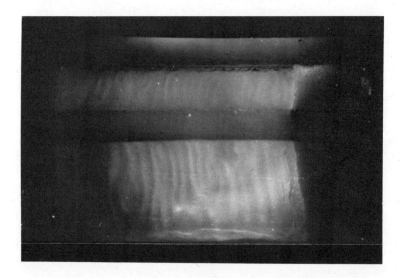

(b)

Figure 2. Electric field effect in Eastman nematic (No 11643).
Electrode separation = 0.5cm,   and electrode dimensions = 1.0x1.7cm
(a) Sample holder and sample with director parallel to both the free
surface and electrode-to-liquid crystal interface. (b) After applying a 50 volt (50 Hz) source for 7 min with no magnetic field
present.

An interesting observation was made using both an electric and a magnetic field. In Figure 3 a 50 volt source had been on 7 min followed by a 2 kG magnetic field perpendicular to the elec- trodes for 0.5 min. After both fields have been applied the sample should be reasonably well aligned, with the director perpendicular to the electrodes, except at the walls (defects) which can be easily seen in the photograph. The walls (not domain spacing) became narrower when the magnetic field was increased. This would be expected if the walls were inversion walls[7] of the splay-bend or twist type. From many observations at much higher electric fields, we believe that the walls tend to form rather quickly. Initially they should be more like splay-bend walls, but energy considera- tions[9] normally favor twist walls. They are probably some combina- tion of the two types and vary with time. When a sheet of polaroid was placed in front of the microscope, the walls were usually fainter for light polarized perpendicular to the walls which indicated that they were more like twist walls than splay-bend walls.

MOLECULAR ALIGNMENT IN HIGH ELECTRIC FIELDS

It was assumed[2] that distortions of the director for Williams domains can be represented by sinusoidal functions, but for patterns at high electric fields we are proposing a model illus- trated in Figure 4. When a high dc (or ac field in conduction

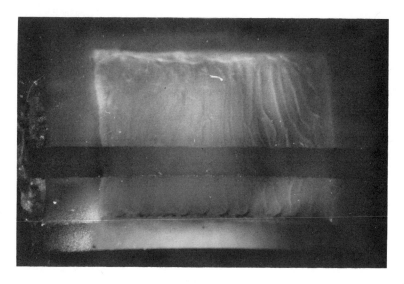

Figure 3. Walls in Eastman Nematic after applying a 50 volt (50 Hz) source for 7 min (Figure 2b) followed by a 2 kG magnetic field perpendicular to electrodes for 0.5 min.

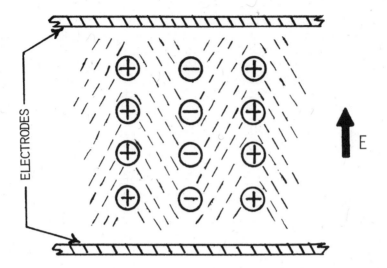

Figure 4. Model for molecular alignment and material flow due to
an external electric field. Charge accumulates at the walls (defects) which are perpendicular to the electrodes and the plane of
paper.

regime) field is turned on, a pattern resembling Williams domains
first appears but this pattern is soon changed to one illustrated
in Figure 4. This time period will normally be much less than a
second for very high fields. When the alignment becomes distorted
due to the electric field, walls tend to form where the fluid is
moving with its greatest velocity. This was shown in earlier work[6]
involving samples 0.15 cm thick. As these walls start forming the
space charge density increases at the walls because of the conductivity anisotropy. Forces due to the interaction of the electric field with the space charge at the walls (defects) tend to
shear the sample. Because of shear flow,[10] the director associated
with the sample between the walls is turned toward the electric
field as illustrated in Figure 4. This model also applies to ac
electric fields providing the frequency is well below the space-
charge relaxation frequency.

    The model in Figure 4 appears to explain some results from
magnetic resonance experiments[11]. When the aligning effect of
the electric field (conduction regime) is much greater than that
due to the magnetic field, these experiments have indicated that
in a large portion of the sample, the director aligns at small
angles with respect to the electric field. These angles vary
with different materials and temperature and may be greater than

20 degrees. The angle that the director makes with the electric
field could be compared with the flow-alignment angle[10] since this
angle can be determined from other experiments. However, it is
difficult to make a serious comparison at the present time because
there appears to be some uncertainty in the literature concerning
the measurements of this quantity.

Flow patterns similar to those discussed here have been
observed[6] at the free surface in samples with electrodes separations of 0.15 cm and 135 microns. For the 135 micron separation,
dynamic scattering was observed when viewing perpendicular to the
electrodes, while flow patterns which appeared to involve walls
were observed at the free surface. This suggests that the model
in Figure 4 may be involved in the dynamic scattering mode.

DOMAINS DUE TO A MAGNETIC FIELD

The photograph in Figure 5 shows domains due to a 500 G magnetic field. The director was initially aligned parallel to the
electrodes before applying the field perpendicular to the electrodes. This photograph was taken 4 min after turning on the
500 G field. Domains, with a spacing of approximately 400 microns,
can be seen at the free surface and the electrode-to-liquid crystal
interface. The domain width is less than that shown in Figure 3,
but the spacing in Figure 3 was due to the 50 volt source.
Domains due to magnetic fields in samples with an electrode

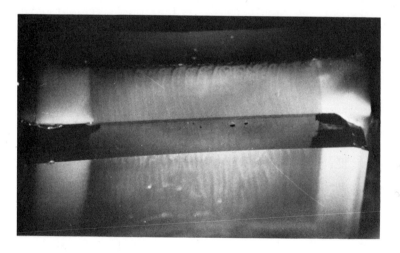

Figure 5. Domains due to a magnetic field. A 500 G Field was
applied perpendicular to the electrodes for 4 min. The director
was initially parallel to both the free surface and electrodes.

separation of 0.15 cm were discussed elsewhere[8]. We believe that
these domains were separated by walls, which are probably a combination of the splay-bend and twist type, but behave more like twist
walls.

FLUID FLOW

Figure 6 gives information on the movement of the liquid
crystal while a 400 volt (50 Hz) source was applied to the electrodes (electrode separation = 0.5 cm). Observations were made
at the free surface. The photograph in Figure 6a was taken after
a 2 kG magnetic field had been applied parallel to the electrodes
for 1 min. The director was initially aligned perpendicular to
the electrodes before applying the 2 kG field. Defects, which we
believe are walls, can be seen parallel to the electrodes. The
formation of these defects will be discussed elsewhere[12]. The
reason for presenting them here is only to gain information about
the movement of the fluid. In Figure 6a the director was parallel
to the electrodes except at the defects. When the 400 volt source
was turned on (magnetic field was zero), the defects start to move
indicating the movement of the fluid. The photograph in Figure 6b
was taken approximately 2 sec after turning on the 400 volt source
and that in 6c was taken 4 sec after turning on the field. The
shape of the defects gives information[6] about the flow cell width,
and the movement of the defects, as determined from successive
photographs, gives information about the average velocity of the
fluid. These photographs show that the distortion of the defects
in the center of the sample tends to look more like a triangular
wave then a sine wave. This suggests shear flow which is consistant with our previous discussion.

Figure 6 shows that the velocity of the fluid increases with
flow cell width. Attempts were made to measure the maximum
velocity of the fluid (near center of sample) as a function of the
electric field intensity. The electric field was varied from 200-
3200 V/cm. There was considerable variation in velocity, but if
the flow cell width was not appreciably changed, the maximum
velocity seemed to increase as the square of the electric field
intensity (E). However, there was too much variation to establish
an E^2 dependence. Efforts are now underway to make measurements
without being involved with changes in molecular alignment. If the
alignment is not changing the space charge density at the walls
(defects) should be roughly proportional to the electric field
intensity making the force on the walls proportional to E^2.

Kai, Yoshitsune and Hirakawa[13] have made flow velocity
measurements in MBBA. They measured average velocities by
observing the movement of particles. Although the size and
geometry of our sample holders are quite different from that of

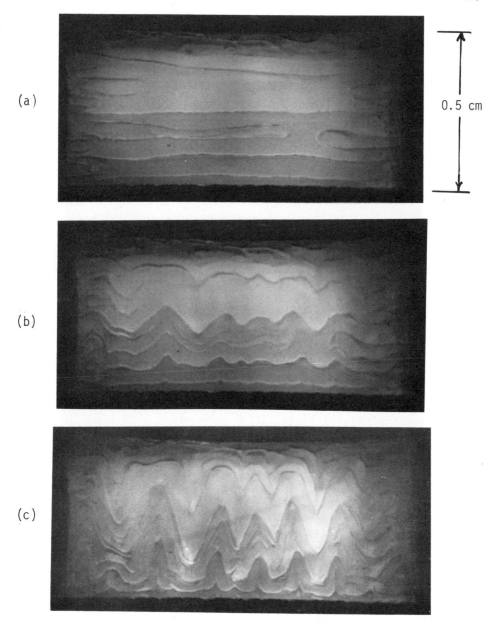

Figure 6.Distortion of walls due to a 50 Hz electric field at the free surface. Electrode separation = 0.5cm. (a) A 2 kG magnetic field applied parallel to the electrodes for 1 min. The director was initially perpendicular to the electrodes (b) Two sec after applying a 400 volt source. (c) Four sec after applying a 400 volt source.

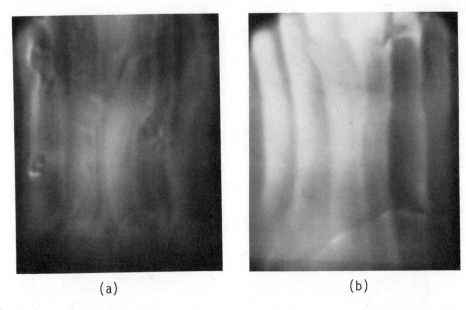

(a)   (b)

Figure 7. Domains due to a 40 Hz electric field at the electrode-to-liquid crystal interface in the Merck Nematic (5A). Electrode separation = 0.15 cm. (a) Forty sec after applying a 50 volt source. (b) Ten min after turning off the 50 volt source which had been on for 40 sec.

Kai et.al., the results appear to be consistant with what they have reported.

DOMAINS AND LIGHT SCATTERING

The results in Figure 7 were obtained using a sample holder with an electrode separation of 0.15 cm. The sample was purchased from E Merck (5A). The patterns were photographed at the electrode-to-liquid crystal interface. The laser beam was directed into the sample through the free surface near the back electrode. The light that was scattered by the sample through the electrode was sufficient for observation of the phenomena.

The photograph in Figure 7a was taken immediately after turning off a 50 volt (50 Hz) source which had been on for 40 sec. The director had been initially aligned parallel to the electrodes. Patterns in Figure 7a are difficult to identify, but Figure 7b, taken 10 min after the electric field was turned off, shows domains. These results imply that the flow cells, which were created by the electric field, caused other distortions in the alignment which relaxed before the alignment associated with the flow cells became destroyed. This implies that distortions which are caused by flow within the flow cells may be responsible for some of the scattering in the dynamic scattering mode. The work presented here involves bulk samples, but flow cells similar to those discussed have been observed[6] in samples with an electrode separation of 135 microns.

CONCLUSION

When electric fields of sufficient magnitude are applied to a nematic liquid crystal, causing large distortions in the align-ment, the mechanism illustrated in Figure 4 is probably involved. This includes the dynamic scattering model[14] in thin samples, and what has often been referred to as anamalous alignment[3] in bulk samples. When the electric field is applied, the initial behavior of the nematic material is similar to that giving rise to Williams domains but walls (defects) start to form quickly. The space charge density should be maximum near where the wall is forming, and eventually most all the space charge should be concentrated on the wall because of the conductivity anisotropy. The forces, due to the interaction of the electric field with the space charge at the walls, tend to shear the sample. Because of shear flow, the director associated with the sample between the walls (defects) is turned toward the electric field. It is interesting to note that Heilmeier[14] et.al. reported that dynamic scattering could be produced by mechanically shearing the sample.

We do not intend to imply that dynamic scattering is completely explained by the mechanism discussed here but rather that his mechanism is very likely to be involved in the dynamic scattering mode. The work discussed here assumes that we start with a well aligned sample and apply the electric field perpendicular to the nematic director. After the field has been on an appreciable length of time many of the walls will have been destroyed but investigations[8,15] involving flow cell width as a function of electric field intensity imply that new walls are continuously being created.

ACKNOWLEDGEMENT

The author wishes to express his appreciation to Mr. R.W.H. Kozlowski for many helpful discussions and to Miss Karen Carr for her efforts in preparing the illustrations.

REFERENCES

1. R. Williams, J. Chem. Phys. 39, 384 (1963).

2. W. Helfrich, J. Chem. Phys. 51, 4092 (1969).

3. E.F. Carr, Mol. Cryst. and Liq. Cryst. 7, 253 (1969).

4. G. Durand, M. Veyssie, F. Rendelez, and W. Leger, C.R. Acad. Sci. B 270, 97 (1970).

5. P.A. Penz, Phys. Rev. Lett. 24, 1405 (1970)

6. E.F. Carr, P.H. Ackroyd and J.K. Newell, Mol. Cryst. and Liq. Cryst. (to be published).

7. W. Helfrich, Phys. Rev. Lett. 21, 1518 (1968).

8. E.F. Carr, Mol. Cryst. and Liq. Cryst. 34, 159 (1977).

9. P.G. deGennes, The Physics of Liquid Crystals, Clarendon Press, Oxford (1974).

10. F.M. Leslie, Arch. Ration. Mech. Anal. 28, 265 (1968); Ch. Gahwiller, Mol. Cryst. and Liq. Cryst. 20, 301 (1973) and Phys. Rev. Lett. 28, 1554 (1972); S. Meiboom and R.C. Hewitt, Phys. Rev. Lett. 30, 261 (1973); P. Pieranski and E. Guyon Phys. Rev. Lett. 32, 924 (1974).

11. J.H. Parker, Thesis, University of Maine (1971); T.E. Kubaska, C.E. Tarr, and T.B. Tripp, Mol. Cryst. Liq. Cryst. 29, 155 (1974); J.C. Rowell, W.D. Phillips, L.R. Melby, and M. Panar, J. Chem. Phys. 43, 3442 (1965); E. Gelerinter, A.L. Berman, G.A. Fryburg, and S.L. Golub, Phys. Rev. A9, 2099 (1974); G.R. Luckhurst, Chem. Phys. Lett., 9, 289 (1971); M. Schara and M. Sentjurc, Solid St. Commun., 8, 593 (1970).

12. R.W.H. Kozlowski and E.F. Carr, unpublished results.

13. S. Kai, K. Yamaguchi and K. Hirakawa, J. Phys. Soc. (Japan) 40, 267 (1976).

14. G.H. Heilmeier, L.A. Zanoni and L.A. Barton, Proc. IEEE 56, 1162 (1968).

15. E.J. Sinclair and E.F. Carr, Mol. Cryst. and Liq. Cryst. 37, 303 (1976).

THE DETERMINATION OF MOMENTS OF THE ORIENTATIONAL DISTRIBUTION FUNCTION IN LIQUID CRYSTALS BY THE DEPOLARIZATION OF FLUORESCENCE OF PROBE MOLECULES

L. Lawrence Chapoy
Instituttet for Kemiindustri
The Technical University of Denmark
DK-2800 Lyngby

Donald B. DuPré
Department of Chemistry
University of Louisville
Louisville, Kentucky 40208 U.S.A.

Edward T. Samulski
Department of Chemistry and
Institute of Materials Science
University of Connecticut
Storrs, Connecticut 06268 U.S.A.

1. INTRODUCTION

The approach to the mathematical formulation of the distribution function that describes long range orientational ordering of molecules in liquid crystals is necessarily an approximation. Some workers propose model forms with one or more adjustable parameters used to fit data of some experimental property sensitive to an average over this unattainable function. These attempts are open to criticism of the reasonableness and uniqueness of the chosen form of the approximation. An alternative procedure, which is mathematically rigorous and unambiguous, is to write the function as a truncated series expansion whose coefficients are experimentally determinable moments of the real distribution. This method is usually limited by the availability of only the second moment. Raman[1-4], electron spin resonance[5], and fluorescent emission spectroscopy[1,2] are capable of supplying both the second and fourth moments, hence extending the expansion and making it a more fiithful representation of the real distribution. The fluorescence technique, however, has inherent complications in that there is usually a significant delay (1-10 nsec)

Acknowledgement: The authors would like to thank Statens teknisk-videnskabelige Forskningsrad, Contract 516-6569. K-339 for partial support of this work.

between the absorption and emission process and emitted radiation may emanate polarized along a different axis in the excited molecule. The effect of rotational Brownian motion of the fluorescent site during the lifetime of the excitation and the nonparallelism of absorption and emission oscillators has been considered[6] and an expression derived for the anisotropy of emission which contains the possibility of a nonrandom distribution of fluorescent sites before photoselection[7,8] (a process that curiously creates a nonrandom population of molecules even in an initially isotropic collection). A nonzero intramolecular energy transfer angle, δ, was included; and rotational relaxation appeared through a reduced variable, τ/τ_R, the ratio of fluorescence lifetime to the rotational relaxation time. It was shown that considerable misinterpretation of experimental data, and hence the form of the orientational distribution function, can occur if the analysis of the experiment is not properly placed in the regime of δ and τ/τ_R. After a short review of some fundamental problems with the use of fluorescence depolarization as a probe to molecular order, new data will be presented for the thermotropic liquid crystal p-methoxybenzylidene-p'-n-butylaniline (MBBA).

2.   THEORY

If we consider a macroscopically aligned liquid crystal situated such that the major symmetry axes of the specimen are parallel to a laboratory baded coordinate frame O-xyz, a representative molecule whose unique symmetry axis (long axis of the liquid crystal molecule) is denoted by $\sim$ can be located by polar and azimuthal angles (θ,ϕ). In discussing the orientational order of a collection of such molecules, it is convenient to introduce a distribution function $f(\theta,\phi)$, relating the vectors $\sim$ to the macroscopic coordinates of the specimen. The function is a measure of the relative density of the molecular vectors $\sim$ on the surface of a unit sphere surrounding the sample. Alternatively $f(\theta,\phi)$ $\sin\theta d\theta d\phi$ may be regarded as the fraction of molecules with their long axes found within the solid angle $\sin\theta d\theta d\phi$ and is thus normalized to unity. In many liquid crystals only uniaxial synmetry about the director is present and we only need consider an abbreviated orientation distribution, $f(\theta)$. In this case the angle ϕ is an unimportant random variable. We will limit our discussion to this symmetry case.

In liquid crystal research it is the function $f(\theta)$ that is sought as it completely describes the long range and anisotropic orientational order that gives these fluids their unique physical properties. The significance of $f(\theta)$ is to liquid crystals as the single particle distribution function is to "normal" liquids. It is equally   difficult to obtain. The function, however, may be reproduced with accuracy sufficient for most purposes by considering the following expansion[9-11] in terms of the complete set of Legendre polynomials, $P_l (\cos\theta)$:

$$f(\theta) = \sum_0^\infty c_l P_l (\cos\theta)$$

(1)

l even

where the coefficients c_l are given by:

$$c_l = \frac{2l+1}{2} \int_{-1}^1 P_l (\cos\theta) f(\theta) d(\cos\theta) = \frac{2l+1}{2} <P_l (\cos\theta)>$$

(2)

(Uniaxial symmetry conditions on $f(\theta)$ eliminate the odd terms in the expansion). The first few averages can be written explicitly in terms of the lowest moments of the distribution $\overline{\cos^2\theta}$ and $\overline{\cos^4\theta}$ as:

$$<P_o> = 1$$

$$<P_2> = \frac{1}{2} (3\overline{\cos^2\theta}-1)$$

$$<P_4> = \frac{1}{8} (35\overline{\cos^4\theta} - 30\overline{\cos^2\theta} + 3)$$

(The bar indicates the same type of statistical average as in Eqn. (2) and will be used interchangeably with the brackets where space may be conserved). The average of the second Legendre polynomial $<P_2>$ is usually reported as S, the liquid crystal order parameter[12], but it is seen that $<P_4>$ contains further information about orientational order present and should in fact be more sensitive to molecular fluctuations because of the higher powers of the deviation angle. $<P_4>$ could, therefore, be termed the hyperorder parameter.

Although X-ray scattering can in principle measure all the moments of the distribution $f(\theta)$, in practice it is limited by the lack of definite repeating crystalline order in liquid crystals. Optical birefrengence; IR, visible, and UV dichroism measurements; and magnetic resonance techniques are the most frequently cited means of determining the degree of liquid crystalline order. The optical methods are inherently limited to only a measure of $\cos^2\theta$. Both the second and fourth moments are available, however, from suitable analysis of ESR[5] and polarized Raman[1-4] and fluorescent emission[1,2] spectra of active sites (embedded probes or sensitive molecular elements) in the liquid crystal. Hence a description of $f(\theta)$ up to the third term of the series

Eqn. (1) is possible.

Fluorescent emission has many things in common with the Raman scattering as it is a two step spectroscopic process modulated by molecular motion. The singular difference is that, on the scale of time, the Raman effect is instantaneous (at least as fast as a molecular vibration, 10^{-12} sec.) whereas fluorescent emission may follow absorptive excitation by many nanoseconds, perhaps even emanating from a different direction in the molecule. These factors lead to experimental and interpretative difficulties particularly when relaxation processes occur. There also exist complications when intermolecular energy transfer processes occur which also depolarize the emission in an (as now) unpredictable manner.

The intensity of light emitted that is observed emerging through a polarization analyser placed directly along a laboratory polarization axis j, resulting from light absorbed from an excitation source polarized directly along laboratory axis i is given by[9,13]:

$$I_{ij} = <M^2_{ai} \ M^2_{ej}> \qquad\qquad (4)$$

where M^2_{ai} and M^2_{ej} are the squares of the absorption and emission oscillator components projected onto the laboratory axes i and j, respectively. The average is over all molecules in the illumination volume and includes all possible angular positions of the absorption and emission oscillators weighted by the appropriate statistical distributions. Instrumental, concentration, and volume factors have been incorporated into the definition of each I_{ij}, which are thus reduced intensities as it is the effect of the geometric disposition of fluorescent molecules that one is seeking to clarify.

The extent of the partially polarized fluorescent emission is frequently quoted in terms of r, the <u>emission anisotropy</u> defined through: $r = (I_{\parallel} - I_{\perp})/(I_{\parallel} + 2I_{\perp})$, where $I_{\parallel}$ and $I_{\perp}$ are the emission intensities measured in a right angle experimental geometry with polars parallel and perpendicular. In this definition of r, the excitation polarization is taken to be normal to the base plane and parallel to the major symmetry axis of the sample. For a randomly oriented collection of immobile molecules whose absorption and emission moments are parallel, $r_o = 2/5$ where the subscript indicates the random case. If the absorption and emission oscillators form a fixed, nonzero intramolecular angle δ, the Polarization of emission is reduced[14] and $r_o = 2/5 \cdot P_2(\cos\delta)$. r_o is therefore an <u>intrinsic</u> polarization anisotropy factor internal to the fluorescent molecule. The quantity may be measured in a random glass of a suitable solvent or in the isotropic phase of the liquid crystal quenched to freeze out thermal motions that

also act to reduce the emission anisotropy. r_o has the range:
$-1/5 \le r_o \le 2/5$ as δ may vary between 0 and $\pi/2$.

If the distribution of molecular orientations is not random
but otherwise static, it can be shown that[6]:

$$r = \left(\frac{\overline{9\cos^4\theta} - \overline{3\cos^2\theta}}{\overline{6\cos^2\theta}} \right) \cdot P_2(\cos\delta) \qquad (5)$$

The term in parenthesis reduces to the constant factor of 2/5 in
the isotropic limit. In anisotropic media this quantity gives
the information sought about the moments of the distribution, $f(\theta)$.
Note should be taken that r could range as high as 1.0 in anisotropic media if perfect order and a zero intramolecular angle δ
obtained.

Another experiment is necessary however to separate $\overline{\cos^2\theta}$ from
$\overline{\cos^4\theta}$. A measurement of the absorption dichroic ratio, D =
$\langle M_{az}^2 \rangle / \langle M_{ax}^2 \rangle$, would be sufficient as: $\overline{\cos^2\theta}$ = D/D+2. Choosing another fluorescence geometry will also provide a second expression
for the moments which can be solved simultaneously with Eqn. (5).
If we define a new polarization anisotropy r', measured in the
right angle geometry with the incident polarization lying now in
the xy plane, i.e., normal to the major symmetry axis of the
sample, then[6]:

$$r' = \frac{(\overline{6\cos^2\theta} - \overline{5\cos^4\theta} - 1)}{(5 - \overline{6\cos^2\theta} + \overline{\cos^4\theta}) + (\overline{6\cos^2\theta} - \overline{3\cos^4\theta} - 3)\cos^2\delta} \cdot P_2(\cos\delta) \qquad (6)$$

The intramolecular vector angle δ does not factor out completely
as in Eqn. (5), but $\cos^2\delta$ is a number intrinsic to the molecule
and can be obtained from measurements on the random "glassy" state
as noted above.

Considerations so far explicitly assumed that the experiment
monitors a statistical average over a large collection of molecules
statically oriented throughout the lifetime of the fluorescence
process. Rotational diffusion of the fluorescent sites will result
in a further degradation of the polarization. Due to the anisotropic nature of the absorption process, molecules in certain
orientations with respect to the incident beam polarization are
more likely to become excited. Polarized illumination therefore
has the effect of producing an oriented population of molecules
within the medium even when the initial distribution of all molecular axes is random. That is, inherent in the physics of the
experiment, we always deal with a subset of all molecules present
in the illuminated volume, those not eliminated through unfavorable

orientations in the initial absorption step. The process may be referred to as photoselection[7]. In the aligned liquid crystal photoselection occurs in an already ordered condition of the molecules so that initial anisotropic and isotropic molecular organizations are still distinguishable.

Diffusion complications may be introduced in terms of a _time dependent orientational distribution function of the photoselected population_, $g(\theta,t)$ which is governed by a rotational diffusion equation of the form[15]:

$$\partial g/\partial t = D[\frac{1}{\sin^2\theta}\frac{\partial^2}{\partial\phi^2} + \frac{1}{\sin\theta}\frac{\partial}{\partial\theta}(\sin\frac{\partial}{\partial\theta})]\, g(\theta,t) \qquad (7)$$

where D is a diffusion constant. A general solution to Eqn. (7) is available which is made subject to the photoselection initial condition that: $g(\theta,0) = M_{ai}^2\, f(\theta)$. The above is simply a statement that from the moment of illumination, the orientational distribution function followed is weighted by the probability of initial absorption, M_{ai}^2. The time evolution of the intensity elements, Eqn. (4), will thus be given by:

$$I_{ij}(t) = \iint M_{ej}^2\, g(\theta,t)\, P(t)\, \sin\theta d\theta d\phi \qquad (8)$$

where $I_{ij}(0) = I_{ij}$, and P(t) is the probability that a molecule emits light at time t during the emission lifetime[8,16]. P(t) may be represented in exponential form as $\tau^{-1}e^{-t/\tau}$, where τ is the emission lifetime.

It follows that[6], for the steady state experiment:

$$r(\tau/\tau_R;\delta) = \frac{\bar{P}_2(\cos\theta)+(\tau/\tau_R+1)^{-1}[\frac{72}{70}\bar{P}_4(\cos\theta)+\frac{4}{7}\bar{P}_2(\cos\theta)+\frac{2}{5}]}{1+2\cdot(\tau/\tau_R+1)^{-1}\cdot\bar{P}_2(\cos\theta)} \times$$

$$P_2(\cos\delta) \qquad (9)$$

where $\tau_R = 1/6D$ is the rotational relaxation time.

If the emission is instantaneous ($\tau=0$) or the rotational diffusion highly hindered ($\tau_R>>\tau$), Eqn. (9) reduces to Eqn. (5) above for static anisotropic media.

On the other hand if rotational motion of the probe relaxes out much quicker than the lifetime of the excitation ($\tau_R<<\tau$), Eqn. (9) becomes:

$$r(\infty;\delta) = \quad P_2(\cos\theta)>\cdot P_2(\cos\delta) \qquad\qquad (10)$$

and no information about the fourth moment of the distribution is
obtainable. In this case the fluorescence experiment supplies no
more information than an absorption measurement but is complicated
by the possible noncoincidence of absorption and emission oscillators of the probe molecules.

In the limit of totally random molecular order, Eqn. (9) reduces to Perrin's equation[17]:

$$r = 2/5\cdot P_2(\cos\delta)\cdot(\tau/\tau_R+1)^{-1} \qquad\qquad (11)$$

While Perrin's Eqn. (11) does not apply in liquid crystal systems,
Eqn. (9) is appropriate in such anisotropic media and contains
more information about the distribution of molecular order. Fig-
ure 1 is a plot of $r(\tau/\tau_R;\delta)$ versus τ/τ_R for various values of the
molecular angle δ with reasonable choices of $\cos^2\theta = 0.6$, $\cos^4\theta =$
0.4 for the liquid crystal. It is clear that a considerable misinterpretation of fluorescent emission data can occur if it is
not certain where the experiment falls in the regime of relaxation times and intramolecular energy transfer angles, δ. A faulty
approximation to the distribution function will result, if esti-
mates of δ and τ/τ_R are substantially in error. A measurement of
the quantity r_0 to obtain the value of δ is therefore necessary.
One possible means of avoiding, or minimizing, the rotational
difficulty would be to rapidly quench the sample at each temperature within the mesomorphic range freezing out thermal motion
while preserving the molecular order of the higher temperature
phase of interest. This would assure that the experiment is
placed on the far left of the curves of Figure 1.

3. EXPERIMENTAL SECTION

The thermotropic liquid crystal MBBA was examined. Two candidates for probe molecules, tetracene and dimethylamino-nitro-
stilbene (DS) were considered. These compounds have been shown[18]
to be soluble in MBBA, with absorption bands shifted to the red of
the liquid crystal primary UV absorption.

r_0 was determined as the limit r in the viscous solvent di-
methylphthalate $[T/\eta \rightarrow o]$ where T and η are the temperature and
solvent viscosity, respectively. Tetracene was rejected due to its
very low value of r_0 (< .02) and its propensity to bleach as
evidenced by rapid and sustained decreases in emission intensity
under conditions of continuing illumination. The value of r_0
of DS, however, was close to the theoretical value for an ideal

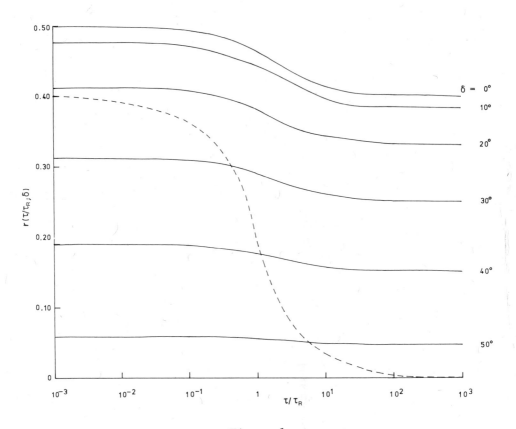

Figure 1

Anisotropy of fluorescent emission in uniaxially ordered system as a function of the reduced relaxation time variable τ/τ_R for various intramolecular energy transfer angles δ. Solid curves were calculated assuming $\overline{\cos^2\theta} = 0.6$, $\overline{\cos^4\theta} = 0.4$. The broken curve is Perrin's equation, for the isotropic condition, with $\delta = 0°$.

probe molecule with intramolecular energy transfer vector angle of $\delta = 17°$.

Temperature and viscosity measurements also indicated that $\tau/\tau_R \sim \tau T/\eta V \ll 1$, where V is the effective molecular volume of a probe molecule. Thus the complications of rotational diffusion can be neglected for this probe.

The liquid crystal was doped at a concentration of 10^{-2} M DS. Differential thermal analysis showed the clearing temperature depressed to $T_c = 38°C$. Monodomain samples were prepared on glass slides by the familiar rubbing technique. Absorption and fluorescent emission intensities were measured with polars parallel and perpendicular to the axis of the director. In the fluorescence experiments, the excitation polarization was also placed parallel and perpendicular to the director in order to obtain both r and r'. The instrumentation consisted of a Beckman DK-2A visible ultraviolet double beam absorption spectrophotometer and a Perkin-Elmer MPF-2A fluorescence spectrophotometer. The latter is equipped with dual slits and monochromators. Excitation was performed at 445 nm and emission was observed at 570 nm with slits set at 10 nm. In spite of extensive light scattering from the liquid crystal, complete resolution was attained between the broad scattering peak and the probe emission. The emission at 570 nm was a broad featureless structure. The emission intensities were obtained in the "ratio mode" which compensates for fluctuations in excitation lamp intensity, thus stabilizing the observed signal. Additional stabilization of the signal is effected by averaging the signal in the real time domain using a Solartron 1860 Computing Digital Voltmeter. This procedure averages out low frequency noise, which cannot be effectively removed by capacitative filtering, according to: $<i> = \frac{1}{t} \int_0^t i(t)dt$, where i(t) is the effective integrated average of this signal. Precision in the determination of emission intensities is essential as the determination of r involves small differences in large numbers.

Dichroic ratios of the preparation were determined as a function of temperature throughout the liquid crystal region and the order parameter $S = <P_2>$ calculated. The determination of $\overline{\cos^4\theta}$, and hence $<P_4>$, proceeded as outlined above with one exception. The crystallization rate of MBBA was found to be too fast to obtain a clear glass that we have suggested as a means to eliminate rotational relaxation complications. The extremely small value of τ/V deduced from Perrin's relation in viscous, random solvents for the DS probe leads us to believe, however, that the experiment falls on the left of Figure 1 and fluorescence depolarization data were analyzed under this assumption. The value of r obtained in the isotropic phase of the liquid crystal mixture is approximately the same as that found in viscous liquids in the limit of

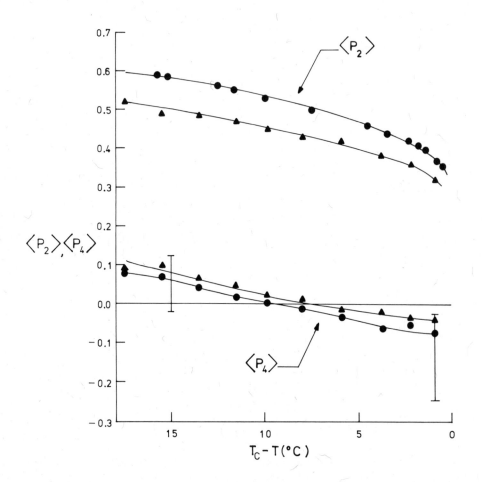

Figure 2

Experimental values of the liquid crystal order parameters $\langle P_2 \rangle$ and $\langle P_4 \rangle$. [Triangles (Δ), from simultaneous solution of Eqns. (5) and (6); circles (0), from combination of fluorescent emission and optical dichroism data.] The error bars are those reported on similar measurements of $\langle P_4 \rangle$ by Raman scattering[4].

TABLE 1

Fluorescent emission anisotropies and moments of the orientational
distribution function for the liquid crystal MBBA.

$T-T_C$ (°C)	r	$\cos^2\theta$*	$\cos^4\theta$	$\langle P_2\rangle$*	$\langle P_4\rangle$	r'	$\cos^2\theta$	$\cos^4\theta$	$\langle P_2\rangle$	$\langle P_4\rangle$
17.5	0.57	0.73	0.56	0.59	0.08	0.41	0.68	0.52	0.52	0.09
17.5	0.56	0.73	0.56	0.59	0.08	0.40	0.66	0.50	0.49	0.10
13.5	0.55	0.71	0.53	0.57	0.04	0.39	0.66	0.49	0.49	0.068
11.5	0.54	0.70	0.52	0.55	0.02	0.38	0.64	0.48	0.47	0.050
9.8	0.53	0.69	0.51	0.54	0.006	0.37	0.63	0.46	0.45	0.025
8.0	0.52	0.67	0.49	0.51	-0.01	0.36	0.62	0.45	0.43	0.014
5.9	0.51	0.66	0.47	0.49	-0.03	0.34	0.61	0.44	0.42	-0.008
3.7	0.49	0.63	0.44	0.45	-0.06	0.33	0.59	0.42	0.38	-0.015
2.2	0.48	0.61	0.43	0.42	-0.05	0.31	0.57	0.40	0.36	-0.030
0.9	0.47	0.59	0.40	0.38	-0.07	0.26	0.54	0.37	0.32	-0.035
iso-trope	0.34	0.33	0.20	0.00	0.00	(0.00)	0.33	0.20	0.00	0.00

*From optical dichroism

$T/\eta \rightarrow o$ [i.e. 2/5 $P_2(\cos\delta)$], further demonstrating that Brownian
motion is not a problem. Values of $\cos^{2,4}\theta$ thus obtained by both
procedures are presented in Table 1 and are plotted in terms of
$<P_2>,<P_4>$ in Figure 2. The $<P_2>$ values from optical dichroism
measurements are in good agreement with those obtained for MBBA
by NMR, Raman, magnetic susceptibility, and optical methods[4].
$<P_2>$ values obtained from the simultaneous solution of fluorescence
depolarization Eqns. (5) and (6) are slightly and consistently
lower than most other literature values. $<P_4>$ values calculated
by either procedure fall within the error bars presented in the
previous Raman measurements[4] of this fourth moment and follow
the same trend. The significance of the low and negative values
of $<P_4>$ in relation to predictions from mean field theory has been
previously discussed[4,5,19]. We note that the discrepancy with
theory cannot be resolved by invoking multiple scattering as our
excitation source is of moderate intensity and the fluorescent
emission band monitored (at 570 nm) is well removed from the
scattering peak.

REFERENCES

1. D. I. Bower, J. Poly. Sci. Poly. Phys. Ed. 10 2135 (1972).
2. D. I. Bower, Structure and Properties of Oriented Polymers,
    I. M. Ward, Ed., Applied Science Publ. Ltd. London 1975.
    Chapter 5.
3. E. B. Priestley and P. S. Pershan, Mol. Cryst. Liq. Cryst.
    23 369 (1973).
4. S. Jen, N. A. Clark, P. S. Pershan and E. B. Priestley, Phys.
    Rev. Lett. 31 1552 (1973).
5. G. R. Luckhurst and R. Poupko, Chem. Phys. Letters 29 191
    (1974).
6. L. L. Chapoy and D. B. DuPré, to be published.
7. A. C. Albrecht, J. Mol. Spectry. 6 84 (1961).
8. J. R. Lombardi, J. W. Raymonda, and A. C. Albrecht, J. Chem.
    Phys. 40 1148 (1964).
9. C. R. Desper and I. Kimura, J. Chem. Phys. 38 4225 (1967).
10. S. Nomura, H. Kawai, I. Kimura, and M. Kagiyama, J. Poly.
    Sci. A-2 8 383 (1970).
11. E. B. Priestley, P. J. Wojtowicz, and P. Sheng, Eds., Intro-
    duction to Liquid Crystals, Plenum Press, New York, 1975
    pp. 72.
12. A. Saupe, Z. Naturforsch. 19a 161 (1964); Angew. Chem. Int.
    Ed. 61 947 (1974).
13. Y. Nishijma, J. Poly. Sci. PtC. 37 353 (1970).
14. G. Weber, "Polarization of the Fluorescence of Solutions",
    chapter 8 in Fluorescence and Phosphorescence Analysis,
    Interscience Publ. New York 1966.
15. K. A. Valiew and L. D. Eskin, Optics and Spectroscopy 12 429
    (1962). (English Transl.).
16. G. Weber, Adv. in Protein Chem. 8 415 (1953).
17. F. Perrin, J. Phys. 7 390 (1926).
18. G. Baur, A. Stieb, and G. Meier, Mol. Cryst. Liq. Cryst.
    22 261 (1973).
19. G. R. Luckhurst and R. N. Yeats, Mol. Cryst. Liq. Cryst.
    Letters, 34, 57 (1976).

LATERAL SPECIES SEPARATION IN THE LIPID BILAYER

P. H. Von Dreele

Northwestern University
Evanston, Illinois

The different lipid species in a system of two different kinds
of lipids arranged in a bilayer can be well mixed with little
lateral species separation or poorly mixed with a large degree of
excess lateral species separation according to the magnitude of the
nonideal intermolecular forces between the two kinds of lipids in
the surface. I shall begin this talk by defining the terms we will
be using to describe the excess lateral species separation;
secondly, comment on the experimental measurements required to
determine the degree of excess species separation; thirdly, discuss
the theory for two dimensional systems which we have developed to
analyze the data; and finally talk about the ways to represent the
results which will give a good physical picture of the membrane.

We will begin by defining a few of the terms which we will be
using. By clusters we mewn the grouping together of lipids of
like species. By domains we mean the grouping together of lipids
of like phase. The arrangement of lipids in the surface of the
bilayer will be represented using a view down on the top of the
bilayer in which each kind of lipid is represented by a capital
letter; e. g. M = dimyristoyl lecithin and S = distearoyl lecithin.
Hence, grid pictures like that shown on the right   -M-M-S-S-
are obtained.   -M-M-S-S-

Having defined the terms to be used to describe the excess
lateral species separation, we looked for a way to measure it.
This requires a physical property of the system which (1) depends
on the kinds of the neighboring lipids and also (2) has different

This talk is based in large part on manuscripts which are in press
in Biochemistry and Biochem. Biophys. Res. Commun.

values for the different kinds of lipid molecules. Two such
properties are the transition temperature, T_{tr}, and the transition
enthalpy, ΔH, which characterize the gel to liquid crystal phase
transition of the lipids. These quantities are measured by
differential scanning calorimetry curves which are used to construct
phase diagrams. It is not possible to inspect the shape of the
phase diagrams and determine whether the mixture is ideal or
nonideal. The equations for calculating the phase diagrams for
ideal mixtures have been derived previously(1). The discussion
in this reference indicates that the line down the center of the
lens (f_A) may or may not be linear with respect to the temperature
depending on the values of ΔS_A and ΔS_B. For small ΔS values,
if $\Delta S_A = \Delta S_B$ then this line is linear. For other values of
$\Delta S_A / \Delta S_B$, the dependence is shown in Figure 1. Even if $\Delta S_A =
\Delta S_B$, a linear relation between T_{tr} and f_A is obtained only for
small ΔS. For larger ΔS values the relation between T_{tr} and f_A
is sigmoidal (Figure 2). The breadth of the lens depends on the
magnitude of S and increases with increasing ΔS (Figure 3).
The calculated phase diagrams for ideal mixtures having the T_{tr}
and ΔH of DML/DPL, DML/DSL, and DPL/DSL mixtures are shown
in Figure 4. Phase diagrams using the data of Shimshick and
McConnell and the theoretical curves for an ideal mixture are
shown in Figure 5. There is a deviation between the theory and the
data indicating the presence of nonideal mixing.

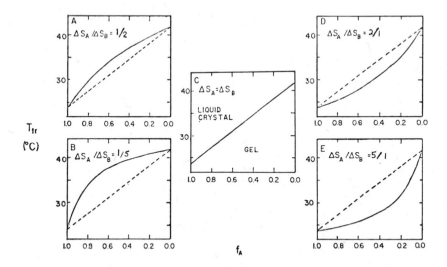

Figure 1   Mole Fraction of A in an Ideal Mixture of A and B
           Lipids versus Temperature

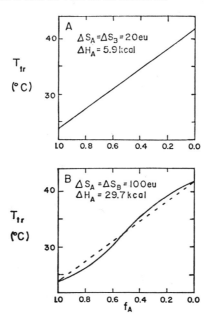

Figure 2   Mole Fraction of Lipid A in the Whole System versus
           Temperature

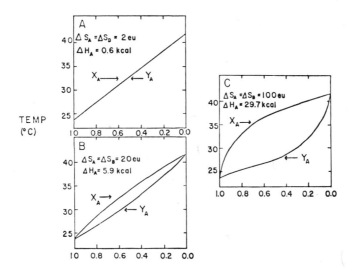

Figure 3   Mole Fraction of A Lipid in the Whole System versus
           Temperature

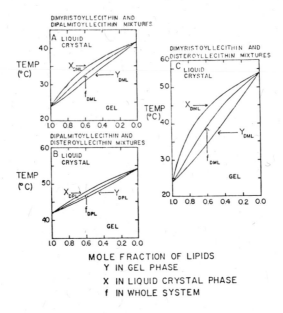

Figure 4   Mole Fraction of A Lipid versus Temperature

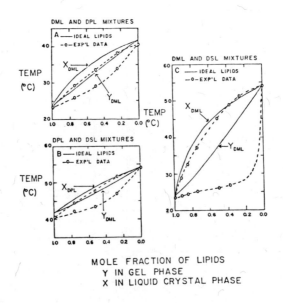

Figure 5   Mole Fraction of A Lipid versus Temperature

This brings us to the third item in this talk. We must develop a theory for nonideal phase diagrams in two dimensions. This theory should contain a nonideality parameter, ν, whose value can be determined by fitting the data. This theory is developed in reference 3.

The final point in this presentation is to find a way to convert the nonideality parameter into a form which will give a good physical picture of the lipid organization of the membrane. This is done in reference 4 and the result is an average cluster size of like lipids in the bilayer surface, $\overline{Cl}_A$. We note that the average cluster size is model dependent.

Z value	$\overline{Cl}_A$	Ideal 1/1 Mixture $\overline{Cl}_A$
2	$1/\nu X_B$	2
4	$(1/\nu X_B)^2$	4
6	$\left(\dfrac{2}{3\nu X_B}\left(1+\sqrt{1-\dfrac{3\nu X_B}{2}}\right)\right)^2$	4

It can be shown that both hexagonal close pack alkyl chains and square planar close pack alkyl chains can be represented by a square planar arrangement of lipid molecules. Therefore the Z = 4 values are likely to be representative of the real lipid mixture. Using the Z=4 model the gel phase of the DML/DSL mixture at 299.8°C is found to have an average cluster size of 19 which means it has an excess lateral species separation of 19/4 or 4.75. Since the ν value is less than unity, the DML-DSL forces are repulsive. In conclusion this series of papers develops a new mathematical tool for use in analyzing membranes and applies this tool to obtain the amount of excess species separation present in the bilayer.

REFERENCES

1. Von Dreele, P. H., Poland, D., and Scheraga, H. A. (1971) Macromolecules 4, 396

2. Shimshick, E. J. and McConnell, H. M. (1973) Biochemistry 12, 2351

3. Von Dreele, P. H. Biochemistry, in press.

4. Von Dreele, P. H. Biochem. Biophys. Res. Commun., in press.

FORMATION OF THE LIQUID CRYSTALS OF POLYRIBONUCLEOTIDE COMPLEXES

Eisaku Iizuka and Jen Tsi Yang

Institute of High Polymer Research, Shinshu University,
Ueda 386, Japan and Cardiovascular Research Institute,
University of California, San Francisco, California 94143

ABSTRACT   Double-stranded helices of poly(A)·poly(U), poly(G)·poly
(C) and poly(C)·poly(I) and triple-stranded helices of poly(A)·2poly
(U) and poly(A)·2poly(I) in concentrated solutions form nematic liquid crystals. Poly(C)·poly(I) also contains some spherulitic structures. The molecular assemblies of the two complexes of poly(A) and
poly(U) are rod-like and align themselves with their long axes parallel to the direction of orientation. On standing for several
weeks these two complexes are converted from the nematic to cholesteric type, which reverses to the oriented nematic crystals under
shearing stresses. All the helical complexes show negative linear
dichroism and strong negative birefringence. In a magnetic field
the pitch of the cholesteric crystals changes with the field
strength; this change also varies with the angle that the helicoidal
axis of the cholesteric structure makes with respect to the magnetic
field. The nematic liquid crystals of the two complexes of poly(A)
and poly(U) appear to orient themselves perpendicular to the direction of a magnetic field of more than ten kilogausses.

Some polyribonucleotide complexes such as poly(A)·poly(U)* and
poly(C)·poly(I) are known to be inducers for interferons that re-
strain cancerous tissues from multiplying.[1] Any structural study
of these complexes might provide information concerning the mechanism of inducing interferons. We report here that the double-
stranded helices, poly(A)·poly(U), poly(G)·poly(C) and poly(C)·poly
(I), and triple-stranded helices, poly(A)·2poly(U) and poly(A)·2poly

*Abbreviations used: poly(A), polyadenylate; poly(U), polyuridylate:
poly(G), polyguanylate; poly(C), polycytidylate; poly(I), poly-
inosinate.

(I), all form liquid crystals in concentrated solutions. In many
aspects these liquid crystals resemble those of some synthetic polypeptides in helix-promoting solvents[2] and of polyglutamic acid[3],
deoxygenated sickle-cell hemoglobin[4] and sodium salt of DNA[5] in
aqueous solutions. We will further show that these liquid crystals
are of the nematic type. (Poly(C)·poly(I) also forms some spherulitic structure.) The structures of poly(A)·poly(U) and poly(A)·2
poly(U) (and possibly poly(G)·poly(C)) are converted to the cholesteric type on standing for several weeks (cf. Figs. 1-1 and 1-4).
The other polyribonucleotide complexes do not appear to change their
liquid crystalline structures with time.

The solutions of polyribonucleotide complexes, except poly(C)·
poly(I), became slightly turbid when the polymers were about 4%
(v/v), above which the solutions became birefringent. The corresponding polarizing micrographs were similar to those of the lyotropic liquid crystals so far observed.[2-5] Figure 1 illustrates the
structural difference of the complexes of poly(A) with poly(U) and
poly(I) at 5.5%. As the polymer concentration was raised to 10.5%,
the solutions of poly(A)·poly(U) and poly(A)·2poly(U) were fairly
viscous but still could flow, but those of poly(G)·poly(C) and poly
(A)·2poly(I) could no longer flow. On the other hand, the poly(C)·
poly(I) solution at less than 25% concentration did not show birefringence; even the 25% solution was not very viscous.

The birefringent solutions of poly(A)·poly(U) and poly(A)·2poly
(U) show the H_v scattering of the $\pm 45°$ type (prior to the cholesteric phase) (Fig. 2-1), suggesting a collection of rod-like assemblies of the helices (molecular clusters) with maximum polarizability
in the direction either parallel or perpendicular to the rod axes.[6,7]
The upper and lower angles of intersection of the cross pattern ($\pm$
45° type) become larger than 90° when the polymer solution is sheared
vertically in an optical cell by moving the spacer up and down inside
the cell (Fig. 2-2). Thus, the molecular clusters align themselves
with their long axes parallel to the direction of shearing stress.[4,7]
Lamellar domain patterns with lines perpendicular to the direction
of shearing stress (see Fig. 1-3) may suggest the presence of smectic
arrangement of the rod-like molecular clusters. The lamellae were
several micrometers thick for 10.5% solutions of poly(A)·poly(U) and
poly(A)·2poly(U). No such distinct lines were observed for the
other complexes tested. The intensity of the V_v scattering was much
weaker than that expected from theory.[7] Since the V_v scattering is
originated from both the orientation fluctuation of the molecular
clusters and density fluctuation of the polymer solution,[6] this discrepancy may be attributed to the molecular clusters dissolved in an
isotropic solution of the polymer molecules (ca. 4%) rather than in
pure solvent. These "free" polymer molecules can exchange the molecules in the clusters and vice versa as in the liquid crystals of
polypeptides.[2,8,9]

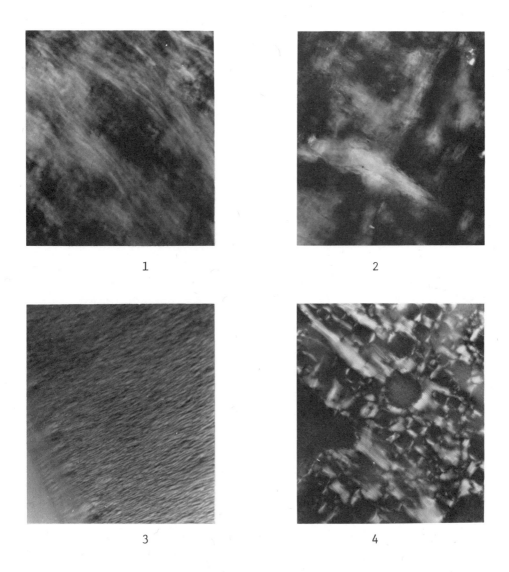

Figure 1. Polarizing micrographs (60X) of 5.5% solutions of poly-
ribonulceotide complexes. 1, poly(A)·poly(U); 2, poly(A)·2poly(I);
3, poly(A)·poly(U) sheared parallel to the edge (lower left) of
the spacer; 4, poly(A)·poly(U) in the cholesteric phase (see text).

         1   2

Figure 2. The H_v scattering patterns of the poly(A)·poly(U) solution. 1, 5.5%(v/v)(without shear); 2, 10.5% (sheared vertical).

     At the absorption band near 260 nm the oriented polyribonucleo-
tide complexes in solution shows a larger absorbance when the electric vector of the incident beam is perpendicular to the direction
of shear than when it is parallel (Fig. 3). The 260-nm band is related to the π-π* transition in the purine and pyrimidine bases of
the nucleotides and the direction of transition is in the planes of
these bases.[10-12] Therefore, the linear dichroism results indicate
that the helical complexes align themselves parallel to the rod-
like axes of the molecular clusters.

     As with negative linear dichroism, the oriented polyribonucleo-
tide complexes in solution showed negative birefringence, whose
magnitude per unit concentration, $-\Delta n/C_v$, increased with polymer
concentration and appeared to approach 0.1 for DNA crystals.[13]
Table I illustrates the results of poly(A) and poly(U) complexes.

     X-ray diffraction patterns of the birefringent solutions of the
polyribonucleotide complexes showed the Debye-Scherrer rings, which
were less broad at high polymer concentration. An example is given
in Figure 4. A film appears to have more rings than a solution.
Thus, the arrangement of the ordered helices is more compact and
stable at higher concentration.

     All the above evidence supports the conclusion that in concen-
trated solutions the double- and triple-stranded helices aggregate

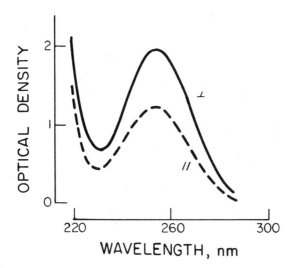

Figure 3.   Linear dichroism of a 10.5% oriented solution of poly(A)·
poly(U).

Table I.   Birefringence of the oriented complexes of poly(A) and
poly(U) in concentrated solution.

Polymer	$-\Delta n/c_v$ [a]		
	5.5%	10.5%	12.8%
Poly(A)·poly(U)	0.009	0.046	0.045
Poly(A)·2poly(U)	0.011	0.073	0.091

[a] Δn, measured at 547 nm; c_v, the volume fraction of the polymer.

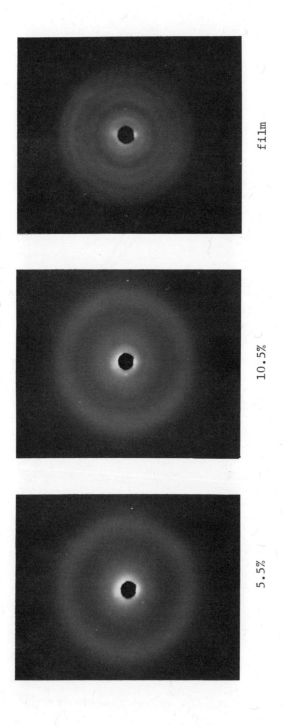

film   10.5%   5.5%

Figure 4. X-ray diffraction patterns of the liquid crystals of poly(A)·poly(U).

to form rod-like molecular clusters and to produce nematic liquid crystals. We also found that at neutral pH single-stranded poly(A), poly(U) and poly(C) did not form liquid crystals in concentrated solutions. Poly(G) seemed to form liquid crystals at concentrations more than 50%, but poly(I) did at about 10 to 20%.

On standing the liquid crystals of poly(A)·poly(U) and poly(A)· 2poly(U) are converted to the cholesteric type. Figure 5a shows that the cholesteric pitch of the two complexes is proportional to $C_v^{-1.1}$, although the range of concentrations (in terms of volume fractions) used are not wide enough to yield a precise slope. The effect of temperature in Fig. 5b first shows a gradual increase in pitch with temperature that is comparable to the coefficient of linear expansion of the polymer solution. This is followed by a sharp increase in pitch, which eventually disappears, as the liquid crystals melt and the solution becomes isotropic; the melting point of the liquid crystals seems to be higher in more concentrated solution.

The cholesteric pitch is known to change under a magnetic field; the change varies with the angle that the helicoidal axis of the cholesteric phase makes to the direction of the field.[14,15] The example shown in Fig. 6 suggests the presence of an anisotropy of the magnetic susceptibility that causes the magnetic field to exert torques within the cholesteric structure.

Films of poly(A)·poly(U) and poly(A)·2poly(U) were cast from their nematic solutions under a magnetic field of 25 kilogausses (the film was parallel to the magnetic field). The X-ray diffraction patterns show two equatorial spacings (3.3 and 3.8 A) and two meridian spacings (4.6 and 5.8 A) when the direction of the magnetic field is vertical (Fig. 7). The 3.3 A spacing is related to the stacked bases nearly normal to the helical axis.[16] Thus, the X-ray results suggest that these polymer complexes are oriented in the direction perpendicular to the magnetic field and, furthermore, their magnetic susceptibilities are stronger in the direction perpendicular than parallel to the helical axes. Such an orientation can occur because the rod-like molecular clusters, whose induced magnetic dipole moments are large, are still able to flow.

The light scattering patterns of the other three complexes, poly(G)·poly(C), poly(C)·poly(I) and poly(A)·2poly(I) did not show distinct molecular clusters that were rod-like. These liquid crystals were not oriented very much under shearing stresses. Nor were they oriented under a magnetic field because the concentrated solution of poly(G)·poly(C) and poly(A)·2poly(I) was too viscous to flow and that of poly(C)·poly(I) contained spherulites.

The rod-like molecular clusters are reported to appear in some

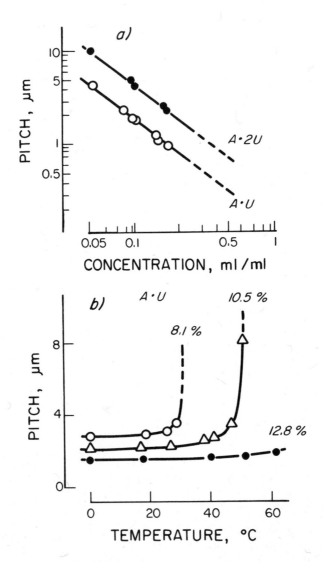

Figure 5. The cholesteric pitch of poly(A) and poly(U) complexes. a) concentration dependence; b) temperature dependence.

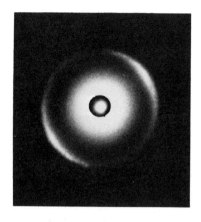
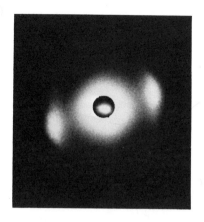

no field   25 kilogausses

Figure 6.   Laser light diffraction patterns of the cholesteric phase
of poly(A)·poly(U) (8.1%) under a magnetic field.   Direction of
the field: horizontal.

poly(A)·poly(U)   poly(A)·2poly(U)

Figure 7.   X-ray diffraction patterns of the magnetically oriented
films of poly(A) and poly(U) complexes.   Direction of the magnetic
field: vertical.

lyotropic liquid crystals by one of the authers.[2-5] The cholesteric structure is found in some polypeptides[17] and DNA[18]. The magnetic-field orientation of liquid crystals has also been observed in certain polypeptides.[2,19]

EXPERIMENTAL

Potassium salts of poly(A) and poly(I), ammonium salt of poly(U) and sodium salt of poly(G) were purchased from the Miles Laboratories and potassium salt of poly(C) was from Schwarz Bioresearch Inc. All polyribonucleotides were dissolved in 0.08 M NaCl and 0.02 M sodium citrate plus citric acid with a pH of 7.0.[20] The preparations of double- and triple-stranded complexes have been described elsewhere.[21,22] The concentrated polymer solutions were introduced into quartz cells of 1 mm pathlength and stored for at least three weeks before measurements. A quartz spacer was inserted into each optical cell to shorten the lightpath to about 0.1 mm or less.

Micrographs of concentrated solutions were taken under crossed polarizers (to increase the contrast). Light scattering (low angle) and diffraction patterns were photographed with a 2-mW He-Ne laser (6328 A) as the light source. For birefringence measurements the solutions were sheared by moving the spacer up and down inside the cell. For linear dichroism measurements the solutions were sheared between two quartz spacers. X-ray diffraction patterns of the solutions were photographed in a glass capillary tube of 1 mm diameter. All experiments were done at room temperature unless stated otherwise.

Acknowledgment. This work was supported in part by the U. S. Public Health Service Program Project grant HL-06285.

REFERENCES

1. C. Colby, Jr., Prog. Nucleic Acid Res. Mol. Biol. 11, 1 (1971).
2. E. Iizuka, Adv. Polymer Sci. 20, 79 (1976).
3. E Iizuka, Y. Kondo and Y. Ukai, Polymer J. 9, 135 (1977).
4. E. Iizuka, Proc. 6th Int. Liq. Cryst. Conf. , Part E, Molecular Crystals and Liquid Crystals, in press.
5. E. Iizuka, Polymer J. 9, 173 (1977).
6. M. B. Rhodes and R. B. Stein, J. Polymer Sci. Part A-2, 7, 1539 (1969).
7. E. Iizuka, T. Keira and A. Wada, Mol. Cryst. Liq. Cryst. 23, 13 (1973)
8. E. Iizuka, Biochim. Biophys. Acta 243, 1 (1971).
9. E. Iizuka, J. Phys. Soc. Japan 34, 1054 (1973).

10. S. F. Mason, J. Chem. Soc. (London) 1240 (1959).
11. S. F. Mason, J. Chem. Soc. (London) 1247 (1959).
12. A. Rich and M. Kasha, J. Am. Chem. Soc. 82, 6197 (1960).
13. G. B. B. M. Sutherland and M. Tsuboi, Proc. Roy. Soc. (London) A239, 446 (1957).
14. R. B. Meyer, Appl. Phys. Letters 12, 281 (1968).
15. P. G. de Gennes, Solid State Commun. 6, 163 (1968).
16. W. T. Astbury, Symp. Soc. Exp. Biol., I. Nucleic Acid, University Press, Cambridge, 1947, p 66.
17. C. Robinson, Trans. Faraday Soc. 52, 571 (1956).
18. C. Robinson, Tetrahedron 13, 219 (1961).
19. S. Sobajima, J. Phys. Soc. Japan 23, 1070 (1967).
20. G. C. Chen and J. T. Yang, Biophys. Chem. 1, 62 (1973).
21. P. K. Sarkar and J. T. Yang, J. Biol. Chem. 240, 2088 (1965).
22. P. K. Sarkar and J. T. Yang, Biochemistry 4, 1238 (1965).

THEORETICAL MODEL FOR THE PHOSPHOLIPID BILAYER SYSTEM: SINGLE

COMPONENT PHASE TRANSITION AND BINARY MIXTURE PHASE DIAGRAM

R. G. Priest and J. P. Sheridan

Naval Research Laboratory

Washington, D. C. 20375

INTRODUCTION

An interesting and important problem in statistical mechanics
is the phospholipid main phase transition and related phenomena.
A phospholipid molecule may be thought of as being composed of a
polar head group molecular subunit to which are attached two alkyl
chain "tails."[1] In aqueous solution, under conditions of excess
water, these molecules are organized into bilayer structures. The
bilayer may be viewed as a planar assembly with thickness approximately twice the length of the chain tails (~ 60 Å). The polar
head groups are confined to the two planes which define the surface of the bilayer and are in contact with the water on the outside of the bilayer. The hydrophobic alkyl chains are confined to
the interior of the bilayer and extend from the polar head group
to which they are attached to the plane which is the mid plane of
the bilayer. Water is almost totally excluded from the interior
of the bilayer.

Bilayers formed from phosphatidyl cholines and phosphatidyl
ethanolomines (and many other lipids) exist in a low temperature
phase called the gel phase and a high temperature phase called the
fluid phase. As the low temperature phase is heated it is transformed into the high temperature phase at a first order phase
transition.[1] The transition temperature depends on the chain
length of the tails and on the nature of the polar head group.
In the gel phase the carbon-carbon bonds in the chains are nearly
all in the trans state. Thus the chains are straight and at
temperatures near the gel-fluid transition, are oriented normal to
the polar head group planes. In the fluid phase there are appreciable numbers of gauche bonds in the chains. The hydrocarbon

interior of the bilayer is disorganized, fluid, in nature although
the chains remain roughly parallel to each other and normal to the
surface planes. The polar head groups remain confined to the surface plane at the phase transition but the area per head group
increases about 20% on going from the gel to the fluid.[2] There is
a small volume change at the transition. The gel-fluid phase
transition manifests itself in binary mixtures of two different
species. In particular, mixtures in which the components differ
only in chain length separate into a long chain rich gel phase
and a short chain rich fluid phase for certain values of temperature and relative composition.[3] The fact that this mixed phase
region is confined to temperatures intermediate between the short
and long chain single component gel-fluid transition temperatures
indicates that the phase separation results from the gel-fluid
phase transition.

The main phase transition is clearly an order-disorder
phenomena mainly involving the degrees of freedom associated with
the alkyl chains. The statistical mechanical problem is essentially to model the behavior of highly parallel alkyl chains. The
difficulties encountered in this effort fall into two main areas.
The first is simply the questions of what types of interactions
are important and of how they can be adequately modeled. There
are contributions to the thermodynamic potentials proportional to
L, the number of carbons in the chains, and independent of L.
Both hydrostatic pressure and lateral surface pressure are relevant intensive variables. In the case of the phosphatidyl cholines
there is an imperfectly understood pretransition[1] which may be
important. Recent evidence suggests that the two molecular
chains are in non-equivalent physical environments in both the
gel and fluid states.[4] There is a question of to what extent the
two leaflets of the bilayer interact at the bilayer midplane.

The second type of problem is a technical difficulty of some
fundamental interest. There are two constraints on the organization of the segmented chains. The individual chain segments
cannot interpenetrate each other and are each members of a specific chain. The first constraint of noninterpenetration of aspher-
ical objects is familiar from liquid crystal research. The second
constraint is, of course, familiar in polymer research. However,
most of the rigorous results in the statistical mechanics of
polymers are obtained for dilute systems, whereas the lipid bilayer is dense.

In this paper we concentrate on the effects of the constraints
and take the interactions into account in a phenomenological manner. As input to the phenomenology we use data on the phase
transition in single component systems. The best set of data is
for the phosphatidyl choline series so we devote our attention to

this class of molecules. We use the results obtained from the data
on the single component systems to calculate the phase diagrams of
binary mixtures in which the components differ only in chain
length.   We have used the Raman technique to measure the relative populations of trans and gauche bonds in these systems. This
provides an experimental determination of the boundaries of the
mixed phase region in the temperature-composition plane. The results show that our theoretical approach can explain the basic
features of this type of system so long as the chain length difference is not too great.

   Various authors[5-14] have introduced a variety of methods for
dealing with the constraints. They range from assumptions about
the statistical consequences of the constraints[7] to numerical
approaches[5,8] to the problem. A very promising line of attack on
this problem was initiated by Nagle[6] who studied an exactly solv-
able artificial two dimensional model having these same types of
constraint. The exact solution brought to light some of the important features of this type of system. The models of Jacobs
et al.,[13] Jackson,[11] Priest,[12] and Wiegel[14] incorporate ideas
generated in part from the study of Nagle's exact solution.

   The models of ref. 11-14 have features of both mathematical
and physical nature which make straightforward application to the
binary mixture problem difficult. In this paper we use the physical insight gained in these papers to develop a Landau phenomenology. We introduce a few simplifying assumptions to deal with
the complexity arising from the many different types of interaction
mentioned above. We do not consider vacancies in the bilayer interior in any detail other than to show how they can be incorporated into the model in principle. We make this approximation in
light of the fact that there is only a 3% change in the volume at
the phase transition. We also assume that the discontinuity in
the density of gauche bonds at the phase transition is the same
for all phospholipids having the same polar head group. In particular that it is the same for dimyristoyl phosphatidyl choline
(DMPC), dipalmitoyl phosphatidyl choline (DPPC) and distearoyl
phosphatidyl choline (DSPC). As will be seen below, this allows
us to ignore the lateral pressure as a variable and considerably
simplifies the treatment in other respects. It will become clear
that this does introduce some quantitative error into the model.
However, the nature and approximate extent of this error can be
clearly understood in the context of the model. In any case the
approximation fails mainly for the shorter chain lengths and would
be rather better if longer chain length molecules could be studied
experimentally.

SINGLE COMPONENT MODEL

A useful concept in crystal physics is that of the dislocation. Like a crystal, the gel phase of lipid bilayers is highly ordered. With the possible exception of the bonds near the mid plane, almost all the C-C bonds are trans. The approach taken in ref. 12 may be thought of as a study of the simplest type of dislocation cores that can exist in a system of highly parallel all trans chains. These dislocations, called strings, were described in relation to the idealized lattice dimer model of Nagle.[6] Here we develop the concept of a string as the underpining of a phenomenology.

An important aspect of the dislocation approach is that we do not consider effects which cause the density of gauche bonds to vary with location with respect to the polar head group plane. This is in line with the observation by deGennes[15] that the density of gauche bonds should not depend on position except to the extent that the chains can fold back on themselves near the bilayer midplane. This feature contrasts with ref. 11, 13.

The starting point for the dislocation analysis is the observation that in a system of tightly packed all trans chains, a single gauche bond can not exist in isolation. A less disruptive combination is a $G \pm T G^+$ kink.[16] This kink is a succession of a right (left) gauche bond, a trans bond and a left (right) gauche bond. If the initial direction of an alkyl chain is imagined to be vertically downward, a GTG kink results in a continued downward vertical propagation after a sidewards jog. The jog, however, requires that neighboring chains on either side of a chain having a kink also have kinks or that empty space, a vacancy, be left in the system. This situation is illustrated in schematic form in Fig. 1. A dislocation in a system of highly parallel chains is shown. The Van der Waals boundaries of the chains are shown by the rectilinear lines. The zig-zag lines represent the backbone of the chain. The dislocation segment shown is comprised of four kinks (1,3,4,5) and a vacancy (2). The vacancy is included to show the generality of the concept. We ignore the role of vacancies in the model solution below. The kink labeled 5 takes the dislocation out of the plane of the paper. In three dimensions the dislocations are not confined to a plane.

From the figure it can be seen that if the chains were infinite in length, there would be no way to terminate the group of kinks and vacancies. The figure shows that the kinks and vacancies in the aggregate can be labeled according to their vertical position. The aggregate is therefore one dimensional and the name string is appropriate. A single string is the simplest type of dislocation that can exist in a system of nearly parallel chains.

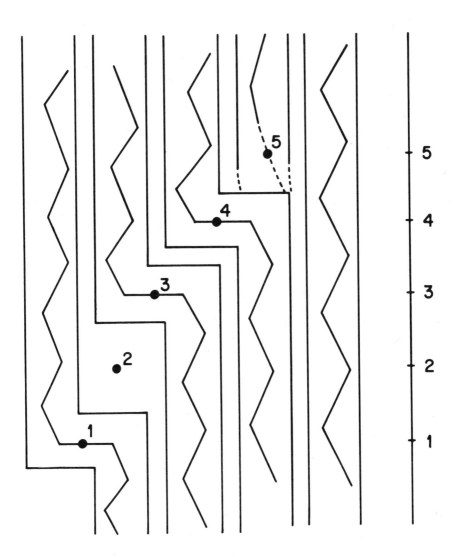

Figure 1 - A string dislocation in a system of nearly parallel
alkyl chains is shown in schematic form. The zig-zag lines
represent the carbon backbones of the chains. The vertical and
horizontal lines outline the space occupied by each chain. The
string segment is composed of four kinks and one vacancy. The
kink labeled 5 takes its chain and string location out of the
plane of the paper. The scale on the right shows that the
vertical displacement of the kinks from each other are all equal.

The horizontal excursions of a string are arbitrary. However, to a certain approximation, the vertical extent of the string is quite regular. The vertical progress of the string is nearly equal for every kink and vacancy. This is illustrated by the scale on the right hand side of Fig. 1.

For chains of finite length such as found in one half a lipid bilayer, strings are of finite length starting at a location in the plane of polar head groups and ending at the surface which is the midplane of the bilayer. For the finite case all possible strings have the same number of kinks (we ignore here the role of vacancies). This number of kinks per string is proportional to L. In the limit of a low density of kinks all the gauche bonds in the system are incorporated into strings. The density of gauche bonds is therefore independent of position in the bilayer. This fulfills the condition mentioned in Ref. 15.

A Landau phenomenology is a model which represents the free energy of a system as a power series in a variable which is the density of an extensive quantity. For the phospholipid bilayer system the appropriate density is the number of strings per unit area of polar head group plane, or equivalently of any parallel plane. We call this density ρ. We notice at once that the energy associated with the kinks is linear in ρ. The number of kinks in a string is αL where space filling models give an estimate of $\alpha \sim .4$. The energy density associated with the kinks is $\varepsilon \alpha \rho L$ where ε is the energy of a single kink. The product of Avogadro's number, N_o, and ε is approximately 1 Kcal.[17]

There is also an entropic contribution to the free energy density proportional to L. This is the case because a string, like a chain, can have many different conformations. The entropy associated with these string conformational degrees of freedom is given by $k_B \ln(\Omega)$ where k_B is Boltzmann's constant and Ω is the number of conformations that a string can assume. The number Ω is calculated in the same way as for a chain in which the bonds can be in any of several states. The result for a chain is $\Omega = a^N$ where a is the number of states that a bond can assume and N is the number of bonds. For alkyl chains a is 3 there being two gauche and one trans states available to a bond. In the case of strings $\Omega \alpha a^{\alpha L}$. In the model of ref. 6, a = 2. In the three dimensional case a > 2, but need not be an integer since neighboring chains can be rotated about their long axes with respect to each other. The quantity a is an average over such relative orientations. The entropy density due to the string conformational degrees of freedom is $k_B \alpha \rho (L \ln a - \hat{K})$. The quantity $\hat{K}$ arises from the proportionality constant in Ω. It is independent of L.

There are a variety of contributions to the free energy which are independent of L. In addition to $\hat{K}$ just mentioned there are: polar head group interactions, lateral pressure terms (the area per head group increases with increasing ρ), and effects due to changes in the leaflet midplane interaction with changing ρ. This last effect may be important if the degree of non-equivalence between the two molecular chains changes at the phase transition.

We discuss one L independent entropic contribution to the free energy in more detail. For small ρ the entropy associated with the distribution of string ends in the head group plane is proportional to $\rho\ln\rho$. This term ensures that the density of strings can never be zero even in the gel state. Numerical studies of the model including the $\rho\ln\rho$ term indicate that the value of ρ in the gel state is of order 10^{-2}. All the other effects of the $\rho\ln\rho$ term can be included in the approach discussed below in which we do not consider this term explicitly. This simplification does result in the gel value of ρ being exactly zero. In view of the small value of the correct value, this is not a serious problem.

Rather than attempting to deduce the detailed forms of the L independent contributions to the free energy we write for the free energy density G_S:

$$\frac{G_S}{k_B T} = \alpha\rho L \left(\frac{\varepsilon}{k_B T} - \ell na\right) - \frac{W'}{k_B T} f(\rho) - K'g(\rho) \tag{1}$$

The functions $f(\rho)$ and $g(\rho)$ are assumed to satisfy $f(0) = g(0) = 0$. Equation (1) represents both energetic and entropic L independent contributions to G. Not included are L dependent contributions not linear in ρ. There are two reasons for this omission. First, the role of the higher order terms in ρ is mainly to fix the location of the minima of G as a function of ρ. Since the fluid state has a lower density of gauche bonds than a hydrocarbon liquid at the same temperature, it must be the L independent terms which fix the minimum. Second, we are going to assume that the discontinuity in ρ at the phase transition is L independent. We can therefore eliminate the nonlinear energetic terms by redefining ρ so that it is always proportional to the gauche bond density. The nonlinear entropic terms can be eliminated by a (slight) redefinition of a, a quantity which is going to be obtained by a fit to the data in any case.

In principle, the equilibrium values of ρ are obtained by minimizing eq. (1) with respect to ρ. Since the phase transition is first order, the coefficient of the ρ^2 term in eq. (1) must be negative and those of the higher order terms positive. The free energy density will then have two minima, one at $\rho = \rho_g = 0$ and one at $\rho = \rho_f$ corresponding to the gel and fluid states. We assume that the value of ρ_f is L independent. The transition temperature can be obtained by setting $G_S(\rho_g) = G_S(\rho_f)$. The resulting equation is:

$$0 = \left(\frac{\varepsilon}{k_B T_c} - \ell na\right) L - \frac{W'}{k_B T_c} \frac{f(\rho_f)}{\rho_f} - K' \frac{g(\rho_f)}{\rho_f} \qquad (2)$$

If we define $W = W' \dfrac{f(\rho_f)}{\rho_f}$ and $K = K' \dfrac{f(\rho_f)}{\rho_f}$, the expression for T_c is

$$\frac{\varepsilon}{k_B T_c} = \frac{K + \ell na}{L - W/\varepsilon} \; L \qquad (3)$$

Since ε is known, this is a three parameter expression for T_c. For this approach to make sense a must be in the range $2 > a > 4$. A fit of a, K and W to the data of Mabrey and Sturtevant[18] taken on phosphatidyl cholines with L = 12, 14, 16, 18 and of Ladebrooke and Chapman[19] for L = 22 gives a = 3.22, K = -1.94 and W/ε = 5.43. The fit is quite good giving a RMS error of on $\varepsilon/k_B T$ as a function of L of less than 10^{-2}. More importantly, a turns out to be in the expected range and K and W are of reasonable magnitude.

The expression for the latent heat per mole of chains is:

$$\Delta H = N_o \varepsilon \; b \; a\rho_f (L - W/\varepsilon) \qquad (4)$$

where b is the area per chain in the fluid phase. According to eq. (4) the latent heat should be linear in L intercepting the L axis at L = 5.43. The experimental results for phosphatidyl cholines[18,19] show that the latent heat is a monotone increasing function of L. However the dependence is not strictly linear and the zero of the latent heat apparently occurs at L = 10. An extrapolation of the values for L = 22 and L = 18 to ΔH = 0 yields L = 8. This is an indication that ρ_f is not absolutely independent of L in the physical range of L, 12 < L < 22. Evidentally ρ_f increases with increasing L starting from zero at L = 10. The limiting behavior of ρ_f independent of L is reached for L $\gtrsim$ 22. Fortunately eq. (3) remains valid under weaker conditions than eq. (4).

If the latent heat did follow the form of eq. (4), the data could be used to find $a\rho_f b$. The slope of the experimental curve giving ΔH versus L varies from .88 Kcal/mole/carbon at L = 12 to .53 Kcal/mole/carbon at L = 22. Since $N_o \varepsilon$ is equal to 1 Kcal, this would imply that $a\rho_f b$ lies in the range .88 < $a\rho_f b$ < .53. However, since the fraction of C-C bonds in the gauche state is $2a\rho_f b$, these numbers are clearly too high. They reflect the fact that ρ_f increases with L. To take account of this increase, we take $a\rho_f b$ = .5 in the calculations below.

BINARY MIXTURES

The extension of the model to binary mixtures of two lipids differing only in chain length is straightforward. We consider a binary mixture of two phosphatidyl cholines one with chain length L_O the other with length $L_O + \Delta L$, $\Delta L > 0$. The molar fraction of L_O molecules is 1-x, that of $L_O + \Delta L$ molecules is x. The free energy density G can be written as:

$$G = G_s + G_m \qquad (5)$$

with G_m given by

$$2b \; \frac{G_m}{k_B T} = x \ln x + (1-x) \ln(1-x) + \Delta \mu x \qquad (6)$$

The first two terms on the RHS of eq. (6) are the entropy of mixing. The last is the difference in chemical potential between the two components. The factor of 2 reflects the fact that there are two chains per molecule. The RHS of eq. (6) is the mixing free energy per molecule. It is convenient to introduce $\Delta \mu$ because we wish to calculate phase separation conditions. We must identify phases with equal free energies and equal values of intensive thermodynamic variables such as $\Delta \mu$ and T. The free energy G_s is formally identical with eq. (1). The change that must be made is to correctly interpret L which appears in this equation.

At first thought it appears that $L_e = L_O + x \Delta L$ is the best choice for L_e. This is not correct however. In a system with a mixture of chain lengths, there can exist strings with different numbers of kinks. The strings with less kinks are energetically favored but longer strings have a greater phase space due to the larger number of conformations available to them. These effects are the same as entered into the L dependent part of eq. (1). If there are two types of strings possible with the effective value of L differing by δL, the energy favoring the shorter strings is $\alpha \delta L \varepsilon$ and the ratio of the number of conformations of the long strings to those of the short strings is $\exp(\alpha \delta L \ln a)$. We assume that the midplane sites for terminating strings are such that a fraction x correspond to $L_O + \Delta L$ strings and a fraction 1-x to L_O strings. According to the Boltzmann distribution the ratio of L_O strings to $L_O + \Delta L$ strings is $\exp(\alpha \Delta L \frac{\varepsilon}{k_B T} - \alpha \Delta L \ln a)$. This leads us to an expression for L_e:

$$L_e = L_O + \Delta L \; \frac{x}{x + (1-x) \exp(\alpha \Delta L \; E\lambda)} \qquad (7)$$

with $E \equiv \frac{\varepsilon}{k_B T} - \ln a$ and $\lambda = 1$.

To solve the model in the mean field approximation it is necessary to simultaneously minimize the free energy with respect to
x and ρ. Due to the nature of the assumptions made above, the
minima with respect to ρ occur at $\rho = 0$ and $\rho = \rho_f$. The free
energy densities at these two minima are now functions of x. They
can be written as:

$$\frac{G_g}{k_B T} = \frac{1}{2b} [x\ln x + (1-x)\ln(1-x) + \Delta\mu x] \qquad (8a)$$

$$\frac{G_f}{k_B T} = \frac{1}{2b} [x\ln x + (1-x)\ln(1-x) + \Delta\mu x]$$
$$+ \alpha\rho_f [(\frac{\varepsilon}{k_B T} - \ln a) L_e - \frac{W}{k_B T} - K] \qquad (8b)$$

The minima of G_g and G_f with respect to x will be called x_g and x_f
respectively. The gel (fluid) is the thermodynamically stable
state if $G_g(x_g)<(>)G_f(x_f)$. If, for a given $\Delta\mu$ and T, $G_g(x_g) \neq$
$G_f(x_f)$ then for that set of $\Delta\mu$ and T the system exists in a single
phase, either gel or fluid whichever has the lower G. The condi-
tion $G_g(x_g) = G_f(x_f)$ defines a relation between $\Delta\mu$ and T, $\Delta\mu=\Delta\mu_c(T)$.
For a given value of T, if $\Delta\mu = \Delta\mu_c(T)$ then the three conditions
for phase coexistence are met by the gel and fluid states: $T_g = T_f$,
$\Delta\mu_g = \Delta\mu_f$ and $G_g = G_f$. For a given value of T the minimum value
of x_g corresponding to a stable gel phase is $x_g = x_g (T, \Delta\mu=\Delta\mu_c(T))$.
For a given value of T the maximum value of x_f corresponding to a
stable fluid phase is $x_f = x_f(T, \Delta\mu=\Delta\mu_c(T))$. Therefore for a given
value of T, the range of x:

$$x_f(T, \Delta\mu = \Delta\mu_c(T)) < x < x_g(T, \Delta\mu = \Delta\mu_c(T)) \qquad (9)$$

is the mixed phase region for that value of T. A physical system
having x in this range will have a fraction of p of its material
in a gel phase with $x_g = x_g(T, \Delta\mu = \Delta\mu_c(T))$, and a fraction 1-p
of its material in a fluid phase with $x_f = x_f(T, \Delta\mu = \Delta\mu_c(T))$. The
fraction p is determined by

$$x = px_g + (1-p) x_f. \qquad (10)$$

The functions of T, $x_g (T, \Delta\mu = \Delta\mu_c(T))$ and $x_f (T, \Delta\mu = \Delta\mu_c(T))$
give the boundaries of the mixed phase region in the x, T plane.

We included a parameter λ in eq. (7) because a closed form
solution for these functions can be given for $\lambda = 0$ which corresponds to $L_e = L_o + x\Delta L$. The closed form solution is of value in
understanding analytic features of the phase diagram. The result
for this case is:

$$x_g(T, \Delta\mu = \Delta\mu_c(T)) =$$

$$\frac{1 - \exp[2\alpha\rho_f b \ (EL_o - W/k_B T - K)]}{1 - \exp(-2\alpha\rho_f E\Delta L)}$$

$$x_f(T, \Delta\mu = \Delta\mu_c(T)) =$$

(11)

$$\frac{\exp[-2\alpha\rho_f b(EL_o - W/k_B T - K)] - 1}{\exp[2\alpha\rho_f E\Delta L] - 1}$$

From eq. (3) we have $EL - \frac{W}{k_B T_c} - K = 0$. Therefore according to eq. (9) $x_g = x_f = 0$ for $T = T_c(L_o)$ and $x_g = x_f = 1$ for $T = T_c(L_o + \Delta L)$. It can be seen from these equations that the mixed phase region is entirely confined to the region $T_c(L_o) < T < T_c(L_o + \Delta L)$. These results also carry over to the physical case $\lambda = 1$.

For $\lambda = 1$ we must rely on numerical solutions for the minima of eq. (8b). The results for the case $L_o = 14$ and $\Delta L = 2$ are presented in Fig. (2). These values correspond to a DMPC/DPPC mixture. The smooth curve is the theoretical result. The individual points are derived from the Raman data. The method of data reduction is discussed in a companion paper in this volume (Sheridan, Schnur and Schoen). The Raman data are in substantial agreement with the calorimetric data.[18] The parameters $\alpha\rho_f b$, a, W and K which enter into the theoretical curve were taken from the single component data as discussed above. An investigation of the systematics of the numerical solution shows that the width of the mixed phase region in T depends primarily on $\alpha\rho_f b$. The fact that the width is correctly given by $\alpha\rho_f b = .5$ indicates that using this value, taken from the latent heat data, correctly accounts for the L dependence of ρ_f. The deviation of the phase diagram from ideality in the sense discussed by Mabrey and Sturtevant[18] is almost entirely due to λ. The closed form solution of eq. (9) (with $\lambda = 0$) exhibits nearly ideal behavior. The fact that the calculation for $\lambda = 1$ agrees with experiment indicates that this model has the correct explanation for the non ideal nature of the phase diagram.

Figure 3 presents the phase diagram for $L_o = 14$, $\Delta L = 4$ which is for DMPC/DSPC mixtures. Again the smooth curve is the theoretical result and the individual points are the Raman data. The general shape of the theoretical phase diagram is correct except that the width in T is too small. Very good argument with experiment can be forced for this case by taking $\alpha\rho_f b = 1$. This indicates that for $\Delta L > 2$, features not considered in this model are important. One such feature is the effect of chain mismatch at the midplane of

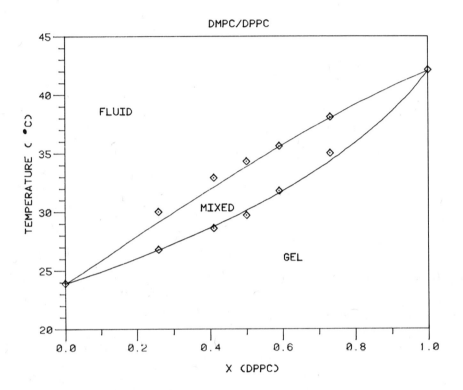

Figure 2 - Phase diagram for DMPC/DPPC mixture in the temperature-
molar fraction of DPPC plane. The fluid, gel and mixed regions
are labeled. The smooth curve is the theoretical result. The
individual points are derived from the Raman data.

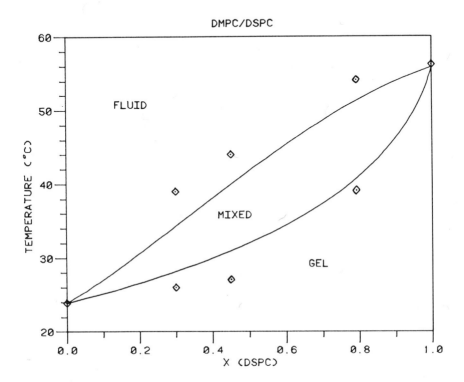

Figure 3 - Phase diagram for DMPC/DSPC mixture in the temperature-
molar fraction of DSPC plane. The fluid, gel and mixed regions
are labeled. The smooth curve is the theoretical result. The
individual points are dervied from the Raman data.

the bilayer.   This mismatch grows in severity with increasing ΔL.
The fluid state will have less "unsatisfied" Van der Waals bonds
than the gel in which the last few segments of the long chains
stick out of the leaflet.   This effect contributes to the difference in free energies between the gel and fluid states and will
give rise to a broadening of the two phase region.   This same
mechanism will, for large enough ΔL, cause a phase separation in
the gel itself at all relevant temperatures.   When this happens
the gel-fluid phase separation discussed above will be affected.
The gel-mixed phase boundary becomes horizontal (independent of
temperature) for $x \lesssim .9$ in this case.

CONCLUSION

The analysis of phospholipid bilayer systems using the concept of string dislocations is seen to be able to account for
the important features of the binary mixture phase diagrams at
least in the case $\Delta L \sim 2$.   The width of the mixed phase region
is determined primarily by the increase in free energy density
with L.   The deviation of the phase diagram from ideal behavior
in the case $\Delta L = 2$ caused by the fact that short strings are
favored energetically over longer ones.   The approximation that
the discontinuity in the gauche bond density is independent of L
fails for $L < 22$.

ACKNOWLEDGEMENTS

This work was supported in part by the Naval Medical Research
and Development Command.

The authors would like to thank Dr. Joel M. Schnur for many
helpful and stimulating discussions.

REFERENCES

1. A description of phospholipid bilayers and their properties may be found in D. Chapman, R. M. Williams and D. L. Ladbrooke, Chem. Phys. Lipids 1, 445 (67); D. Chapman Biological Membranes, Vol. 1, D. Chapman editor, (Academic Press, New York, 1968).

2. M. C. Phillips, D. E. Graham, H. Hauser, Nature 254, 154 (75).

3. E. Shimshick and H. M. McConnell, Biochemistry 12, 2351 (73).

4. P. B. Hitchcock, R. Mason, K. M. Thomas and G. G. Shipley, Proc. Nat. Acad. Sci (USA) 71, 3036 (74).

5. P. Bothorel, J. Belle, B. Lemaire, Chemistry and Physics of Lipids 12, 96 (74).

6. J. F. Nagle, J. Chem. Physics 58, 252 (72); 63, 1255 (75).

7. S. Marcelja, Biochim. Biophys. Acta 367, 165 (74).

8. H. L. Scott, J. Theor. Biol. 46, 241 (74); J. Chem. Physics 62, 1347 (75).

9. S. White, Biophysical Journal 15, 95 (75).

10. J. A. McCammon and J. M. Deutch, J. Amer. Chem. Soc. 97, 6675 (75).

11. M. B. Jackson, Biochemistry 15, 2555 (76).

12. R. G. Priest, J. Chem. Physics 66, 722 (77).

13. R. E. Jacobs, B. Hudson, H. C. Andersen, Proc. Natl. Acad. Sci. USA 72, 3993 (75).

14. F. W. Wiegel, Physics Letters 57A, 393 (76); J. Stat. Phys. 13, 515 (75).

15. P. deGennes, Phys. Lett. A47, 123 (74).

16. H. Trauble, Biomembranes, F. Kreuzer and J.F.G. Slegers, editors (Plenum, New York, 1972) Vol. 3, p.197.

17. P. J. Flory, Statistical Mechanics of Chain Molecules, (Interscience, New York, 1969).

18. S. Mabrey and J. M. Sturtevant, Proc. Natl. Acad. Sci. USA 73, 3862 (76).

19. B. D. Ladbrooke and D. Chapman, Chem. Phys. Lipids 3, 304 (69).

THE ORDERING OF CHOLESTERIC MESOPHASES BY LINEAR CELLULOSIC POLYMERS

T. Sarada and P. F. Waters

Department of Chemistry, The American University

Washington, D. C. 20016

Studies of the effects of macromolecules on the optical properties cf mesophases can delineate those features of the macromolecules which are important for stabilizing or destabilizing mesophases. Measurements of the variations of the cohesive energies of mesophases induced by added macromolecules can quantify the energetics of the intermolecular interactions and thereby direct the choice of additives appropriate to either process. Since the optical measurements are easier to carry out they can be used to screen candidates for thermodynamic evaluations when the latter are required.

The stabilization of mesophases is important for their use in display or monitoring devices. The destabilization of the mesophases known to be present in the atherosclerotic plaque may induce a benign resorption of the complex plaque components into the bloodstream.

Fergason characterized the effects of small molecule additives on cholesteric mesophases by measuring the shift in the wavelength of maximum intensity of reflected light.[1] The research reported in this paper describes optical methods which can be used to observe the effects of added macromolecules on the organization of cholesteric mesophases. The optical properties which were measured are: the birefringence at 589 nm, the bandwidth at half height of the peak of minimum transmission, and the intensities of reflected light at several angles.

Mesophase organizing macromolecules are those which enhance the birefringence and the intensity of light reflected from the

system, they narrow the half intensity width of the reflection
spectrum and they do not increase the half-bandwidth of the
transmission peak. A decrease in the birefringence widens the
half intensity width of the reflection spectrum. Cellulose esters
were selected as the additives because of 1) the possibility of
configurational accommodations amongst the six-membered rings of
the cellulose backbone units and those of the cholesteryl framework, 2) the likelihood of mutual compatibility attending the
similar van der Waals interactions of the ester moeities in both
the macromolecules and the mesophases and 3) the availability of
cellulose esters with rigid rod, comb and random coil structures
with identical backbones. The structures of a cellulose repeat
unit and of cholesterol are shown in Fig. 1. The repeat unit of
PMMA is illustrated in Fig. 2. By examining these, the influence
of macromolecular configuration and its weighting vis-a-vis
functional group similarity, on the mesophase organizing ability
of macromolecules, can be assessed. This work is an exploratory
rather than a comprehensive study of the problem.

EXPERIMENTAL

Materials

     The materials used are listed in Table 1. They were used
without further purification.

Preparation of Samples

     A mixture of 1.00 part OCC and 0.79 parts EE by weight was
prepared by heating and mixing in a 40°C oven for 10 minutes. It
formed a cholesteric mesophase between 8° and 18°C. It was stored
at 4°C. The additives were incorporated into the mixture as
follows: the polymers were weighed and dispersed by heating in
1,4-dioxane and these were then mixed thoroughly with the mesophase.
Two fused silica plates, 1 in. square by 1/16 in. thick were used as
the cell. Each was rubbed several times, unidirectionally, with
lens paper. A drop of the mixture was placed on one plate. The
other plate was placed upon the drop and the mesophase was made
uniform by rubbing the plate back and forth in a top-to-bottom
direction. Each sample was examined with a magnifying lens for
evidence of air bubbles or discontinuities in texture and with
crossed polarizers for single domain structure. The rubbed
direction of the cell was always mounted in the vertical direction
in the photometers. The compositions of the samples and the number
average molecular weights, $\overline{M}_n$, which are available are recorded in
Table 2.

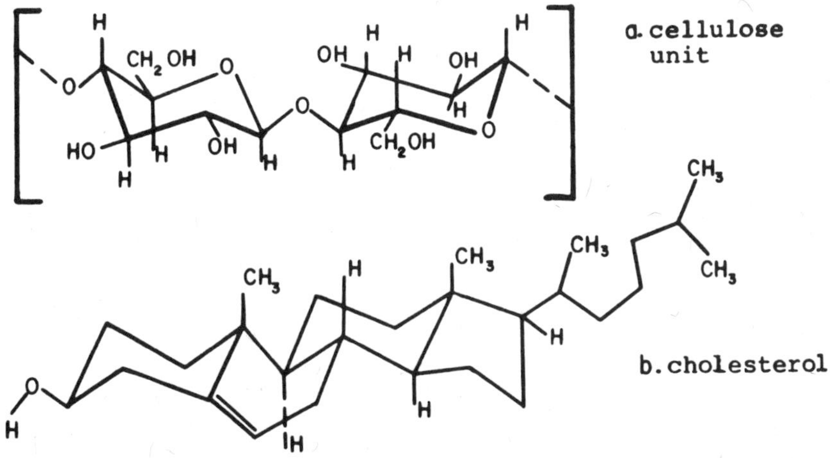

Figure 1. Configuration of cellulose unit and cholesterol.

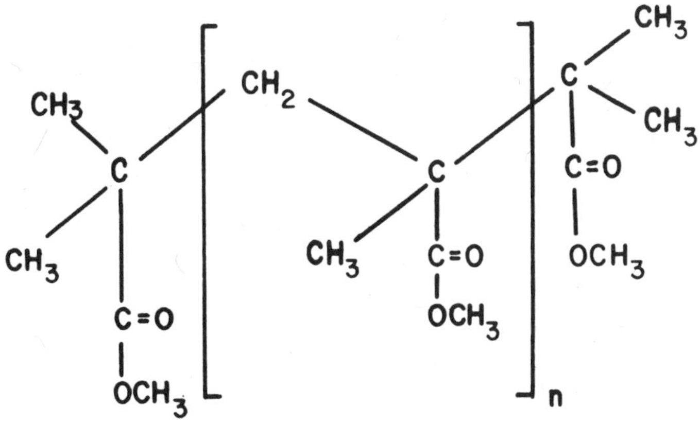

Figure 2. Polymethyl Methacrylate.

TABLE 1

Materials

Substances	Eastman Catalogue No. [a]
Cholesteryl oleyl carbonate (OCC)	10053
Cholesteryl 2-(2-ethoxy ethoxy)ethyl carbonate (EE)	10016
Cellulose acetate, 39.8% acetyl (2.45 acetyls per glucose unit) ASTM viscosity 3 (CA-3)	4644
Cellulose acetate, 40.0% acetyl (2.47 acetyls per glucose unit) ASTM viscosity 25 (CA-25)	4650
Cellulose acetate, 39.4% acetyl (2.41 acetyls per glucose unit) ASTM viscosity 45 (CA-45)	4655
Cellulose triacetate, 43.2% acetyl (2.81 acetyls per glucose unit)	2314
Cellulose acetate butyrate, 17% butyl (CAB)	4623
Cellulose tridecanoate (CTD)	P-7137
Polymethylmethacrylate (Plexiglas, PMMA) [b]	
1,4-Dioxane [c] (D)	

[a] Eastman Organic Chemicals Co.

[b] Rohm & Haas Co.

[c] Fisher Scientific Co.

TABLE 2

Weight Per Cent and Mole Per Cent

of Polymers Added to a Mixture of

1.00 Part OCC and 0.79 Parts EE

Polymer	$\overline{M}_n{}^a$	Weight Per Cent Polymer & Dioxane	Weight Per Cent Polymer	Mole Per Cent Polymer X 10^3
CA-3	27,000	3.89	0.42	8.0
CA-25	48,000	4.24	0.24	2.5
CA-45	58,000	15.54	0.51	2.9
CTD	–	2.41	0.89	–
CTA	110,000	5.16	0.29	1.3
CAB	32,000^b	5.80	0.12	1.8
PMMA	–	7.20	0.22	–

^{a}From Eastman Kodak Co.

^{b}Estimated from viscosity data.

Apparatus

The refractive indices were measured in an Abbe refractometer.[2]
A schematic of the divergent optical paths through the refractometer
which result when light is incident perpendicular to the optic
(helical) axis of the cholesteric mesophase is depicted in Fig. 3.
Therein, a is the illuminating prism, b is the mesophase, c is the
refracting prism, d is an Amici prism, e is the field viewed through
the ocular of the refractometer, r_e, is the angle of refraction
associated with the extraordinary ray and r_o is the angle of refraction associated with the ordinary ray. Polaroid film, HN38
(0.03 in. thick, 38% transmission), properly oriented, was placed
on the ocular to read n_e and n_o.

The transmission spectra were measured in a Cary 14 spectrophotometer. The reflection intensities were measured in a Brice-
Phoenix light scattering photometer modified by replacing the
mercury lamp with a voltage-regulated 150 watt tungsten lamp and
the filter turret with a Schoeffel Model QPM 30 quartz prism mono-
chrometer.

A cell holder was fabricated to allow both transmission and reflection measurements to be made on the same sample. A side view of the holder is shown in Fig. 4, where a is the mounted fused silica cell and b is a dull black shutter which is dropped behind the cell for reflection measurements. The window opening is 7/8 in. square. The holder is placed inside a metal cylinder through which a thermostatic fluid is circulated. The cylinder is keyed to bases for alignment in either the spectrophotometer or the light scattering photometer.

RESULTS

Birefringence

Values of the extraordinary refractive index, n_e, the ordinary index, n_o, the birefringence, $\Delta n = n_e - n_o$, and the changes in n_e and n_o induced by the additives are presented in Table 3.

Transmission Measurements

The transmission spectra were measured at 14.0° ±0.5°C with unpolarized light (UP), left-circularly polarized light (LCP) and right-circularly polarized light (RCP). Polaroid film HNCP37 (0.03 in. thick, 37% transmission) was used for circular (elliptical) polarization. This consists of a linear polarizing filter and a quarter wave retardation of 140 ±20 nm oriented at an angle 45° to the transmission direction of the polarizer. The temperature was monitored with an iron constantan thermocouple. The optical density, O.D., the wavelength of minimum transmission, $\lambda_{min}(T)$, and the half-bandwidth of the transmission peak, hbw, measured for each sample are entered in Table 4. Since the O.D. depends on the sample thickness and since molecular optical absorption is not a property of the esters of cholesterol or any of the additives, the O.D. can be used as an analogue of sample thickness. The half-bandwidth values entered in Table 4 are those measured at the O.D. cited. The changes in half-bandwidth, Δhbw, were calculated from the half-bandwidths measured on each spectrum and the half-bandwidth of the unadulterated mesophase of the same O.D. derived from a plot of O.D. vs. hbw for the latter. Fig. 5 shows spectra which are typical of those which are recorded using UP, LCP and RCP light. Fig. 6 illustrates spectra typical of a sample containing a mesophase-organizing macromolecular additive.

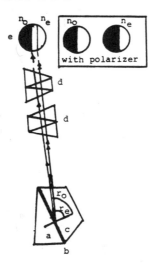

Figure 3. Schematic of the refraction by a birefringent material
in a Abbe refractometer.

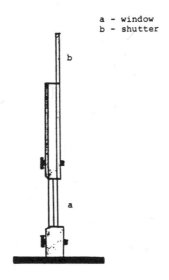

Figure 4. Cell holder.

TABLE 3

Refractive Indices of Mixtures of a Mesophase of

1.00 Part OCC and 0.79 Parts EE and Additives

Measured at 589 nm and 14.0° ±0.1°C

Sample	Additive	n_e	Δn_e	n_o	Δn_o	$-\Delta n$
OCC/EE (1.00/0.79)		1.4946	–	1.5100	–	0.0154
	CA-3	1.4945	-0.0001	1.5103	+0.0003	0.0158
	CA-25	1.4952	+0.0006	1.5066	-0.0004	0.0114
	CA-45	1.4948	+0.0002	1.5081	-0.0019	0.0133
	CTD	1.4944	-0.0002	1.5106	+0.0006	0.0162
	CTA	1.4951	+0.0005	1.5105	+0.0005	0.0154
	CAB	1.4948	+0.0002	1.5098	-0.0002	0.0150
	PMMA	1.4945	-0.0001	1.5106	+0.0006	0.0161

TABLE 4

Transmission Characteristics of Mixtures of a Mesophase of 1.00 Part OCC
and 0.79 Parts EE and Additives Measured at 14.0° ±0.5°C

Sample	Additive	O.D.	UP $\lambda_{min}(T)$ nm	hbw nm	Δhbw nm	O.D.	LCP $\lambda_{min}(T)$ nm	hbw nm	Δhbw nm
OCC/EE (1.00/0.79)		0.090	575	50	–	0.250	575	55	–
		0.100	582	43	–	0.258	583	50	–
		0.170	583	43	–	0.400	583	43	–
		0.350	585	65	–	0.650	585	35	–
	CA-3	0.265	585	30	-23	1.170	585	21	-16
	CA-25	0.140	588	50	+ 8	0.338	588	40	– 5
	CA-45	0.090	578	50	0	0.225	575	41	-14
	CTD	0.100	575	43	0	0.250	575	44	– 9
	CTA	0.100	585	45	+ 2	0.220	585	41	-12
	CAB	o.100	583	41	– 2	0.263	583	45	– 8
	PMMA	0.095	630	27	-18	0.250	630	29	-22
	D	0.235	593	23	-26	0.765	590	19	-19

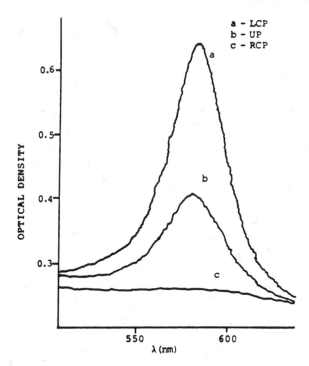

Figure 5.   Transmission spectra of the mixture, OCC/EE = 1.00/0.79.

Reflection Measurements

The intensities of UP and LCP light reflected from the samples were measured relative to the empty cell at 14° ±0.5°C at angles of 22.5°, 30°, 37.5°, and 45°. The data for LCP at selected angles of incidence, i, and reflection, r, are presented here. Plots of the reflection intensities of LCP vs. λ for different thicknesses of samples of the mesophase alone are given in Fig. 7. Inspection of these reveals that the relative intensity of the reflected light and the wavelength at which that intensity is greatest, as

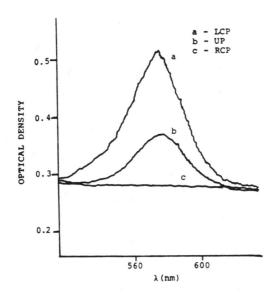

Figure 6.   Transmission spectra of the mixture, OCC/EE = 1.00/0.79
with CTD.

predicted by Fergason,[3] depend upon the sample thickness. Consequently, comparisons of reflection intensities at a given angle,
of samples with and without additives, were made with the former
and the latter at the same O.D. values. Plots of the maximum
intensity of LCP light, $I_{R_{max}}$, reflected from the mesophase alone
vs. the O.D. for two angles are shown in Fig. 8. Plots of I_R(LCP)
vs. λ for the mesophase with additives CAB and CTA, and a plot of
I_R(LCP) for the mesophase without any additive and at the same
O.D. values are given in Fig. 9. In Fig. 10, I_R(LCP) measurements
recorded at two angles, are plotted vs. λ for the mesophase containing CA-3. The relative maximum intensities of LCP light
reflected from mesophases containing additives normalized to the
mesophase alone are entered in Table 5.

     The contributions of the different additives to the normalized
$I_{R_{max}}$ are shown in Fig. 11. The data are for the samples, as
compounded, compared with the unadulterated mesophase at the same
O.D. They are, therefore, not weighted according to relative mole
fraction contributions.

TABLE 5

Normalized Maximum Intensities of UP and LCP Light Reflected

From Mixtures of a Mesophase of 1.00 Part OCC and 0.79 Parts

EE and Additives Measured at 14.0° ±0.5°C

Additive	$I_{R_{max}}$ $i = r = 30°$		$I_{R_{max}}$ $i = r = 45°$	
	UP	LCP	UP	LCP
CA-3	1.73	2.16	1.48	2.05
CA-25	1.35	1.44	1.46	1.62
CA-45	1.22	1.09	1.14	1.01
CTD	1.72	1.64	1.74	1.74
CTA	1.23	1.14	1.36	1.32
CAB	1.51	1.42	1.69	1.37
PMMA	0.49	0.54	0.52	0.59
D	0.98	1.26	0.84	1.08

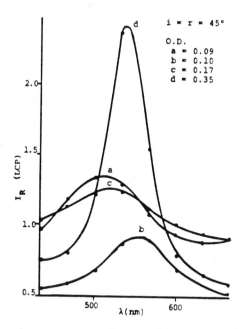

Figure 7. Reflection spectra of the mixture, OCC/EE = 1.00/0.79.

These exploratory measurements suggested that CA-3 and CTD should be examined as additives, at concentrations greater than trace quantities, in a mesophase of the same constituents. A new mesophase containing 1 part OCC to 0.60 parts EE was prepared. The polymers were dispersed, as before, in 1,4-dioxane and they were then mixed into the mesophase. The normalized maximum intensities of reflected UP and LCP light are given in Table 6.

DISCUSSION

The cellulose polymers used in this study have different molecular skeletal configurations. The cellobiose units in the polymer backbone have an oxygen-oxygen repeat distance of 1.03 nm. To account for this distance it is necessary to assume that the rings are angled and twisted with respect to one another. Cellulose acetate fractions with weight average molecular weights, $\overline{M}_w$, below about 80,000 exhibit rod-like

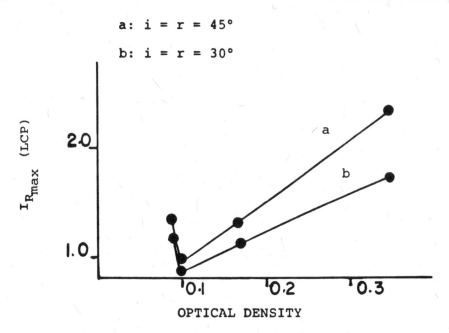

Figure 8. Reflection intensities of the mixture OCC/EE = 1.00/ 0.79 for different optical densities.

properties in dilute solution.[4,5,6] The value of $\overline{M}_w$ for CA-3 is 46,000.[7] The value of $\overline{M}_w/\overline{M}_n \gtrsim 1.7$. While there is no assurance that the other polymers have the same degree of polydispersity, if we assume that they do then the following are approximate values of $\overline{M}_w$ for the polymers used in this study: CA-25 $\gtrsim$ 82,000, CA-45 $\gtrsim$ 100,000, CAB $\gtrsim$ 55,000, CTA $\gtrsim$ 190,000. The molecular weight of CTD is not known. However, we interpret its ready solubility in 1,4-dioxane to mean that the heavy loading with decanoate moeities is on a relatively short backbone. The molecular weight of PMMA is unknown.

Within the confines of this analysis we conclude that CA-3 and CAB are rod-like molecules; CTD is probably a short chain molecule with a degree of esterification approaching that of CTA (which has 2.81 acetyl groups per glucose unit), i.e., it is a comb-like molecule; CA-25 probably has incipient coiling but it is predominantly rod-like; CA-45, CTA, and PMMA are random coil polymers.

The polymers were introduced as dispersions in 1,4-dioxane. As Fig. 11 shows, the 1,4-dioxane has no measurable effect on $I_{R_{max}}$. Thus the enhancement of the reflection characteristics of the mesophases containing the additives can be attributed, in a first approximation, primarily to the added macromolecules. The order

TABLE 6

Normalized Maximum Intensities of UP and LCP Light Reflected

From Mixtures of a Mesophase of 1.00 Part OCC and 0.60 Parts

EE and Additives Measured at 15.0° ± 0.5°C

Additive	Weight Per Cent	IR_{max} $i = r = 30°$		IR_{max} $i = r = 45°$	
		UP	LCP	UP	LCP
CA-3	2.83 (15.27)*	1.91	2.63	2.30	2.27
	5.82 (31.35)	1.31	1.74	2.19	2.05
	7.83 (39.17)	1.60	1.61	1.84	2.00
CTD	5.77 (15.97)	1.53	1.07	2.11	1.47
	10.00 (30.39)	1.85	1.08	2.05	1.45
	15.90 (44.06)	1.13	0.75[+] and 0.71	1.64 and 1.58	1.02 and 1.02

*values in parentheses are dioxane and polymer weight per cent.

[+]twin peaks were observed even in the transmission spectrum.

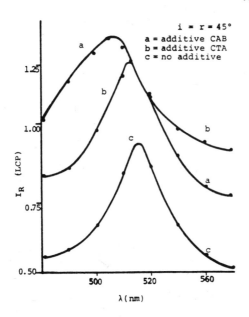

Figure 9. Reflection spectra of the mixture OCC/EE = 1.00/0.79
with and without additives.

of the normalized intensities of the samples containing CA-3,
CA-25, CAB and CTA is in the order of their mole fractions in the
mesophases. (The mole fraction of CTD is not known.) If the
effect were one of molar concentration of polymer alone, CA-45
should be more effective than CA-25, assuming no synergistic or
abrogating effect due to 1,4-dioxane, either in the same or in
different concentrations.

The method by which the samples were prepared mechanically by
a thorough mixing of the dioxane dispersions of the macromolecules
with the mesophase and the subsequent reciprocative shearing of
the mixture between the fused silica plates, is most conducive to
associative alignments of rod-like and comb-like molecules with
the cholesteryl esters within the planes of the mesophase.

The macromolecules are likely to be accommodated alongside
the cholesteryl esters within and/or between the planes of the
mesophase. The substantial effect of CA-3 and CTD on the normalized intensities of maximum reflection in such minute quantities

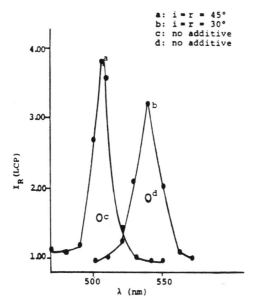

Figure 10.   Reflection spectra of the mixture, OCC/EE = 1.00/0.79 with CA-3.

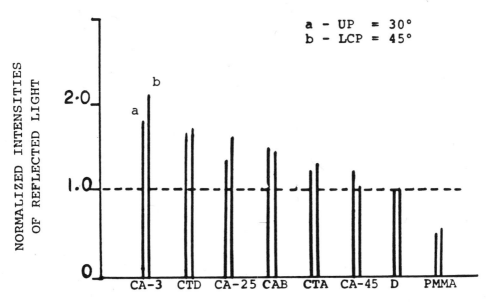

Figure 11.   Summary of the reflection intensities of the mixture, OCC/EE = 1.00/0.79, with the additives.

implies that these additives function to organize the mesophase
during the preparation of the samples. These same additives
increase the birefringence of the mesophase while dioxane, in
itself, decreases the birefringence and the effect can, therefore,
be attributed to the polymers. The longer chain esters, CA-25,
CAB, CTA, and CA-45 increase the normalized intensities and
decrease the birefringence. CTA does not alter the birefringence.
The change in Δn influences the half intensity width for non-
normal incidence. This is seen in the reflection spectrum. The
reflection spectra of the mesophase with the additives CA-3, CTD
and PMMA have narrow and well-defined peaks and those with CA-25
CA-45 and CAB are broad humps. The reflection spectra of the
mesophase with CTA have the same half intensity width as the
mesophase without any additive. Futhermore, the much higher
concentrations of CA-3 and CTD added to the mesophase, as noted
in Table 6, show that the effect is still beneficial for concentrations of these polymers which are about an order of magnitude
greater than those of the samples catalogued in Fig. 11. We
interpret this to mean that the mesophase organizing potential
of short chain linear cellulosic polymers is substantial.

Samples prepared with trace amounts of CA-3 in this and other
cholesteric mesophases have retained about the same reflectivity
as they exhibited when first prepared, one year ago.

ACKNOWLEDGEMENT

The authors gratefully acknowledge the financial support given
by the Camille and Henry Dreyfus Foundation. T. Sarada has been
a Fellow of the Epworth Fund, N.Y.

REFERENCES

1. J. L. Fergason, U.S. Patent No. 3409404, Nov. 5, 1968.
2. P. F. Waters, and T. Sarada, Mol. Cryst. Liq. Cryst., 25,
       1 (1974).
3. J. L. Fergason, Appl. Opt., 7, 1729 (1968).
4. P. Doty and H. S. Kaufman, J. Phys, Chem., 49, 583 (1945).
5. R. S. Stein and P. Doty, J. Am. Chem. Soc., 68, 159 (1946).
6. W. J. Badgley and H. Mark, J. Phys. & Coll. Sci., 51, 58
       (1947).
7. D. W. Tanner, J. Borch, C. G. Berry and T. G. Fox, Quarterly
       Summary Report, Office of Saline Water, U. S. Dept. of the
       Interior, Grant No. 14-01-001-1648, Aug. 1 - Oct. 31, 1969.

EFFECT OF LIQUID CRYSTALLINE MATERIAL ON STABILITY OF NONAQUEOUS FOAM

Syed Iftikhar Ahmad

Technicon Instruments Corporation

Tarrytown, New York   10591

ABSTRACT

Stability of foam in nonaqueous solution using phase equilibria of nonionic surfactant (EMU 050), water, and hexadecane has been described. Liquid crystalline phase in the presence of liquid phase was found to enhance the stability of foam. Foam drainage and breakage were studied in mixtures having different ratios of liquid crystal and isotropic solution, separated from compositions with different amounts of nonionic surfactant, water, and hexadecane. An empirical equation: $V_1 = at/(1 + bt)$ was found applicable to the data of foam drainage. Foam breakage was treated as a chemical reaction which stopped before final thermodynamic equilibrium was reached.

A model for foam stability in presence of lamellar liquid crystal and isotropic solution is suggested.

INTRODUCTION

Foams and emulsions have attracted great industrial attention due to their immense applications in technological processes (1). The successful application of foams in industrial and related processes completely depends on a better understanding of the laws of physics and chemistry governing their formation and destruction.

The stability of foams and emulsions is dependent on the stability of thin films. Kitchener (2) stated four factors responsible for the stability of thin films, viz., 1) the Laplace capillary suction pressure, 2) the electrical double layer

243

repulsion, 3) the long range Van der Waal's pressure, and 4) the "steric hindrance" of closely packed monolayers.

Considerable attention has been given to the factors 1, 2, and 3 in the study of thin films by Mysels et al., (3,4) Derjaguin et al., (5), Scheludko (6), and Tien (7). The steric factor has not yet received any theoretical or experimental treatment. But it is referred to as the molecular obstruction offered by monolayers in which molecules are closely packed together and molecular cohesion prevents their separation when subjected to considerable pressure (2). Some general principles which govern this type of stabilization are slowly emerging (8).

Ordered packing conditions are present in the liquid crystals formed by combinations of surface active agents with water and amphiphilic substances (9,10,11). Friberg et al., (12) have shown that liquid crystals play a decisive role in the stability of emulsions. Ahmad and Friberg (13) have shown that the foam stability in a hydrocarbon is completely determined by the presence of a liquid crystalline phase in equilibrium with the liquid phase having reversed micellar structure.

The results - stabilization of foams with liquid crystalline phase - appeared to provide new possibilities for obtaining information about the "steric factor" by using liquid crystals as model. The new rules for foam formation in hydrocarbons using phase equilibria which give well defined boundaries between different phases formed with surfactants, water and amphiphilic compounds appeared to be of technical value. Further investigations were conducted in nonaqueous solutions.

The present paper describes the kinetic studies on the stability of foam using a phase equilibria of non-ionic surfactant (EMU 050), hexadecane and water. The effect of liquid crystalline phase on the foam drainage and foam breakage has been investigated.

EXPERIMENTAL

Materials

Analytical reagent, Hexadecane (E. Marck) was used without further purification. The water was doubly distilled.

Nonionic surface active agent, Polyoxyethylene (5) nonyl phenol ether, EMU 050, (Berol, Sweden) was purified and made free from long chain C_{12-14} alcohols. EMU 050 100 ml, was mixed with sodium chloride, 50 ml (5 M/1). Nonionic surfactant was extracted by dry ether, 500 ml (A.R.). The surfactant was made free from ether by water bath distillation. The nonionic surfactant was

Drainage

    Results for a typical run of a solution with compositions
EMU 050 70%, water 17%, and hexadecane 13% (W/W) from which phases
N and L are separated and remixed in different proportions, are
shown in Figures 3 and 4.

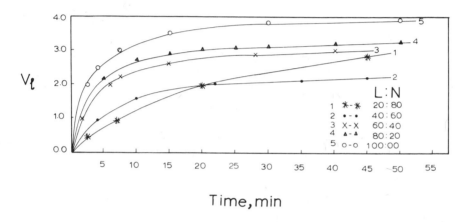

Figure 3.   Volume of liquid (ml) drained from the foam with time.
Solutions having different ratios of phase L and N prepared from
the solution having EMU 050 70%, water 17% and hexadecane 13%.

    In Fig. 3, the volume of the liquid drained (V_1) from the
foam is plotted against time. An initial steady increase in the
volume of liquid drained, slowed down at longer time. The
empirical equation: $V_1 = at/(1 + bt)$ was found to fit the data.
In the equation, "a" is a constant for initial drainage rate at
t = 0, which could be obtained by differentiating volume with respect to time (dV/dt). The constant "b" describes the change of
rate which is time dependent. At infinite time (t = ∞), the final
volume of drainage is a/b. In Fig. 4 the reciprocal of the volume
($1/V_1$) is plotted against reciprocal of time ($1/t$). The data is
found to conform to the equation: $V_1 = at/(1+bt)$ and a deviation
of less than 5% is observed between experimental and calculated
values of V_1.

    In Table I, the calculated values of various constants a,
b and dV/dt, are given. The values of the constants were calculated using the least square method.

TABLE I

Values of the constants a,b, final drainage (a/b), initial drainage rate (a), slope (1/a) intercept (b/a) and foam breakage constant K for different compositions with phase L and N.

Compositions	1/a	b/a	a	b	a/b	K	Remarks
EMU 050: H_2O:Hexadecane							
85 : 15:0	–	–	–	–	–	–	Does not foam
Phases							
L : N							
0 : 100	–	–	–	–	–	–	Does not foam
20 : 80	–	–	–	–	–	–	Does not foam
40 : 60	2.525	0.272	0.396	0.108	3.666	1.281	Foam
60 : 40	5.681	0.249	0.176	0.043	4.016	1.611	Foam
80 : 20	4.400	0.278	0.227	0.063	3.597	2.025	Foam
100 : 0	1.590	0.269	0.628	0.168	3.711	2.071	Foam
55 : 21 : 24	3.275	0.281	0.305	0.084	3.558	–	Foam
L : N							
0 : 100	–	–	–	–	–	–	Does not foam
20 : 80	2.458	0.247	0.406	0.111	3.649	0.414	Foam
40 : 60	10.000	0.284	0.100	0.028	3.510	0.119	Foam
60 : 40	6.217	0.311	0.160	0.049	3.213	0.188	Foam
80 : 20	3.720	0.257	0.268	0.069	3.891	0.349	Foam
100 : 0	2.760	0.257	0.360	0.092	3.891	0.667	Foam
41 : 18 : 41	1.502	0.257	0.665	0.169	3.890	–	Foam
L : N							
0 : 100	–	–	–	–	–	–	Does not foam
20 : 80	0.848	0.326	1.178	0.348	3.067	0.230	Foam
40 : 60	1.545	0.290	0.647	0.187	3.444	0.368	Foam
60 : 40	1.735	0.307	0.576	0.177	3.257	0.736	Foam
80 : 20	1.425	0.289	0.701	0.203	3.460	1.104	Foam
100 : 0	1.073	0.286	0.931	0.266	3.496	1.243	Foam
21 : 10 : 69	0.257	0.250	3.891	0.972	4.000	–	Foam
L : N							
0 : 100	–	–	–	–	–	–	Does not foam
20 : 80	0.150	0.276	6.666	0.183	3.162	3.453	Foam
40 : 60	0.504	0.286	1.984	0.143	3.504	4.604	Foam
60 : 40	0.823	0.315	1.214	0.383	3.166	13.812	Foam
80 : 20	0.572	0.291	1.747	0.508	3.434	18.416	Foam
100 : 0	0.200	0.296	4.982	1.477	3.370	20.718	Foam

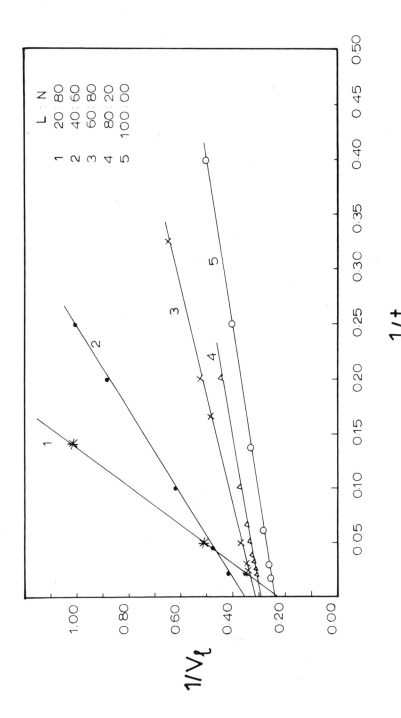

Figure 4. 1/V (ml⁻¹) versus 1/t (minute⁻¹) curve for solutions
having different ratios of phase L and N, prepared from solution
having EMU 050 70%, water 17% and hexadecane 13%.

In Fig. 5 the value of the slope (1/a) is shown for runs having different compositions of phases N and L separated from the mixtures having EMU 050 70%, water 17%, and hexadecane 13% and EMU 050 85%, water 15%, and hexadecane 0%. The Fig. 5 and Table 1 show the final volume (a/b) remained almost constant for all the compositions studied while initial rate "a" of the liquid varied due to presence of the liquid crystalline phase in the solution.

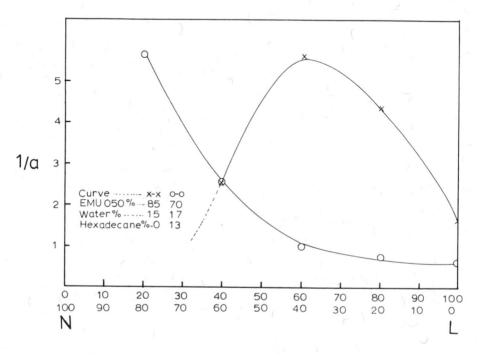

Figure 5.   Effect of different ratios of phase L and N on initial drainage (a). Phase L and N prepared from solution having EMU 050 70%, water 17% and hexadecane 13%.

Foam Breakage

In Fig. 6, the breakage of foam is shown for a run having different compositions of phases N and L separated from a mixture containing EMU 050 70%, water 17%, and hexadecane 13%. Initially foam remained unbroken for a certain period of time before breaking starts. In certain compositions, where liquid crystalline phase 60-80% is present, the foam stops breaking after some time and remains unbroken for a very long period. The unbroken foam either suddenly breaks up or stays unbroken.

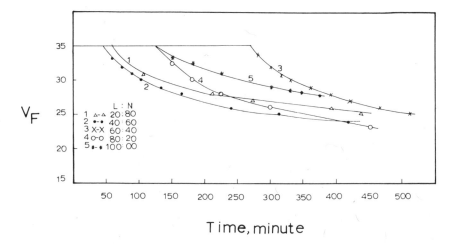

V_F

Time, minute

Figure 6. Volume of foam (V_F,ml) broken with time (minute) for
solutions having different ratios of phase L and N, prepared from
solutions having different compositions.

In stable foam, when a large amount of liquid drains out, a
small amount of liquid remains in the foam and foam collapse
begins. The film collapse rate may be slowed down by accumulation
of stabilizing material that falls on the remaining liquid films
from the above collapsed foam. Under these conditions the foam
may stop breaking short of thermodynamic equilibrium. As an
anology, consider a block which is sliding down a sloped surface.
If the sliding is frictionless, the block will end up at the lowest point, the equilibrium position; but if there is friction
between block and surface, the block may end up short of equilibrium (14).

Analogously, the breaking of foam which stops short of
thermodynamic equilibrium will have a rate equation, with a
friction like or resistance term. If we consider breaking of
foam irreversible, the rate equation for foam breakage which
does not go to completion is:

$$\frac{dV_f}{dt} = K\,V_f - Q \qquad \ldots \ldots (A)$$

where V_f is volume of foam left in the column at time t.

When $dV_f/dt = 0$ we find that $Q = KV_e$, where V_e is the foam
volume where breaking stops before reaching the thermodynamic

equilibrium.

Therefore, $-\dfrac{dV_f}{dt} = K(V_f - V_e)$ . . . . . (B)

on integrating we get:

$$- \ln \frac{V_f - V_e}{V_0 - V_e} = K(t - t_0) \quad . . . . . \text{(C)}$$

where t_0 is the time when foam starts breaking and V_0 is volume of initial foam.

The above equation could be used for foams which are very stable and stop breaking after certain time due to accumulation of stabilizing material. A similar equation is given by Levenspiel (14) for chemical reaction which stops short of equilibrium.

The value of the constant K is calculated from the slope of the curve log $\dfrac{V_t - V_e}{V_0 - V_e}$ against $t-t_0$ for different compositions (Fig. 7). The constant K for foam breaking is affected by the presence of liquid crystalline phase in the solutions.

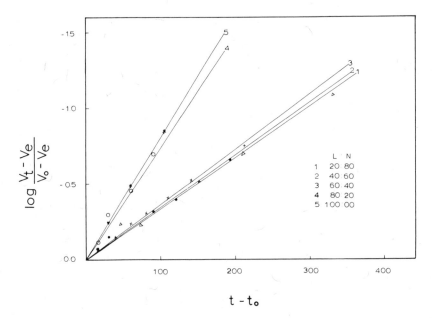

Figure 7. Log $\dfrac{V_t - V_e}{V_0 - V_e}$ versus time (minute) for solutions having different ratios of L and N, prepared from solution having composition EMU 70%, water 17% and hexadecane 13%.

DISCUSSION

The equation, $V = V_0(1 - e^{-kt})$ is found to describe the drainage of foam under the influence of gravity (15). But the equation is inapplicable for the losw draining foams in their earlier and final stages. For our system where foams are formed by a non-ionic surface active agent, the equation $V_1 = at/(1+bt)$ is applicable and fit the experimental data well (Figs. 2 and 3). A similar equation is given by Kruglyakov (1) for the drainage of stationary foams.

The value of "a", the intial drainage rate, should be low for the stable foams (15). From Table I it is evident that for foams having liquid crystalline phase 40-60% in the solutions, the values of "a" are quite low. The fractional decrease in drainage rate with increasing time, on which constant "b" depends, is low for the foams having an optimum amount of liquid crysalline phase, 40-60%. The low values of the constant "b" indicate a very small drainage from the foam even after a moderate time.

The results for the foam drainage are explained simply in terms of a lubrication theory (6,15). According to this theory, the constant "a" should vary with the bulk density and viscosity of the experimental solution (15). Foams from the solutions having lower shear viscosity, for example, will have higher initial drainage rate "a" and less stable foam. The results given in Table I and Fig. 5 show that after an optimum L/N ratio, the solutions which should have higher shear viscosity and low initial drainage rate "a", give higher values of "a" and consequently less stable foams. The exceptions are for the solutions foamed from a composition of EMU 050 70%, water 17% and hexadecane 13%.

The high stability of foam at the optimum L/N ratio shows that the presence of liquid crystalline phase in combinations with micellar solution offer a possibility to provide information about the "steric factor" in stabilizing the foam.

In film of non-polar solvents, the electric forces have a minor influence on the stability of films. The forces remaining responsible for the stability of non-aqueous foam are attractive forces due to long range Van der Waals' interaction and "steric hindrance". The steric hindrance effect has not yet received any attention but is vaguely regarded as the molecular obstruction offered by monolayers in which the adsorbed molecules are so closely packed together that molecular cohesion (e.g., between the chains of hydrocarbons) prevents their separation when subjected to a considerable pressure (2).

From the results, it is also evident that properties of foams
in the presence of liquid crystalline phase are affected in the
initial stages (the value of "a" is changing), when properties of
the foam lamellae are subjected to some changes due to the action
of gravity and Laplace capillary suction. In the final stages
these effects are minimized and the value of the final drainage
rate a/b remains constant (Table I).

The final drainage rate is found to be very much dependent
on the expansion of foam (15). In the present study the foam
expansion value 8.7 (volume of foam formed/volume of liquid) was
kept constant and the final drainage rate a/b was not affected.

In the emulsion stability, the structure conferred by
emulsifier molecules on the thin films of the continuous phase
separating the flocculated globules play an important role (16,17).
Now it appears equally worthwhile to study the structural organization within these liquid films. Recent studies by Friberg et
al., (12,18) on the emulsion stability, using phase equilibria,
showed the properties of the thin liquid films between the
globules in the emulsion are affected due to the presence of
liquid crystalline phase. Friberg et al., (12,18) have demonstrated that the liquid crystalline phase (in presence of micellar
phase)., which covers the globules in the emulsion, is responsible
for the increased stability of emulsion. A theoretical treatment
of the Van der Waal interaction between the emulsified droplets
covered with a liquid crystalline phase is given by Friberg (19).

Considering the work of Friberg et al., (10,12,19) on the
stability of emulsion, a possible model for stable foaming in a
two phase region (liquid crystalline and micellar phase), where
long range Van der Waals' forces and steric hindrance are present,
is shown in Fig. 8.

The proposed model for the stability of foam gives a
reasonable explanation as to why micellar solutions give less
stable foams. Micelles do not form layered structures, and
maximum cohesion between the molecules is not experienced. Stable
foam formed at optimum L and N ratio is due to dispersion of liquid
crystalline phase around nitrogen bubbles by micellar solution.
The liquid crystal monolayers adsorbed at the surface of the gas
bubbles offer molecular cohesion and prevent their separation under
the considerable pressure from Van der Waals' forces over a thin
layer between the bubbles (19).

The breakage of foam is treated as a chemical reaction which
stops before the final thermodynamic equilibrium is reached (14).
The value of the constant "K" showed a similar trend as the final
drainage rate "b". For very stable foam, the value of the constant
is less compared to the value of the unstable foam. The value of

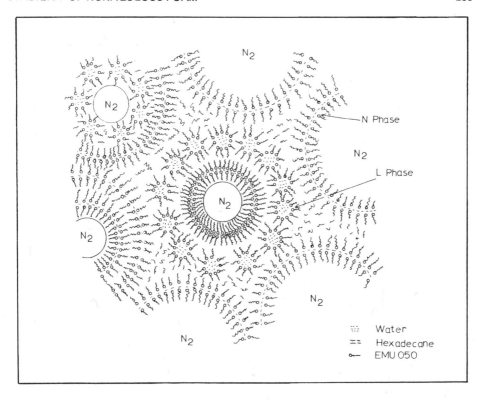

Figure 8. A schematic representation of the foam stability in presence of liquid crystalline phase (N) and reversed micelles (L).

the foam breakage constant "K" increases as the amount of the liquid crystalline phase in the solution decreases. The value of "K" showed that the breakage of the foam follows a similar mechanism as the final drainage constant "b" which is time dependent.

The investigations are continued with the determination of shear viscosity, interfacial tension, thickness of the thin films, and foam bubble size distribution in the mixtures containing liquid crystalline phase (N) and micellar solution (L).

CONCLUSIONS

In phase equilibria on nonionic surfactants, water and hexadecane, a liquid crystalline phase (N) and an isotropic liquid containing hydrocarbon and emulsifier with dissolved water

were found.   Stable foams were found in the two-phase region L + N.
Stable foams were found for mixtures having 40-60% liquid crystalline phase.   Foam drainage could be described by an empirical
equation: $V_1 = at/(1 + bt)$.   Foam breakage was treated as a
chemical reaction which stops before final thermodynamic equilibrium was reached.

A foam model, stabilized by the presence of a liquid crystal
is suggested.

Nomenclature:

V = volume, cm^3
t = time, minute
a = initial drainage rate, cm^3 min^{-1}
b = final drainage rate, min^{-1}
Q = friction constant
K = foam breakage constant, min^{-1}

Subscripts:

1 = liquid
e = equilibrium
f = foam

ACKNOWLEDGEMENTS

The author is thankful to Professor Stig Friberg for his
valuable suggestions in preparing this article and the facilities
provided by him to work at the Swedish Institute for Surface
Chemistry, Stockholm, Sweden, where this work was carried out.
The investigations were supported by the Swedish Board for Technical Development.

LITERATURE CITED

(1) J.J. Bikerman, "Foams" Springer-Verlag, New York, 1973.
(2) J.A. Kitchener, Recent Progress Surface Sci., 1, 51, (1964).
(3) K.J. Mysels, K. Shinoda and S.P. Frankel, "Soap Films"
    studies of their thinning and bibliography, Pergamon, Oxford
    (1959).
(4) K.J. Mysels, M.C. Cox and J.D. Skewis, J. Phys. Chem., 65,
    1106 (1961).
(5) B.V. Derjaguin and A.S. Titievskays, Kolloid Zhur., 15, 416
    (1953).
(6) A. Scheludko, Adv. Colloid and Interface Sci., 1, 391 (1967).
(7) H.T. Tien, Advan. Exp. Biol., 7, 135 (1970).
(8) D.H. Napper, Ind. Eng. Chem. Proc. Res. Dev., 9, 467 (1970).
(9) P. Ekwall, Mol. Cryst., 8, 157 (1969).

(10) P. Ekwall, in Liquid Crystal, Eds. Johnson, J.F. and
     R.S. Porter, Plenum Press, New York, pp. 177, 1974.
(11) S.I. Ahmad and S. Friberg, Acta Polytechnica Scand., 102
     (1971)
(12) S.I. Friberg, Mandell and M. Larsson, J. Colloid and Interface
     Sci., 29 155 (1969).
(13) S. Friberg and S. I. Ahmad, J. Colloid and Interface Sci.,
     35, 175 (1971).
(14) O. Levenspiel, "Chemical Reaction Engineering", John Wiley,
     N.Y., 1962.
(15) J.T. Davies and E.K. Rideal, Interfacial Phenomena, Acadamic
     Press, N.Y. 1963.
(16) J. Boyd, C. Parkinson and P. Sherman, J. Colloid & Interface
     Sci., 41, 359 (1972).
(17) S. Friberg and L. Mandell, J. Pharm. Sci., 59, 1001 (1970).
(18) S. Friberg and P. Solyom, Kolloid Z. Polym., 236, 173 (1970).
(19) S. Friberg, J. Colloid & Interface Sci., 37, 291 (1971).

SPECTROSCOPIC STUDIES OF DYES IN LIQUID CRYSTAL IMPREGNATED

MICROPOROUS POLYPROPYLENE FILMS

E. J. Poziomek and T. J. Novak

Chemical Systems Laboratories
Abeerdeen Proving Ground, MD 21010

R. A. Mackay

Department of Chemistry
Drexel University
Philadelphia, PA 19104

ABSTRACT

Microporous polypropylene films such as Celgard 2400 (Celanese Plastics Co., Summit, N. J.) contain an ultra-fine pore structure resulting in a high film surface area on the order of $50m^2/g$. The pores are about 0.2μ long and 0.02μ wide, forming slightly tortuous channels extending from one surface to the other. We have found that when this film is impregnated with a nematic liquid crystal containing a dissolved dye, the absorption of polarized light by the dye is strongly dependent upon the orientation of the film with respect to the plane of polarization. The liquid crystal is therefore spontaneously oriented by the micropores, and this orientation is not affected by subsequent directional application of a static charge or mechanical deformation. Further, the results are consistent with the slit-like nature of the voids since a cylindrical pore would not be expected to exhibit polarization effects. A potentially important application of this observation lies in the simplicity of the technique. Quantitative as well as qualitative results are readily obtained, and the matrix can be rendered reversibly isotropic by heat or vapor absorption. The results of these and other measurements are discussed.

INTRODUCTION

There has been much interest in recent years on the spectroscopy of liquid crystals (1), and on the measurement of the spectra of solutes in oriented liquid crystals. In this regard, various infrared, ultraviolet-visible, nuclear magnetic resonance and electron spin resonance spectra have been reported (1-3).

A number of techniques have been employed to orient liquid
crystals and include the use of electric fields (2,4), rubbing glass
plates (5), and the use of specially prepared surfaces. The latter
may consist of mechanical and chemical cleaning, deposition of compounds such as silicones or surfactants (6), or the use of resins as
aligning agents (7). We report here a simple technique involving
self-orientation of liquid crystals. The effect is achieved by
impregnating microporous polypropylene film with the liquid crystal
(host) which may contain added solute (guest) molecules. The use of
these liquid crystal-impregnated films as a detection device for
vapors was first reported by Poziomek et al. (8). For suitably
chosen solutes, the guest-host interaction will also produce some
degree of alignment in the guest molecules. Visible absorption
spectra of dyes and electronic spin resonance spectra of a nitroxide
radical are given as examples of the wide scope of potential applications.

EXPERIMENTAL

Microporous polypropylene film (Celgard 2400) was obtained from
the Celanese Plastics Co., Summit, N. J. The pores are very uniform
and are rectangular slits about 0.2μ long and 0.02μ wide, and the
film thickness is about 25μ . A detailed discussion of the preparation and properties of microporous films is given by Bierenbaum et
al. (9).

The impregnation procedure consisted of placing a solution of
the solute in the liquid crystal on the microporous polypropylene
film for 30 seconds, then wiping off the excess with a tissue. The
film was then placed on a glass microscope slide, rubbed smooth with
a tissue and mounted on a holder which kept the slide rigid and the
film tight. The direction in which the film is taken off the roll
is the machine direction and is parallel to the long axis of the
rectangular pores. The orientation of the pore with respect to the
glass slide is thus known. The visible absorption spectra were
obtained on a Cary 14 spectrophotometer, and the electron spin
resonance (esr) spectra on a Varian E-12 spectrometer system operating at 9.5 GHz with 100 kHz field modulation. To obtain the
polarized absorption spectra, polaroid sheet of known orientation
with respect to the film was placed in front of the sample. Impregnated film (no solute) and polarizer were employed as the reference.

The liquid crystals were obtained from Eastman, as were the dyes
Sudan IV, DAPN and PAN (vide infra). The nitroxide TAPN is from
Aldrich, and the PDDA was synthesized according to the procedure of
Wilson et al. (10).

RESULTS AND DISCUSSION

Screening

A number of dyes were examined visually in the liquid crystal
N-(p-methoxybenzylidene)-p-butylaniline (MBBA). The orientation
was marked on the film which was then impregnated and rotated in
front of a source of plane polarized light. The results are given
in Table 1. It may be noted that all of the dyes examined, with
the exception of Sudan II and possibly NAR, exhibits some degree of
orientation, and that the principal component of their transition
moments is parallel to the long axis of the pores.

In addition to MBBA, several nematic mixtures as well as two
cholesterics, 2-(2-butoxyethoxy)ethyl cholesteryl carbonate and
cholesteryl oleyl carbonate, were screened with the dye PDDA on the
film. All of the nematics worked, some stronger and some weaker than

Table 1. Visual Polarization Effects of Dyes in MBBA on Microporous
Film (vide text).

Dye[a]	Solubility in MBAA	Color	Visual effect[b]
PDDA	good	red-brown	strong
DAPN	good	royal blue	strong
NAR	moderate	yellow	none(weak)?
Methyl Red	mod/low	orange	moderate
Sudan II	good	orange-red	none
Sudan Blue	good	blue	mod/weak
Sudan Green	good	green	mod/weak
Sudan IV	mod/good	red	mod/strong

a. PDAA = N-(5-pyrrolidinopenta-2,4-dienylidene)-2,4-dinitroaniline;
   DAPN = N-(p-dimethylaminophenyl)-1,4-naphthoquinone-4-imine;
   NAR = 4-(p-nitrophenylazo)-resorcinol.

b. Change from dark to light. In all cases, maximum absorption
   (dark) is for light polarized parallel to the long axis of pores.

MBBA. The cholesterics gave no effect, but they were viscous, did
not dissolve the dye well, and did not impregnate the film readily.
It is probable that a nematic-cholesteric mixture could be successfully used.

If the impregnated film was heated by placing it near a tungsten
bulb, the effect was eliminated, but returned upon cooling. A film
with a thickness greater than 25µ also works, but the visual effect
may be somewhat weaker.

Absorption Spectra of Oriented Dyes.

MBBA exhibits its absorption maximum in the ultraviolet, but
has a tail which extends into the visible to give a yellow color.
At room temperature, using absorption data on the tail at 500, 600
and 700 nm, the order parameter S is estimated to be about 0.63.
In all of the cases to be discussed below, the order parameter is
calculated with respect of the axis of the transition moment of the
electronic absorption band. Thus, S is given by $(R-1)/(R+2)$, where
the dichroic ratio $R = \epsilon_{\parallel}/\epsilon_{\perp}$. The quantities $\epsilon_{\parallel}$ and $\epsilon_{\perp}$ are the extinction coefficients for light polarized parallel and perpendicular
to the optic axis. If the transition moment axis coincides with
the long molecular axis, then S is also the order parameter of the
molecule. For MBBA at 25°C, the order parameter is reported to be
0.59(11). This value has been corrected for the strong anisotropy
of the polarization field in the medium (11,12), but none of the S
values reported here have been so corrected. This correction is
often on the order of $\pm 20\%$, depending upon the sign of the birefringence.

We may therefore conclude that the liquid crystal is spontaneously oriented by impregnation in the film. From the orientation
of the polarizer with respect to the film, we may also conclude that
the optic axis is parallel to the long axis of the pore.

The spectrum of DAPN in MBBA on the film is shown in figure 1.
It is clear that the dye is oriented with respect to the film since
the absorption depends on the direction of polarization, in accord
with the visual observations. Thus, the dye is oriented by the liquid crystal, which in turn is oriented by the pores. Various concentrations of DAPN, Sudan IV and Chlorophyll a in the range 7-35
mg/ml were examined, all giving substantially the same results. These
are summarized in Table 2. The three different guest dyes appear
less ordered than the host liquid crystal and all have S values
around 0.25. However, it cannot necessarily be concluded at this
point that this is more than coincidence since an R value of 2.0
will be obtained if S = 0.6, and the angle between the long molecular axis of the dye and the transition moment is about 23° (1).

If the liquid crystal containing the dye is placed between
rubbed glass plates, the liquid crystal lines up the respect to the
direction of rubbing and the orientation effect is achieved. However, the effect is weaker than on the film. In contrast, rubbing

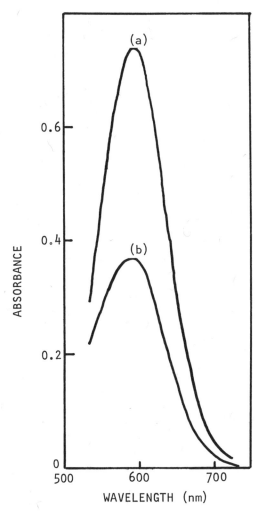

Figure 1.   DAPN in MBBA (34mg/ml)
on film.   Spectra (a) and (b)
are for axis of polarization parallel and perpendicular to the
long axis of the pore, respectively.

Table 2.   Dichroic Ratios for Various Dyes in MBBA on Microporous
Polypropylene Film.

Dye	S^a	$\varepsilon_{\shortparallel}/\varepsilon_{\perp}{}^b$
DAPN	0.28	2.17
Chlorophyll a	0.23	1.92
Sudan IV	0.25	2.01
None	0.63	6.11^c

a. Order parameter (S=0, isotropic; S=1, completely ordered).
    Values given are averages of several runs with an average
    deviation of ± 0.02.

b. Ratio of extinction coefficients for light polarized parallel
    ($\varepsilon_{\shortparallel}$) and perpendicular ($\varepsilon_{\perp}$) to long axis of pore.

c. Obtained from tail of absorption band at 500, 600, and 700 nm.

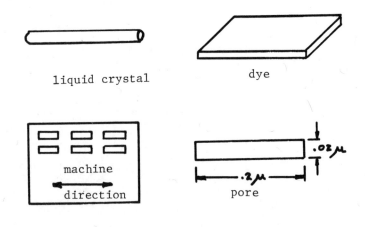

liquid crystal   dye

machine direction   pore

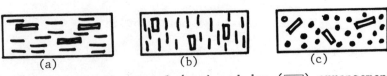

(a)   (b)   (c)

Figure 2.   Possible liquid crystal (——) and dye (⟹) arrangements
in pore: (a) parallel to long axis of pore; (b) perpendicular to
long axis of pore; (c) perpendicular to plane of film. Drawing is
not to scale.

the film vigorously with pressure or sandwiching the film between rubbed glass plates does not affect the orientation. A schematic illustrating the principal features of the situation is shown in figure 2. The long axis of the rod-like liquid crystal tends to line up parallel to the long axis of the slit-like micropore. The dye molecules are roughly approximated by rectangles. The $\pi \rightarrow \pi^*$ transitions are polarized in the plane, and the dye molecules are expected to line up with their long axis parallel to that of the liquid crystal consistent with the arrangement in figure 2(a). The alignment in figure 2(b) would give opposite polarization results, and that in figure 2(c) would result in no dye polarization.

Finally, it is possible to obtain quantitative as well as qualitative results. Although the technique we employed to impregnate and mount the films was moderately simple, the results were satisfactorily reproducible as shown in figure 3. In this regard it may also be noted that two films impregnated with DAPN in MBBA and "stacked" gave an absorbance twice that of a single film.

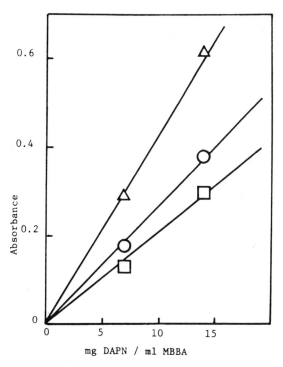

Figure 3. Absorbance <u>vs</u> concentration at 600 nm (circles, unpolarized) and 590 nm (triangles and squares have polarization parallel and perpendicular to long axis of pore respectively).

Effect of Temperature and Absorbed Vapors.

If the liquid crystal becomes isotropic, the effects described above should disappear. The visual observation on film heating (vida supra) is in accord. Also, absorbed vapor is known to cause a nematic → isotropic transition which can serve as a basis for vapor detection systems (8). The effect of chloroform vapor on the spectrum of DAPN in MBBA on the film was examined. When placed in the spectrophotometer in a sealed can with windows and exposed to chloroform at its own vapor pressure at room temperature the polarization effects vanished. When the vapor was allowed to desorb, the effect returned. No effects were observed for DAPN in an isotropic fluid such as cyclohexanol.

When a solute is dissolved in a liquid crystal, the mesomorphic transition temperature is lowered, the magnitude of the depression depending upon the solute concentration and the specific solute and liquid crystal employed (13). The bulk MBBA transition temperature is reported to be around 46°C (our sample gave 43°C), and bulk MBBA containing 3% (w/w) DAPN gave the transition at 37°C. However, rather different results are obtained on the film.

The behavior of the impregnated film as a function of temperature was monitored in the spectrometer employing a previously described technique (14). The results are clearly shown in figure 4. For both MBBA and MBBA containing 3% (w/w) DAPN, the transition occurs at about 35°C. A visual observation of MBBA on the film between crossed polars showed the transition at 34°C in substantial agreement with the spectral results. It is interesting that the film seems to equalize the transition temperature with and without solute, and that it is lower than in the bulk irrespective of the fact that the film spontaneously orients the liquid crystal. When a thin film of nematic liquid crystal is oriented by an electric field, the transition temperature is increased (15). It is not inconceivable that the transition at 35°C on the film represents a change in alignment direction from parallel to the long axis of pore to perpindicular to the film surface. However, the DAPN/MBBA trace in figure 4 was continued to 65°C with no observable change.

Since the film is used as received with no prior treatment such as washing, it is possible that impurities present in the pores are responsible, in whole or in part, for the lowering.

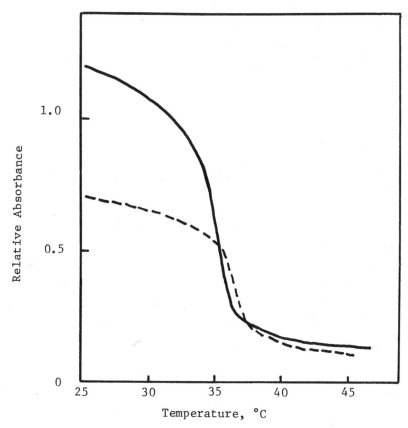

Figure 4. MBAA in microporous film. Solic line is for no added dye (500 nm) and dashed line is for MBBA containing DAPN (610 nm). In both cases, the axis of polarization is parallel to the long dimension of the pores.

Electron Spin Resonance Spectra.

It should also be possible to employ these films to obtain an-isotropic esr spectra, since the film can be placed in a known orientation with respect to the magnetic field. An initial study was performed using the nitroxide free radical 2,2,6,6-tetramethy-4-aminopiperidine-N-oxide (TAPN). This particular molecule is far from ideal since it is neither rod-like nor large and planar. None-theless, if any effects are observed, then more favorable molecules should yield good results. Some difference in the esr spectrum as a function of film orientation has been obtained as shown in figure 5. The quantitative data are given in Table 3.

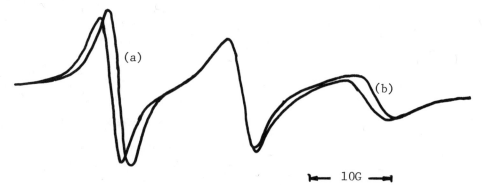

Figure 5. Esr spectrum of the 4-Amino-2,2,6,6-tetramethylpiperidino-oxy free radical in MBAA on the microporous film at room temperature. (a) long axis of pore and film surface parallel to magnetic field; (b) long axis of pore $\perp$ (vertical) to field; film surface perpendicular to field.

Table 3. ESR Parameters for TAPN in MBBA on Microporous Film.

Orientation w.r.t. field		hyperfine		line width		
long axis of pore	film surface	a_{-10}	a_{01}	γ_{-1}	γ_o	γ_1
(bulk)	(bulk)	15.1	15.4	2.6	2.7	3.8
$\perp$ (vert)	$\parallel$	16.3	16.6	2.8	3.4	5.2
$\perp$ (vert)	$\perp$	16.1	16.6	3.0	3.6	5.5
$\parallel$	$\parallel$	15.1	15.0	3.0	3.6	5.5
$\perp$ (hor.)	$\perp$	15.4	15.7	2.5	3.4	4.8

The g value is 2.001. The hyperfine coupling constants (a) and line widths (γ) are in gauss (+0.2). The subscripts refer to the M_z values.

The concentration of the nitroxide was high (ca. 5×10^{-2}M) so that electron spin exchange interactions are not totally absent. Nonetheless, the variation in the hyperfine coupling constants (a values) indicate that some degree of ordering is present. The smaller a values for the parallel-parallel orientation (Table 3), comparable to those in the bulk, are in accord with the expected pattern since $\Delta a \stackrel{\sim}{\sim} 26$ (16) and the bulk sample should tend to line up with the 3.4kG magnetic field (parallel to field). In fact, it is interesting that the liquid crystal in the microporous film appears to maintain its alignment in the presence of the magnetic field of the esr spectrometer since electric fields on the order of several kV cm^{-1} are generally necessary to overcome this orientation (17).

The increased line widths (γ values) may indicate a decreased rate of tumbling of the solute in the oriented liquid crystal relative to that in the bulk. In any event, much lower concentrations of spin label can be employed, and if necessary the films can easily be stacked.

Summary

1. Nematic liquid crystals are spontaneously oriented by microporous polypropylene films. The long axis of the liquid crystal tends to align parallel to the long axis of the pore.

2. Guest molecules of appropriate geometry can be aligned in the liquid crystal host, providing a convenient means of attaining polarized ultraviolet-visible spectra and anisotropic esr spectra.

3. The liquid crystal on the film can be rendered reversibly isotropic by means of heat and vapor absorption. The transition temperature on the film is lower than in the bulk.

ACKNOWLEDGEMENT

Assistance by members of the 400th and 402nd Chemical Laboratories, U.S. Army Reserve, in synthesis of PDDA is gratefully acknowledged.

REFERENCES

1. S. Chandrasekhar and N.V. Madhusudana, Applied Spectroscopy Reviews, 6, 189 (1972).

2. G.H. Meilmeier, J.A. Castellano and L.A. Zanoni, Mol. Cryst. Liq. Cryst., 8, 293 (1969).

3. C.F. Polnaszek and J.H. Freed, J. Phys. Chem., 79, 2283 (1975).

4. L.M. Blinov, G.G. Dyadyusha, F. A. Mikhailenko, I.L. Mushkalo and V.G. Rumyantsev, Doklady Akademii Nauk. SSSR, 220, 860 (1975).

5. G.P. Ceasar and H.B. Gray, J. Am. Chem. Soc., 91, 191 (1969).

6. F.J. Kahn, G.N. Taylor and H.Schonhorn, Proc. of the IEEE, 61, 823 (1973).

7. W.E.L. Haas, J.E. Adams, B. Mechlowitz, U.S. 3, 871, 904 (Cl. 117-72; Go2f), 18, Mar. 1975 (CA 81; 12 9922d).

8. E.J. Poziomek, T.J. Novak and R.A. Mackay, Mol. Cryst. Liq. Cryst. 27, 175 (1974).

9. H.S. Bierenbaum, R.B. Isaacson, M.L. Druin and S.G. Plovan, Ind. Eng. Prod. Res. Develop. , 13, 2 (1974).

10. R.M. Wilson, E. J. Gardener and R.H. Squire, J. Chem. Ed., 50, 94 (1973).

11. R.Chang, Mol. Cryst. Liq. Cryst., 30, 155 (1975).

12. A. Saupe and W. Maier, Z. Naturforsch., 16A, 816 (1961).

13. T.J. Novak, E.J. Poziomek and R.A. Mackay, Mol. Cryst. Liq. Cryst., 20, 203 (1973).

14. T.J. Novak, E.J. Poziomek and R.A. Mackay, Rev. Sci. Instr., 42, 124 (1971).

15. R.Williams, Ch. 5 in "Liquid Crystals and Plastic Crystals", Vol. 2, G.W. Gray and P.A. Winsor, Eds., John Wiley and Sons Inc., N.Y. 1974.

16. O.H. Griffith and A.S. Waggoner, Accts. Chem. Res., 2, 17 (1969).

17. G.R. Luckhurst, Ch. 7, in reference 15.

THERMAL AND CHEMICAL PERTURBATIONS OF MOLECULAR ORDER AT THE INTERFACE AND IN THE ALKYL CHAIN REGION OF A LYOTROPIC LIQUID CRYSTAL

Robert C. Long, Jr. and J. H. Goldstein

Depts. of Medicine and Chemistry

Emory University, Atlanta, Georgia 30322

INTRODUCTION

Magnetic field-orientable lyotropic mesophases are useful both as ordering solvents in NMR spectroscopy and as models for membranes and membrane processes. In NMR studies of the latter category, detailed information concerning the behavior of the alkyl chain of the amphiphilic component of the system is obtainable from the quadrupole splittings of the specifically deuterated chain positions (1). Although this is a highly useful and widely applicable approach it is unable to provide the signs of the order parameters in the general case (unless $|S| > .5$) where the quadrupole splitting is much greater than the direct and indirect couplings (2). We have now measured the ^{13}C shift anisotropies for positions in the alkyl chain for the laurate amphiphile over a range of temperatures. Not only does this eliminate sign ambiguities in the order parameters for the alkyl chain, it also provides a measure of order and its sign for the carboxyl group. The latter information, which is not otherwise available, makes it possible to characterize the situation at the lipid-aqueous interface, an important consideration for model membrane studies. Conventional high-resolution ^{13}C NMR techniques are not ordinarily suitable for shift anisotropy determination, in ordered systems but the solid-state proton-enhanced cross-polarization methods (3) have been successfully applied to thermotropic systems and lecithin dispersions (4-7). In the present study, however, motional averaging of the dipolar couplings brings them within range of the decoupling capabilities of standard high-resolution spectrometers.

Finally, we have followed and partially characterized the phase behavior of a previously reported mesophase (8,9) as a function of the decanol concentration with ratios of other constituents remaining unchanged.

EXPERIMENTAL

The deuterium magnetic resonance spectra were obtained on a Bruker HFX-90 spectrometer at 13.8 MHz, in the continuous wave mode, with proton stabilization. The spectra were recorded using the upfield sideband (4KHz) with centerband suppressed. The parameters were determined from spectra obtained in 5 mm tubes after ∿500 accumulations. The D_2O concentration was ∿.5% in order to minimize overlapping sidebands. Temperature was controlled by a standard Bruker unit which was calibrated by comparison to a Mettler FP-52 hot stage previously calibrated by thermometric standards. The space above the mesophase was sealed with a Kel-F piston and 2 inert o-rings.

Carbon-13 data were acquired with a Bruker WH-90 Fourier transform spectrometer. The sweep width was normally 6KHz with 16K data. A delay of ∿4 sec was employed before each data acquisition. The proton spectrum of the laurate mesophase extends over ∿5-6KHz. Sufficient decoupling was obtained by centering the decoupling frequency and applying noise modulation over ∿1-2KHz at a power level of 20 watts. Six thousand pulses were required to achieve sufficient carbonyl sensitivity. The samples were sealed in Wilmad cylindrical cells (10 mm in height) designed to fit into a standard 10 mm tube. Deuterium lock was provided by D_2O in the external tube. Temperature was controlled in the same manner as ^2H studies to $\pm0.5°K$.

Several samples were used in these studies. The pentobarbituric acid and p-cresol samples (1,2) contained 2.97×10^{-5} and 5.62×10^{-5} moles/gm of mesophase, respectively. The mesophase contained 4.99, 29.92, 5.28, 3.11 and 56.70% by weight of KCl, perdeuterated K laurate, decanol-OD, D_2O, and H_2O, respectively. Sample (3) for deuterium temperature studies contained 5.1% KCl, 5.2% decanol-OD, 33.7% perdeuterated K laurate and 56.0% of a D_2O/H_2O mixture in the ratio of 11gms/200gms. For carbon studies the sample composition was 4.0% KCl, 6.0% decanol-OD, 30.0%K laurate and 60% D_2O. The effect of decanol-OD composition on the carbon-13 shielding anisotropies was obtained from samples prepared from a mixture of 31.9% K laurate, 4.26% KCl and 63.8% D_2O by adding the appropriate weight of decanol-OD.

RESULTS AND DISCUSSION

Our previous studies show that the mesophase (4% KCl, 6% dec-anol-OD, 30% K laurate, 60% D_2O) orients perpendicular to the magnetic field direction (8,9) and spinning the sample does not destroy the order. Although the phase is stable up to $342^{\circ}K$, for the above composition the upper transition temperature depends considerably on sample composition. The deuterium NMR spectrum of D_2O exhibits a quadrupolar split doublet of $\sim(300-500 \text{ Hz})(9)$ which drops discontinuously to zero at the upper transition. Above this temperature a two phase mixture of isotropic material and neat or lamellar mesomorphic material is present. The lamellar material exhibits a powder pattern for D_2O and does not orient spontaneously in the magnetic field. The D spectrum of the perdeuteriolaurate amphiphile in the spinable mesophase consists of six distinguishable doublets centered about a common frequency (1). The linewidths vary from $\sim100Hz$ for the $-CD_3$ group to $\sim300Hz$ for $\alpha-CD_2-$ group. Since $q_{CD} \equiv e^2qQ/h$ is equal to $\sim170KHz$ (10) and is much greater than the D-D direct coupling or indirect coupling we are justified in using the same expression for the doublet splitting, $\Delta\nu_Q$, for $-CD_2-$ and $-CD_3$ groups (2).

$$\Delta\nu_Q = 3/2 \, q_{CD} \, S \qquad (1)$$

S is defined as $<(3\cos^2\theta-1)/2>$ where θ is angle between the applied field and principal axis of the electric field gradient, assumed collinear with the C-D bond. We assume the asymmetry parameter for the electric field gradient tensor vanishes. If α is equal to the angle between some vector $\bar{n}$ and the applied field direction S_{CD} is now $<(3\cos^2\beta-1)/2>$, defined as order of the principal axis of the

$$\Delta\nu_Q = 3/2q_{CD} \, S_{CD}<(3\cos^2\alpha-1)/2> \qquad (2)$$

electric field gradient tensor with respect to $\bar{n}$. In the above transformation it is assumed that cylindrical symmetry exists about $\bar{n}$ (11,12). The order is highest at position 2 and constant from positions 3-8, but drops off for the remainder of the chain as seen in Figure 1. The order gradient compares favorably with that of Charvolin (13) for a K laurate lamellar phase and Na octanoate in the hexagonal phase (14). The alternating effect predicted by Marcelja (15,16) and reported for the nematic-like deuterated decyl sulfate mesophase (17) was not observed in this case. Although the tentative assignments at this single temperature were not confirmed by specific deuteration, the temperature studies of both deuterated chains and carbon-13 shift measurements confirm them.

To have a region of constant order parameter requires the same angular fluctuations for all segments involved. Seelig and Nieder-berger (18) and Seelig and Seelig (19) introduced the arguments of

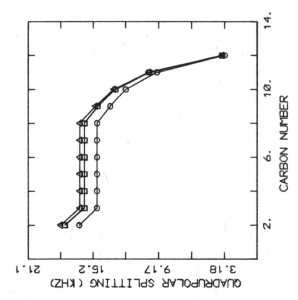

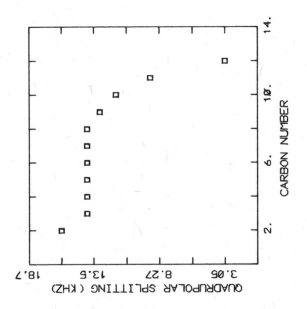

Figure 1.   Deuterium Quadrupolar Splitting versus carbon number. The carboxyl carbon is position 1. For sample (3).

Figure 2.   Effect of solutes on quadrupolar splittings. Triangles: p-Cresol (sample 2). Squares: pentobarbituric acid (sample 1). Circles: pure mesophase. At 298°K.

Pechhold (20) concerning "kink" formation to explain the region of
constant order parameter. If these multiples of gauche states diffuse freely with equal probability for the segments involved, a
region of constant order parameter is predicted. Introduction of
single gauche states near the chain end results in the rapid decrease
in order. In these respects the magnetic field orientable mesophases
exhibit an alkyl chain order picture consistent with pure lipid bilayers as a defective ordered structure (18). Furthermore, the
order gradient obtained for the laurate mesophase is in good agreement with the model calculations of Marcelja (15).

To illustrate the effect of solutes on alkyl chain order, we
prepared sample (1) and sample (2) which contained 2.97×10^{-5} and
5.62×10^{-5} moles/gm of mesophase. Figure 2 compares the S values
for these two samples with those for the pure mesophase at 298°K.
Addition of pentobarbituric acid and p-cresol cause dramatic
increases in the order. One might postulate that the effect is one
involving the interface. Since both molecules have acidic protons,
a change in the effective surface charge of the interface might
lead to a decreased area per head group and an increase in hydrocarbon order. This is equivalent to increasing the lateral pressure in the interface as calculated by Marcelja (15). As one
increases the lateral pressure the order increases in the model
calculations as in Figure 2. This is contrary to the accepted view
that phenol anesthetics cause membrane expansion (21). It should
be remembered that the order parameters may contain a contribution
due to rapid diffusion of the surfactant over the surface of the
superstructure (11). Addition of solutes could possibly alter the
shape of the aggregate or the diffusion rate.

The temperature dependence of $|S_{axis}|$ for each segment is given
in figure 3. S_{axis} is defined as follows for methylene groups:

$$S_{axis} = S_{CD_2}/[1/2(3\cos^2 90^{\circ}-1)] \tag{3}$$

and methyl groups

$$S_{axis} = S_{CD_3}/[1/4(3\cos^2 90^{\circ}-1) \times (3\cos^2 35.25-1)] = -6S_{CD_3} \tag{4}$$

S_{CD_2} and S_{CD_3} represents the order of the C-D bond with respect to

H_o. These order parameters have the property that they are identi-
cal in the all-trans form. In general $|S_{axis}|$ decreases with
increasing temperature for each segment and drop to zero at 342°K.
The most striking feature in the deuterium spectrum concerns the
doublet of relative intensity 6 corresponding to C3-8. At $\sim318^{\circ}$K
the single component splits into 3 components of intensity 4, 1, and
1. Thus two methylene segments are removed from the region of constant order parameter. This is reasonable since an increase in

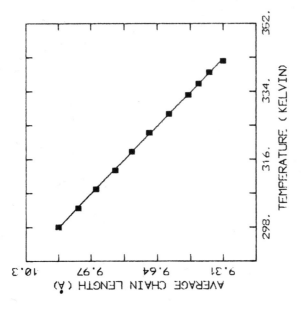

Figure 4.   Temperature dependence of <L>.

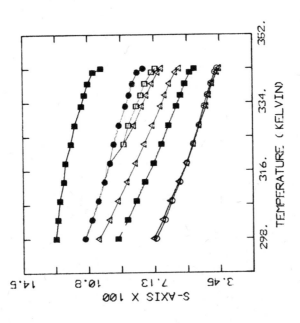

Figure 3.   Temperature dependence of |S_axis| for
each segment. From top to bottom C2,
C3-8, C9, C10, C11, C12.

temperature should increase the number of gauche states in the chain. The curve representing the doublet of relative intensity 4 remains essentially parallel to the curve for the 2 position while the lower two decrease at a comparable rate to the remaining segments. Notice that $|S_{axis}|$ for C11 and C12 are nearly the same. This is consistent with the terminal methylene and methyl moving as a unit with their segment axes parallel.

Consider the nematic phase in this study to be of cylindrical structures with the axes of the cylinders aligned along the spinning axis perpendicular to the field. Such a structure has been proposed in the past for Type II mesophases, Type I phases being aligned with the director along the applied field (22). Diffusion about the cylinder reduces $|S_{axis}|$ by the factor of 1/4 (11). Taking this factor into account gives $|S_{mol}|$ which represents the order of the previously defined segment axes with respect to the normal to the aggregate surface. If all rigid body motion of the surfactant as a whole is contained in the above factor, $|S_{mol}|$ is reduced from unity by internal conformational motion. Seelig developed an approach to calculate the average chain length <L> from a knowledge of $|S_{mol}^i|$ for the ith segment (19). The total length of the hydrocarbon chain for potassium laurate is given at each temperature by

$$<L> = 1.25[11 - 0.5 \sum_{i=1}^{11} (1-|S_{mol}^i|)/1.125] \qquad (5)$$

The results are shown in figure 4. The dependence of <L> on temperature is linear over the temperature range with a slope corresponding to a linear expansion coefficient, $\alpha = \frac{\Delta L}{L\Delta T}$, of $\sim$-1.33 x 10^{-3} $^{\circ}K^{-1}$. The extrapolated length at $0^{\circ}K$ is 15.58 Å. This should be compared with α of -2 x 10^{-3} $^{\circ}K^{-1}$ for egg lecithin bilayers (23) and -3 x 10^{-3} $^{\circ}K^{-1}$ for dipalmitoyllecithin (DPL) bilayers (25) from x-ray diffraction experiments. Deuterium NMR results yield values of -2.5 x 10^{-3} $^{\circ}K^{-1}$ (19) and -1.5 x 10^{-3} K^{-1} for DPL and egg lecithin, respectively.

On the basis of the substituent effect of a carboxyl group on the ^{13}C chemical shifts of dodecane one predicts the following shift assignments for the alkyl carbons in K laurate to be: C2(24.44), C3(12.63), C4-9(16.15), C10(18.33), C11(8.84), and C12(0.0). The above are relative to C12, the terminal methyl taken in aqueous solution. C4-9 are unresolved in solution while the predicted spread is from 15.7 to 16.9 ppm. The K laurate mesophase above the upper transition has values over the temperature range from $344^{\circ}K$ to $363^{\circ}K$ of C2(24.7+0.1), C3(12.6+.1), C4-9(16.3+0.7), C10(18.4+.1), C11(8.9+.1) and C12(0.0). The proton decoupled ^{13}C spectra of the oriented mesophase clearly shows the six alkane resonances mentioned above. On the basis of the deuterium work one

expects C4-9 to move as a unit since the ^{13}C tensor elements for
the interior methylenes should not be too different. Experimentally
this is the case. Thus we can measure the anisotropies of 5 separate
alkyl carbons and the group C4-9. Figure 5 presents the alkyl
chain anisotropies in ppm (downfield from the isotropic value for
the end methyl) as a function of temperature. We use this as our
reference throughout. It is clear that all alkyl carbons are de-
shielded with respect to the isotropic values. The discontinuity
at the phase transition increases in the order C12<C11<C10<C3-9<C2.
This fact confirms the deuterium assignments and shows that the C2
position is most ordered for the alkyl chain. For all lines the
deshielding decreases with temperature and drops discontinuously at
the phase transition to the isotropic value. If these data are used
as a semi-quantitative measure of order gradient down the chain a
picture analagous to the deuterium work is obtained (figure 6).
Figure 7 shows the dependence of the shift anisotropy for the carboxyl group on temperature. The discontinuity at the transition
point is the largest of the ^{13}C nuclei in K laurate. The large
jump is due both to the high order at the interface and to the
shielding tensor elements which are different from those of the
alkyl group.

We use the average principal shielding values determined by
Pines _et al_ (26) for the carboxyl group of solid ammonium tartrate
as representative of the carboxyl group. The principal values are
σ_{11} = -110.5, σ_{22} = -61.7, σ_{33} = 21.6 ppm relative to benzene the
1-axis is in the carboxyl plane bisecting the C-O bonds, and the
3-axis is perpendicular to the carboxyl plane. The ^{13}C of the carboxyl group is most shielded when the applied field is perpendicular
to the carboxyl plane. The chemical shift tensor for the methylene
group has been measured in amorphous polyethylene (7) in ppm upfield
of benzene. The principal elements are: σ_{11} = 78.3, σ_{22} = 92.0 and
σ_{33} = 115.9. VanderHart (27) has determined the principal values
for n - $C_{20}H_{42}$ to be: σ_{11} = 142.6+2.0, σ_{22} = 154.6+2 and σ_{33} =
175.0+2.0. The principal axes are such that the 3-axis is along the
chain axis, the 2-axis bisects the CH_2 group and the 1-axis is along
the H-H vector of the methylene group.

Both shielding tensors are non-axially symmetric and in the
principal axis system three order parameters are necessary to describe the orientation of a given segment with the applied field.
If we consider the molecules to be reorienting rapidly about the
vector $\bar{D}$ (not to be confused with the director) normal to the surface of the aggregate then the average tensor element $\bar{\sigma}_{ii}$ with
components $\bar{\sigma}_{\parallel}$ along $\bar{D}$ and $\bar{\sigma}_{\perp}$ perpendicular to D should be obtained.
Choose $\bar{D}$ to be along the segment z axis defined perpendicular to
the plane containing the atoms of the methylene group. In the all-
trans form $\bar{D}$ would be along the chain parallel to each methylene
segment axis. D makes an angle $\emptyset$ with the 1-axis of the carboxyl

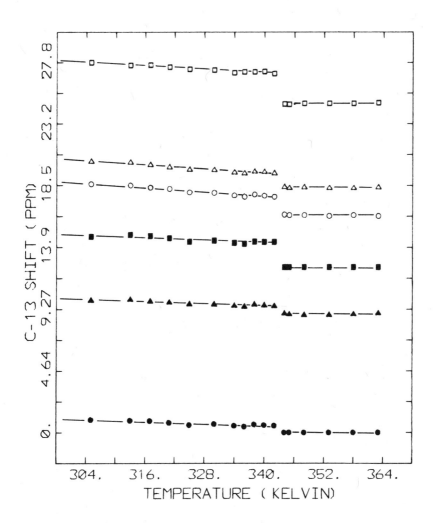

Figure 5. Alkyl chain carbon shifts versus temperature. From top to bottom C2, C10, C4-9, C3, C11 and C12. In ppm downfield from the isotropic methyl shift.

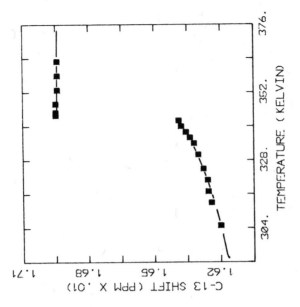

Figure 7.   Temperature dependence of carboxyl
C-13 shift. In ppm downfield from
the isotropic methyl shift.

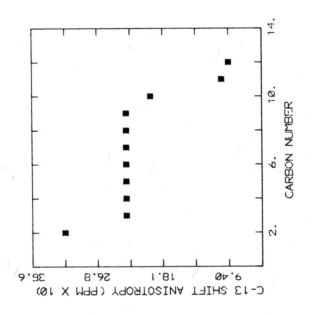

Figure 6.   Alkyl chain anisotropies versus carbon
number at 305°K. Relative to the
isotropic shift of the given carbon
in ppm.

group. Rapid reorientation about D gives average tensor elements of $\bar{\sigma}_{\|}$ = 175.0 and $\bar{\sigma}_{\perp}$ = 148.6 ppm for a methylene group and

$$\bar{\sigma}_{\|} = \cos^2\phi\sigma_{11} + \sin^2\phi\sigma_{22}$$

$$\bar{\sigma}_{\perp} = 1/2[\sin^2\phi\sigma_{11} + \cos^2\phi\sigma_{22} + \sigma_{33}]$$

for the carboxyl group.

For $\phi = 35.25°$, $\bar{\sigma}_{11}$ = -94.24 and $\sigma_{\perp}$ = -28.18. The order of the D vector is then described by a single order element representing the average orientation of D with the applied field. S is given by $3(\sigma-\sigma_i)/2(\bar{\sigma}_{\|}-\bar{\sigma}_{\perp})$. Applying the average shielding elements above shows at once that the tendency is for D to be perpendicular to the applied field for the observed shift anisotropies of K laurate to have the correct sign. Thus S is negative. The above result is not affected by minor changes in the orientations of the principal axis systems.

If D is considered to be oriented with respect to a coordinate frame fixed at the surface of the superstructure then the following equation can be used to calculate the expected shieldings corrected for diffusion over the surface.

$$\delta - \delta(\text{isotropic}) = -\{(8\pi/15)^{1/2} \sigma \sum_{m'} Y_{2,-m'}(\theta,\phi)$$

$$\times [(3/2)^{1/2} D^2_{om'}(\Omega) + \eta/2(D^2_{2m'}(\Omega) + D^2_{-2m'}(\Omega))]\} \tag{6}$$

In the above $\sigma = (\sigma_{zz} - \sigma^i)$, the Y_{2m} are the spherical harmonics of order 2, η the asymmetry parameter of shielding tensor in the principal axis system, and $D^2_{mm'}$, are the elements of the Wigner rotation matrix of order 2. The angles θ and ϕ give the orientation of the applied field in the surface coordinate system. The argument Ω represent the three Euler angles by which the principal axes of the shielding tensor can be brought into coincidence with the surface coordinate system. The dependence of diffusion over the surface is contained in $Y_{2,-m}(\theta,\phi)$. Calculations for several well known structures were carried out to get an estimate of the shielding anisotropies. In these calculations D is normal to the surface. Applying equation (6) for a cylinder perpendicular to the field direction and a lamellar structure with D perpendicular to H_o one obtains the following anisotropies.

	Cylinder	Lamellar	Exp.	(in ppm)
$COO^-\delta - \delta(\text{isotropic})$	+11.0	-22.0	-7	$(305°K)$
$(CH_2)\delta - \delta(\text{isotropic})$	- 4.5	+ 9.0	+2	C3-8

Clearly these represent rigid models and motion reduces them. However, it appears that the cylinder model is at odds with the experimental results.

The utility of using the ^{13}C shift anisotropies as a measure of order at the interface and to follow effects of solutes on this property is illustrated in Figure 8. Shown is the carboxyl anisotropy as a function of decanol-OD concentration. At zero concentration of decanol-OD we have a Type I (27) phase which orients along the field direction. At 305°K this phase contains a small amount of isotropic phase. As the decanol concentration increase to 2.66% the type I phase is converted into an isotropic phase of high viscosity. Further addition of decanol brings on the formation of a Type II phase at ∿4%. It is important to note that the type II mesophase contains no isotropic material. As the concentration increases the order increases up to ∿8% where phase separation occurs. It is interesting to note here that the order of the carboxyl group in the type II material is extremely dependent on decanol concentration. The most interesting point, as reflected in the sign of the anisotropic shift of both Type I and Type II phases, is that K laurate is oriented with the $\bar{D}$ vector preferentially perpendicular to the applied H_o. Both phases have ^{13}C shielding anisotropies at odds with the expected values for a cylindrical superstructure. The use of the cylindrical structure in calculating the values of <L> and α is illustrative of the averaging process required. Since results provided by the case of other structures are not known, the agreement obtained with independent evidence does not constitute proof for the uniqueness of the cylindrical structure.

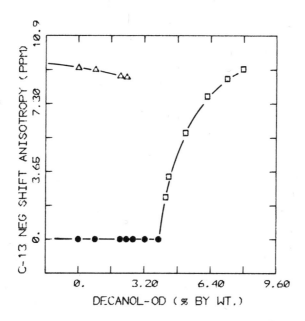

Figure 8. Effect of Decanol-OD concentration (% by wgt.) on the carboxyl group shift anisotropy. Measured relative to the isotropic shift in ppm. Triangles are Type I, circles isotropic and squares Type II mesophases.

ACKNOWLEDGMENT

This work was supported in part by the National Institutes of Health.

REFERENCES

1. R. C. Long, Jr. and J. H. Goldstein, J. Magn. Resonance, 23, 519 (1976). and references sited therein.

2. J. W. Emsley and J. C. Lindon, "NMR Spectroscopy Using Liquid Crystal Solvents," Pergamon Press, New York, 1975.

3. A. Pines, M. G. Gibbey and J. S. Waugh, J. Chem. Phys., 59, 569 (1973).

4. A. Pines, and J. J. Chang, Phys. Rev. A, 10, 946 (1974).

5.   A. Pines, D. J. Ruben and S. Allison, Phys. Rev. Lett., $\underline{33}$, 1002 (1974).

6.   S. J. Opella, J. P. Yerinowski and J. S. Waugh, Proc. Natl. Acad. Sci., $\underline{73}$, 3812 (1976).

7.   Julio Urbina and J. S. Waugh, Proc. Natl. Acad. Sci., $\underline{71}$, 5062 (1974).

8.   R. C. Long, Jr., J. Magn. Resonance, $\underline{12}$, 216 (1973).

9.   R. C. Long, Jr. and J. H. Goldstein, in "Liquid Crystals and Ordered Fluids," Vol. 2 (J. F. Johnson and R. S. Porter, Eds.), Plenum Press, New York, 1974.

10. J. R. Hoyland, J. Amer. Chem. Soc., $\underline{90}$, 2227 (1968).

11. H. Wennerstrom, G. Linblom and B. Lindman, Chemica Scripta, $\underline{6}$, 97 (1974).

12. G. R. Luckhurst, in "Liquid Crystals and Plastic Crystals," (ed. G. W. Gray and P. A. Winsor), Vol. 2. Ellis Horwood Publ., Chichester, Chapter 8.

13. J. Charvolin, P. Manneville and B. Deloche, Chem. Phys. Letters, $\underline{23}$, 345 (1973).

14. A. Henrikson, L. Odberg, and J. C. Eriksson, Mol. Cryst. Liq. Cryst., $\underline{30}$, 73 (1975).

15. S. Marcelja, Biochem. Biophys. Acta, $\underline{367}$, 165 (1974).

16. S. Marcelja, J. Chem. Phys., $\underline{60}$, 3599 (1974).

17. L. W. Reeves and A. S. Tracey, J. Amer. Chem. Soc., $\underline{97}$, 5729 (1975).

18. J. Seelig and W. Niederberger, Biochemistry, $\underline{13}$, 1585 (1974).

19. A. Seelig and J. Seelig, Biochemistry, $\underline{13}$, 4839 (1974).

20. W. Pechhold, Kolloid-2.2. Polym., $\underline{228}$, 1 (1968).

21. H. Machleidt, S. Roth and P. Seeman, Biochim. Biophys. Acta, $\underline{255}$, 178 (1972).

22. D. M. Chen, L. W. Reeves, A. S. Tracey and M. M. Tracey, J. Amer. Chem. Soc., $\underline{96}$, 5349 (1974).

23. R. P. Rand and W. A. Pangborn, Biochim. Biophys. Acta, 318, 299 (1973).

24. G. W. Stockton, C. F. Polnaszek, A. P. Tulloch, F. Hasan and I. C. P. Smith, Biochemistry, 15, 954 (1976).

25. V. Luzzati and F. Husson, J. Cell Biol., 12, 207 (1962).

26. A. Pines, J. J. Chang, and R. G. Griffin, J. Chem. Phys., 61, 1021 (1974).

27. D. M. Chen, F. Y. Fujiwara, and L. W. Reeves, Can. J. Chem., 55, 2396 (1977) have observed a Type I phase for K laurate (32.7), D_2O (65.1), and KCl (2.2) in percent by weight at 35°C.

THE ACOUSTO-OPTIC EFFECT FOR A NEMATIC LIQUID CRYSTAL

IN THE PRESENCE OF AN APPLIED ELECTRIC FIELD

C.F. Hayes

Department of Physics and Astronomy

University of Hawaii,Honolulu,Hawaii 96822

A. INTRODUCTION

It has been over forty years since Frederiks and Zolin[1] observed the acousto-optic effect. They used tuning forks from 200 to 600 Hz and acoustically excited a nematic. The cause of the effect was attributed to a rotation of the optic axis although no details for the mechanism were given. In the last eight years there has been a renewed interest in the effect with a number of quantitative studies performed.[1-8] Most of the experimental reports refer to a threshold of acoustic intensity required before the effect occurs. The first two theoretical explanations for the effect also predict such a threshold.[6,9] In 1976 experimental results were presented which indicated there was no threshold for the effect.[10-12] Two of these reports[10,11] indicate the mechanism responsible for the effect is acoustic streaming, although the formulation in each case is quite different.

In this paper we will extend one of these formulations[10] to include the effects of an applied electric field. In the next section the theory is presented which shows how the previously derived equations need to be modified for inclusion of the field. Although our results are for an electric field they are equally valid for a magnetic field as well. In the last section the results of experimental evaluation of the theory are presented.

B.THEORY

In our sample the nematic is sandwiched between two glass plates. The lateral edges of the nematic are open to a free air boundary. As the ultrasonic wave passes from one glass plate to the nematic to the other glass plate it produces a lateral wave. This lateral wave must be produced since in the nematic there is an ultrasonic wave and hence a pressure and in the adjacent air there is much less intensity for the ultrasonic wave and hence less pressure. The difference in acoustic intensity is the result of the difference in acoustic impedance change for the wave traveling from glass to nematic compared with from glass to air. The difference in pressure in these laterally adjacent regions oscillates with the frequency of the applied wave causing a lateral wave which reflects back and forth from free edge to free edge.

We use the hydrodynamic formulation of Stephen[13] to describe the subsequent motion of the nematic including the director's resulting orientation. With the exception of the electric field terms our equations will be the same as we have previously reported.[10] To first order we have:

$$0 = \frac{\partial n_{x1}}{\partial t} - \frac{\partial v_{x1}}{\partial z} \tag{1}$$

$$0 = \rho_0 \frac{\partial v_{x1}}{\partial t} + C^2 \frac{\partial \rho_1}{\partial x} - (\mu_2 + \mu_9) \frac{\partial^2 v_{x1}}{\partial x^2}$$

$$- \mu_8 \frac{\partial^2 n_x}{\partial t \, \partial z} - \left(\mu_2 + \mu_3 + \frac{\mu_8 + \mu_9 + \mu_{11}}{2}\right) \frac{\partial^2 v_{z1}}{\partial x \, \partial z}$$

$$- \left(\frac{-\mu_8 + \mu_9 + \mu_{11}}{2}\right) \frac{\partial^2 v_{x1}}{\partial z^2} \tag{2}$$

$$0 = \rho_0 \frac{\partial v_{z1}}{\partial t} + C^2 \frac{\partial \rho_1}{\partial z} - \left(\mu_2 + \mu_3 + \mu_5 + \mu_6 + \mu_9 + \mu_{10} + \mu_{11}\right) \frac{\partial^2 v_{z1}}{\partial z^2}$$

$$- \left(\frac{\mu_9 + \mu_{10}}{2}\right) \frac{\partial^2 v_{z1}}{\partial x^2} - \left(\mu_2 + \mu_5 + \frac{\mu_9 + \mu_{10}}{2}\right) \frac{\partial^2 v_{x1}}{\partial z \, \partial x} \tag{3}$$

$$0 = \frac{\partial \rho_1}{\partial t} + \rho_0 \frac{\partial v_{x1}}{\partial x} + \rho_0 \frac{\partial v_{z1}}{\partial z} \tag{4}$$

We are assuming the z direction to be perpendicular to the glass boundaries and the lateral wave to travel along the x axis. The equilibrium orientation of the director we have taken to be along the z axis. We use the subscript 1 to refer to variables oscillating with the frequency of the acoustic wave, and the subscript 2 to refer to the second order solutions. Then t is the time; ρ , the density; v, the fluid speed; μ_i , the i th Leslie viscosity and c, the wave speed. We omit the elastic and electric field terms from Eq. 1 since they may be shown to be small. Taking the approximation $k^2 \ll w \rho_c / \mu_i$ we obtain the following solutions to Eq's. 1-4:

$$n_{x1} = \frac{A}{w} \cos kx \left\{ e^{-\alpha_2 z} [\alpha_2 \sin(wt - \gamma_2 z) + \gamma_2 \cos(wt - \gamma_2 z)] \right.$$

$$\left. - e^{-\alpha_1 z} [\alpha_1 \sin(wt - \gamma_1 z) + \gamma_1 \cos(wt - \gamma_1 z)] \right\} \tag{5}$$

$$v_{x1} = A \cos kx \left\{ - e^{-\alpha_2 z} \cos(wt - \gamma_2 z) + e^{-\alpha_1 z} \cos(wt - \gamma_1 z) \right\} \tag{6}$$

$$v_{z1} = \frac{- k A \sin kx}{\alpha_1 (1 + \gamma_1^2/\alpha_1^2)} \left\{ - e^{-\alpha_2 z} [\cos(wt - \gamma_2 z) \right.$$

$$+ \frac{\gamma_1}{\alpha_1} \sin(wt - \gamma_2 z)]$$

$$+ e^{-\alpha_1 z} [\cos(wt - \gamma_1 z) + \frac{\gamma_1}{\alpha_1} \sin(wt - \gamma_1 z)] \right\} \tag{7}$$

where

$$\alpha_1 + i \gamma_1 = \sqrt{\frac{w \rho_0}{(\mu_8 + \mu_9 + \mu_{11})}} \left\{ [1 + \frac{(2\mu_6 - 2\mu_8 + \mu_9 + \mu_{10}) k^2}{2 w \rho_0}] \right.$$

$$+ i [1 - \frac{(2\mu_6 - 2\mu_8 + \mu_9 + \mu_{10}) k^2}{2 w \rho_0}] \right\} \tag{8}$$

and

$$\alpha_2 + i\gamma_2 = \sqrt{\frac{w(\mu_2 + \mu_q)\hbar^2}{2\rho_o c^2}} \, (1 + i) \tag{9}$$

We expand α_i and γ_i to powers of $\mu_i \hbar^2 / w\rho_o$. To lowest order we have $\alpha_2 = \gamma_2 = 0$ and

$$\alpha_i = \gamma_i = \sqrt{w\rho_o / (\mu_8 + \mu_q + \mu_{11})} \equiv \beta$$

In taking the hydrodynamic equations to first order we have omitted terms involving products of the variables which are oscillating with the frequency of the imposed acoustic wave. If we take the equations to second order we must take these terms into account. Since the variation is sinusoidal these terms involve a constant term plus one which varies with twice the acoustic frequency. We will be interested only in the former type and therefore take a time average of our equations to second order to eliminate terms having twice the imposed frequency. Eq. 2 to second order is then:

$$0 = \left\langle -2\rho_o \, v_{x1} \frac{\partial v_{x1}}{\partial x} - \rho_o \, v_{z1} \frac{\partial v_{x1}}{\partial z} - \rho_o \, v_{x1} \frac{\partial v_{z1}}{\partial z} \right\rangle$$

$$- \frac{\partial P_2}{\partial x} + \left\langle (\mu_5 + \mu_{11}) \frac{\partial}{\partial z} \left(n_{x1} \frac{\partial v_{x1}}{\partial x} \right) \right\rangle$$

$$+ (\mu_5 + \mu_6 + \mu_{10}) \left\langle \frac{\partial}{\partial z} \left(n_{x1} \frac{\partial v_{z1}}{\partial z} \right) \right\rangle$$

$$+ \left\langle \mu_8 \frac{\partial}{\partial z} \left(v_{x1} \frac{\partial n_{x1}}{\partial x} + v_{z1} \frac{\partial n_{x1}}{\partial z} \right) \right\rangle$$

$$- \frac{(\mu_8 - \mu_{11} - \mu_q)}{2} \frac{\partial^2 v_{x2}}{\partial z^2} \tag{10}$$

where the bracket, $\langle \, \rangle$, represents the time average.

In a similar manner we may find an equation for the director neglecting those terms which are zero in the nematic bulk:

$$0 = K_{33} \frac{\partial^2 n_{x2}}{\partial z^2} + \frac{\epsilon_a E^2}{4\pi} n_{x2} - \mu_8 \frac{\partial v_{x2}}{\partial z} \tag{11}$$

We are here assuming that the nematic is a perfect
insulator and also omitting any piezoelectric effects.

We assume the glass boundaries are at $z = 0$ and
$z = h$ which results in the boundary conditions:
$v_{x2} = v_{z2} = 0$ at $z = 0$ and $\partial v_{2x}/\partial z = v_{z2} = 0$ at
$z = h/2$. We obtain:

$$v_{x2} = \frac{3 \hbar A^2 \sin 2 \hbar x \; (2\mu_6 - \mu_8 + 3\mu_9 + 2\mu_{10} + \mu_{11})}{4 \omega (-\mu_8 + \mu_9 + \mu_{11})} \left[c_1 - \frac{z}{h} + \frac{z^2}{h^2} \right] \quad (12)$$

For the director we assume $n_{x2} = 0$ both at $z = 0$ and
$z = h$. We obtain:

$$n_{x2} = \frac{9 \hbar A^2 \sin(2 \hbar x) \; \mu_8 (\mu_8 + \mu_9 + \mu_{11})}{4 \omega K_{33}(-\mu_8 + \mu_9 + \mu_{11}) a^2 h} \left\{ -1 + \frac{2z}{h} \right.$$
$$\left. + \cos(az) - [\; ctn(ah) + csc(ah)] \sin(az) \right\} \quad (13)$$

where

$$a = \sqrt{\epsilon_a E^2 / 4\pi K_{33}} \quad (14)$$

There are also terms in Eq. 14 involving $e^{-\beta z}$ which
we have omitted since they contribute nothing to the
equation in the bulk of the nematic. This assumption
may be stated as $h \gg 1/\beta$ where for our experiment the
ratio of these quantities is of the order of 10 to 1.

If light is now passed through the sample in a
direction perpendicular to the glass plates the intensity
of the light which is transmitted if the sample is
placed between crossed polarizers is

$$I = I_0 \sin^2(2\phi) \sin^2(\delta/2) \quad (15)$$

where ϕ is the angle between the plane formed by the
optic axis and the direction of light propagation and the
axis of polarization and δ is the phase difference
between the ordinary ray and the extraordinary ray

$$\delta = \frac{2\pi h (n_e - n_o) \sin^2 \theta}{\lambda_0} \quad (16)$$

where n_e and n_o are the refractive indicies for the extraordinary and ordinary rays respectively, Θ is the angle the direction of light makes with the optic axis and λ_o is the wavelength of light. We have

$$\sin\Theta = n_{x1} + n_{x2} \tag{17}$$

However, since n_{x1} is zero in the bulk of the nematic we will make the approximation that

$$h \sin^2\Theta = \int_0^h n_{x2}^2 \, dz \tag{18}$$

We are also again assuming $\beta \gg 1/h$

Combining Eq's. 13-18 we find

$$I = I_o \sin^2(2\phi) \sin^2\left[\frac{3^4(n_e - n_o) h \ell A^4 \sin^2(2\ell x) \mu_A^2 F}{2^4 \lambda_o w^2 K_{33}^2 \pi^5}\right] \tag{19}$$

$$\mu_A = \mu_8(\mu_8 + \mu_9 + \mu_{11}) / (-\mu_8 + \mu_9 + \mu_{11})$$

$$F = \frac{1}{s^6 \sin^2(\pi s)}\left[3\pi^2 s^2 \cos(\pi s) + \pi^2 s^2 \sin^2(\pi s)\right.$$

$$\tag{20}$$

$$+ 3\pi^2 s^2 + 3^2 \pi s \sin(\pi s)$$

$$\left. + 3^2 \pi s \sin(\pi s)\cos(\pi s) - 2 \times 3 \sin^2(\pi s)\right]$$

where s is the ratio of the electric field applied perpendicular to the glass plates to the critical value of electric field for a Frederiks transition:

$$S = E/E_c \tag{21}$$

At a lateral position where $2kx = \pi/2$ we would expect the light intensity to vary as $\sin^2(BFV^4)$ where B is a constant, since V, the voltage applied to the transducer, is proportional to A. This dependence has been tested experimentally[10] for the case where no electric field was applied. As V was increased a series of maxima and minima occurred just as the $\sin^2(BFV^4)$ function predicts. However, the purpose of this present study is to examine how the light intensity changes with electric field. As the field increases so does F thereby moving the maxima and minima so they occur at lower values of V. One way to examine the field dependence is to measure the shift in the maxima and minima as a function of electric field. However, the equations involved would still include several unknown constants. One way of eliminating these constants is by setting V and hence A for the first maxima in the absence of an applied field and measuring the intensity as the electric field is increased.

For A fixed so the light intensity is at the first maximum with s = 0 we have:

$$\frac{I}{I_0} = \sin^2\left(2^2 \times 3^2 \times 5 \times 7\, F / \pi^5\right) \tag{22}$$

The only unknown in Eq. 22 is s the ratio of applied electric field to critical electric field. This equation will be the basis for comparison with experimental results.

C. EXPERIMENT

The sample cells were made of a thin layer of N-P-Methoxbenzylidene-P-N-Butylaniline (MBBA) sandwiched between either a conducting glass plate and a front surface mirror or two glass plates. The experiment was performed at a room temperature of 24°C. The plates were held apart by two thin spacers making the MBBA thickness 80 μm. The cross section of the MBBA was circular in shape with a diameter of about 2 mm. The sample was disk shaped with the flat sides in contact with the glass plates and the circular edges open to the air.

The cell was mounted with vacuum grease to a piezoelectric transducer oscillating in the thickness mode causing a continuous compression wave to be propagated through the sample having a frequency of 270.9 kHz. Therefore, the acoustic wavelength in the MBBA was about 5.5 mm, a value much larger than the thickness of the MBBA disk. The value of wavelength is close to the value for a standing wave laterally across the MBBA with the center of the disk a node and the edges antinodes. It would be worthwhile to test the theory we are proposing by studying the effects of variation of the radius of the MBBA disk. We have not as yet made

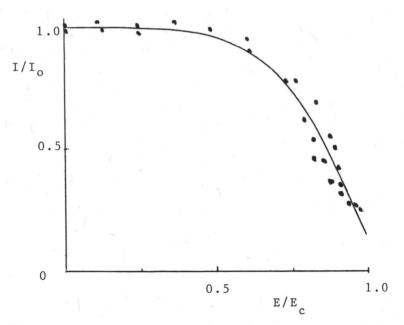

Figure 1. A graph of the ratio for the light intensity
transmitted through the cell with and without an applied
electric field. The data were taken while the cell was
being excited acoustically by an ultrasonic wave of
magnitude equal to that required for the first light
intensity maxima in the absence of an electric field.
The horizontal axis is the ratio of the applied electric
field to that required for a Frederiks transition in
the absence of the acoustic wave. The solid line is for
Equation 22.

such a study. The pattern which appears in the microscope
when the cell is acoustically excited is a dark cross
centered about the middle of the sample extending to the
edges with an orientation parallel to that of the
polarizer and analyzer. Both the center and edges of the
sample are dark. Between them is a ring of light which
appears white at lower transducer voltages and then
colored as the voltage is increased. These observations
are in keeping with the physical model presented in the
previous   section. For higher frequencies more
complicated patterns appear.

Since the effect of acoustic excitation which we
are considering is not sensitive to impurities the

MBBA as supplied by the manufacturer was used without further purification. A thin coating of lecithin on the inner glass surfaces was used to achieve homeotropic alignment. The electric field was maintained across the sample by connecting electrodes to the inner conducting surfaces of the sample cell.

A polarizing microscope in the reflecting mode held the sample cell for study. The white light source of the microscope was used with a filter to allow light of wavelength $0.63 \pm 0.04\ \mu m$ to pass through the sample to the bottom plate or mirror and on to the photomultiplier. The method of detection is described in more detail elsewhere.[10] A x20 objective was used to limit the field of view to a small region (0.2 mm diameter), thereby giving more uniformity.

The procedure was to increase the voltage to the transducer until the first maxima in light intensity was reached while holding the applied electric field to zero. The voltage applied to the transducer for this point was 13.2 Volts. With I the light intensity and I_o for the case of no electric field measurements were taken for I as a function of applied electric field. The results are shown in Fig. 1 where the solid line is a graph of Eq. 22. Within the accuracy of the experiment the data and theory agree.

The agreement strengthens the validity of the physical model we are proposing. However, it would appear that as far as practical applications are concerned the sensitivity of the acousto-optic effect in nematics can not be greatly increased by the application of an applied field. Finally, it should be noted that whereas we have been concerned with an electric field Eq. 22 is equally valid for a magnetic field where $s = H/H_c$ with H_c the critical magnetic field for a Frederiks transition.

REFERENCES

1. V.V.Zolina,Trudy Lomonosov Inst. Akad. Nauk SSSR 8 ,
   11 (1936).
2. L.W.Kessler and S.P.Sawyer, Appl.Phys.Lett. 17,440
   (1970).
3. H.Mailer,K.L.Likins,T.R.Taylor and J.L.Fergason,Appl.
   Phys.Lett. 18, 105 (1971).
4. M.Bertolotti, S.Martellucci,F.Scudieri and D.Sette,
   Appl.Phys.Lett. 21, 74 (1972).
5. P.Greguss, Acustica 29, 52 (1973).
6. S.Nagai and K.Iizuka, Japan J.Appl.Phys. 13, 189 (1974).
7. Y.Kagawa, T.Hatakeyama and Y.Tanaka, J.Sound Vibration
   36, 407 (1974).
8. R.Bartolino,M.Bertolotti,F.Scudieri,D.Sette and
   A.Sliwinski, J.Appl.Phys. 46, 1928 (1975).
9. W.Helfrich, Phys.Rev.Lett. 29, 1583 (1972).
10.C.Sripaipan, C.Hayes and G.T.Fang, Phys.Rev. 15A,
   1297 (1977).
11.S.Candau,A.Peters and S.Nagai, Sixth International
   Liquid Crystal Conference J-30 (1976).
12.J.L.Dion and F.De Forest, Sixth International Liquid
   Crystal Conference K-18 (1976).
13.M.J.Stephen, Phys.Rev.A,2 1558 (1970).

MICROCONTINUUM THEORY OF HEAT-CONDUCTING SMECTIC LIQUID CRYSTALS

M.N.L. Narasimhan and T.E. Kelley

Oregon State University, Department of Mathematics

Corvallis, Oregon 97331

ABSTRACT

Balance laws governing the flow of heat-conducting smectic liquid crystals are presented. Appropriate constitutive equations are formulated taking into account thermal gradients on the basis of the micropolar theory of A.C.Eringen. Full material symmetry considerations of the smectic phase are applied to effect simplifications in the constitutive equations. A generalized form of the Clausius-Duhem inequality governing these constitutive equations is obtained and includes the effects of heat-conduction. A complete set of thermodynamical restrictions governing the material coefficients are derived. The resulting constitutive theory is then specialized to the linear case to facilitate its use in practical applications.

1. INTRODUCTION

During the last two decades there has been a rapid revival of research interest in liquid crystals due to the recognition of their fast developing applications in rather diverse fields such as display technology, nondestructive testing of materials, detection of flaws and cracks in materials, detection of presence of chemical vapor in a room, medical diagnostic devices and biophysical studies. Liquid crystals possess some unusual properties resulting from their unique molecular architecture coupled with their rheological behavior. Among these properties, light scattering into vivid colors in liquid crystals when subjected to external influences such as temperature fluctuations, electromagnetic fields, shear, chemical vapor and boundary influences is mostly what makes them extremely useful for potential applications.

297

X-ray and nuclear magnetic resonance studies have revealed that this iridescent color display is due to changes that occur in the orientational patterns of their molecules.

Liquid crystalline state is considered as an independent thermodynamic state of a certain class of organic compounds, separated from the crystalline state and from the normal liquid by first-order thermal transitions. In this state, liquid crystals are molecularly ordered yet possess mechanical properties such as flow and are essentially anistropic in many of their properties such as thermal and electrical conductivities, viscosity and optical activity including birefringence and dichroism. Liquid crystals have been classified by Friedel [1] based on their molecular structure into three major types — the nematic (threadlike, nonlayered structure, for example, p-azoxyanisole), the smectic (greaselike or soapy, layered structure, for example, p-azoxybenzoate), and the cholesteric (derivatives of cholesterol but not cholesterol itself; double helical, or enantiomorphic structure, for example, cholesteryl benzoate).

The study of mesophase rheology of liquid crystals is relatively of recent origin. It is well documented that rheological measurements represent a simple and definitive method of distinguishing smectic and cholesteric mesophases from nematic ones, which is important from the point of view of applications. Numerous books, reviews and symposia have been devoted in the past to enhance our understanding of the liquid crystal behavior. The best and also the most recent publications on the general properties of liquid crystals are by Brown [2], Fergason [3], the Orsay group [4], and Gray [5]. Fergason and Brown [6] have made an extensive investigation of twist, splay and bend waves in liquid crystal biosystems as such waves may be important for signal detection and propagation in living systems. These are waves unique to liquid crystals and differ from the mechanical waves in solids and in liquids since no motion of the molecular centers is necessarily connected with them. These are waves in which only the rotational motion of the molecules occurs. The most outstanding and highly sophisticated experimental works on mesophase rheology of liquid crystals are by Porter and Johnson [7-13], which have contributed to a proper understanding of the rheological behavior of liquid crystals. Their experimental studies have also stimulated many theoretical investigations.

Detailed works on Molecular theories of liquid crystals have been published by Gray [5], Maier and Saupe [14], Kobayashi [15], and Chandrasekhar [16]. Fairly complete surveys as well as additional contributions to the field were presented recently by Saupe [17,18] and Chistyakov [19].

It appears that continuum theories of liquid crystals were first initiated by Oseen [20], Zocher [21] and Anzelius [22] and further developed by Frank [23]. In regard to more recent developments of continuum theories we mention the works of Groupes d'Etudes des Cristaux liquids (the Orsay Group) [4], Martin, et al [24], Stephen [25], Lubensky [26], Forster, et al [27], Huang [28], and Schmidt and Jahnig [29].

The best and most exhaustive microcontinuum theories of liquid crystals are due to Eringen and Lee [30-34], Ericksen [35-44], and Leslie [45-48]. The unique feature of these theories is that the intrinsic motions of aggregates or ordered domains of liquid crystal media, involving both rotations and deformations are fully taken into account within the framework of continuum mechanics thus essentially differing from the theories of classical continuum and statistical mechanics. Of particular importance and interest are the works of Lee and Eringen [30-34], which allow for a complete and definitive treatment of liquid crystal rheology based on the microcontinuum theory initiated by Eringen [49,50]. It has been well documented in the past that liquid crystals consist of large, relatively rigid molecules, one of whose dimensions is large compared to the others. In some temperature range or, in the case of solutions, in some range of concentrations, these molecules tend to line up more or less parallel to each other, being relatively free to move as long as they remain parallel. In smectic mesophases, molecules are arranged side by side in a series of stratified layers. Molecules within a layer may be further arranged into rows and columns or at random. The long axes of molecules are all parallel to one another and essentially perpendicular to the plane of the layer which can be one or more molecules thick. The smectics have greater ordering than the nematics. Fergason's [3] work reveals that if a smectic sheet could be suspended in space, free from gravity, it would take the form of a perfectly flat surface, since the side-to-side attraction of the molecules in the sheet would be the strongest force acting on it. If such a sheet were bent, it would eventually return to a flat configuration. This indicates the elastic property of smectic liquid crystals. Furthermore, in the smectic phase the temperature is just high enough to ease the bonds between sheets but not high enough to break up the sheets themselves. This accounts for the layered structure in the smectics. Mesophase rheological studies by Porter and Johnson [13] have revealed that the smectic mesophase has viscosities several times higher than those of the nematics. This may be due to the fact that the molecules in the smectic phase are not only parallel to each other but also their ends are constrained to lie in the same plane unlike the nematic case.

The present paper is devoted to the study of certain aspects of mesophase rheology of smectic liquid crystals based on the

micropolar theory initiated by Eringen [49]. Our present work is
an extension of a theory proposed by Narasimhan and Eringen
[51,52] for the mechanics of heat-conducting nematic liquid crystals, to characterize the smectics, and is a direct extension of
the more recent work of Narasimhan [53] on heat-conducting smectic
liquid crystals.

Taking advantage of the viscoelastic nature of the smectic
liquid crystals and by introducing the temperature gradient as an
additional argument in the constitutive functionals, proposed by
Lee and Eringen [33] and subjecting these functionals to the invariance requirements of principle of objectivity and material
symmetry restrictions of the smectic mesophase, we obtain a simplified set of constitutive equations. Next we derive a general
thermodynamic inequality involving heat-conduction effects governing the constitutive equations. We then obtain a complete set of
thermodynamic restrictions on the material moduli which will not
only serve as an invaluable aid in determining the extent to which
the theory would be applicable to real physical problems involving
the smectics but also in the actual determination of the material
moduli. Also, we show that the present theory can be specialized
to the linear case without losing relevance to liquid crystal
phenomena which renders it extremely useful for practical applications. The present work is found to reveal the importance and
necessity of including heat-conduction in the theory. It is also
found that there exists, in general, a contribution to heat conduction due to the micropolar nature of the flow whether or not
an externally imposed temperature gradient is present.

The applications of our theory developed here to the rheological study of smectic liquid crystal in rotational viscometric
flows as well as comparison of our theoretical predictions with
available experimental results are scheduled for another paper
appearing in the proceedings of this conference.

2. BASIC EQUATIONS

We concern ourselves with the viscoelastic nature of the
smectic liquid crystals which are anisotropic due to their molecular arrangements. Thus we regard smectic liquid crystals as
micropolar viscoelastic materials (Eringen [33]).

The basic laws of motion of a micropolar continuum are,
(Eringen [33]):

Conservation of Mass:

$$\dot{\rho} + \rho v_{k,k} = 0 \; ; \qquad\qquad (2.1)$$

Conservation of Microinertia:

$$i_{k\ell} = I_{KL} \chi_{kK} \chi_{\ell L} \; ; \qquad (2.2)$$

Balance of Linear Momentum:

$$t_{k\ell,\ell} + \rho f_\ell = \rho \dot{v}_\ell \; ; \qquad (2.3)$$

Balance of Moment of Momentum:

$$t_{rs,r} + \varepsilon_{spq} t_{pq} + \rho \ell_s = \rho \frac{D}{Dt} [(i_{pp} \delta_{sr} - i_{rs}) \nu_r] \; ; \; (2.4)$$

Conservation of Energy:

$$\rho \dot{\varepsilon} = t_{k\ell}(v_{\ell,k} - \varepsilon_{k\ell m} \nu_m) + m_{k\ell} \nu_{\ell,k} + q_{k,k} + \rho h \; ; \qquad (2.5)$$

Entropy Inequality:

$$\rho \dot{\eta} - (q_k/T),_k - \rho h/T \geq 0 \; . \qquad (2.6)$$

In relations (2.1) to (2.6) we have ρ = mass density, v_k = velocity, $i_{k\ell} = i_{\ell k}$ = microinertia tensor in deformed state, $t_{k\ell}$ = stress tensor, f_k = body force, $m_{k\ell}$ = couple stress, ℓ_k = body couple, ν_k = gyration velocity, ε = internal energy density, q_k = heat flux, h = heat source, η = entropy density, T = absolute temperature, $\varepsilon_{k\ell m}$ = permutation tensor, $\delta_{k\ell}$ = Kronecker delta, and χ_{kK} represents the microrotation with the constraints

$$(\chi_{kK})^{-1} = (\chi_{kK})^T . \qquad (2.7)$$

We shall employ a rectangular Cartesian coordinate system and assume the summation convention over repeated indices. Indices following a comma represent partial differentiation and a super-posed dot represents material differentiation, e.g.

$$v_{k,\ell} \equiv \partial v_k / \partial x_\ell, \; x_{k,K} \equiv \partial x_k / \partial X_K \; , \; \dot{f} \equiv \frac{\partial f}{\partial t} + f,_k v_k \; ,$$

where the x_k denote the spatial position of a material point X_K at time t, both being referred to a rectangular frame of reference.

The kinematic tensors of deformation and their rates are given by [33]:

$$\nu_k = -\frac{1}{2} \varepsilon_{k\ell m} \dot{X}_{\ell K} X_{mK}, \quad C_{KL} = x_{k,K} X_{kL} \, ,$$

$$\Gamma_{KL} = \frac{1}{2} \varepsilon_{KMN} X_{kM,L} X_{kN} \, , \quad \dot{C}_{KL} = (\nu_{\ell,k} - \varepsilon_{k1m}\nu_m) x_{k,K} X_{\ell L} \, ,$$

$$\dot{\Gamma}_{KL} = \nu_{k,\ell} x_{\ell,L} X_{kK} \, . \hspace{4cm} (2.8)$$

3. CONSTITUTIVE EQUATIONS

Eringen and Lee [33] derive the constitutive equations for micropolar viscoelastic smectic liquid crystals to which we shall introduce the variable $T_{,K}$, the temperature gradient, since we are considering heat-conducting smectic liquid crystals. The constitutive equations become:

$$\psi = \psi(C_{KL}, \dot{C}_{KL}, \Gamma_{KL}, \dot{\Gamma}_{KL}, T, T_{,K}) \, ,$$

$$\eta = \eta(C_{KL}, \dot{C}_{KL}, \Gamma_{KL}, \dot{\Gamma}_{KL}, T, T_{,K}) \, ,$$

$$t_{k\ell} = \frac{\rho}{\rho_0} T_{KL}(C_{KL}, \dot{C}_{KL}, \Gamma_{KL}, \dot{\Gamma}_{KL}, T, T_{,K}) x_{k,K} X_{\ell L} \, ,$$

$$m_{k\ell} = \frac{\rho}{\rho_0} M_{KL}(C_{KL}, \dot{C}_{KL}, \Gamma_{KL}, \dot{\Gamma}_{KL}, T, T_{,K}) x_{k,L} X_{\ell L}$$

$$q_k = \frac{\rho}{\rho_0} Q_K(C_{KL}, \dot{C}_{KL}, \Gamma_{KL}, \dot{\Gamma}_{KL}, T, T_{,K}) x_{k,K} \, , \hspace{1cm} (3.1)$$

where $\psi = \varepsilon - T\eta \equiv$ free energy density and q_k = heat flux vector.

We now eliminate h between (2.5) and (2.6) in order to obtain a generalized form of the Clausius-Duhem inequality:

$$-\rho(\dot{\psi}+\dot{T}\eta)+t_{k\ell}(\nu_{\ell,k} - \varepsilon_{k\ell m}\nu_m)+m_{k\ell}\nu_{\ell,k}+T^{-1}q_k T_{,k} \geq 0 \, . \hspace{0.5cm} (3.2)$$

Using (3.1) in (3.2) we obtain

$$-\rho_0[(\frac{\partial\psi}{\partial T} + \eta) \, \dot{T} + \frac{\partial\psi}{\partial T_{,K}} \, \dot{T}_{,K} + \frac{\partial\psi}{\partial C_{KL}} \, \dot{C}_{KL} + \frac{\partial\psi}{\partial \dot{C}_{KL}} \, \ddot{C}_{KL} + \frac{\partial\psi}{\partial \Gamma_{KL}} \, \dot{\Gamma}_{KL} +$$

$$+ \frac{\partial \psi}{\partial \dot{\Gamma}_{KL}} \ddot{\Gamma}_{KL}] + T_{KL} \dot{C}_{KL} + M_{KL} \dot{\Gamma}_{KL} + T^{-1} Q_K T,_K \geq 0 . \quad (3.3)$$

This inequality is postulated to be valid for all independent variations of T, $\dot{T}$, $T,_K$, $\dot{T},_K$, $\dot{C}_{KL}$, $\dot{\Gamma}_{KL}$ and $\ddot{\Gamma}_{KL}$. Since (3.3) is linear in $\dot{T}$, $\dot{T},_K$, $\ddot{C}_{KL}$, and $\ddot{\Gamma}_{KL}$ it cannot be maintained for all values of these quantities unless their coefficients vanish. Thus, we obtain

$$\eta = - \frac{\partial \psi}{\partial T} , \quad \psi = \psi(C_{KL}, \Gamma_{KL}, T) , \quad (3.4)$$

and (3.3) may be rewritten as

$$_d T_{KL} \dot{C}_{KL} + _d M_{KL} \dot{\Gamma}_{KL} + T^{-1} Q_K T,_K \geq 0 \quad (3.5)$$

where

$$_d T_{KL} = {}_T T_{KL} - {}_e T_{KL} = T_{KL} - \rho_0 \frac{\partial \psi}{\partial C_{KL}} , \quad (3.6)$$

$$_d M_{KL} = M_{KL} - {}_e M_{KL} = M_{KL} - \rho_0 \frac{\partial \psi}{\partial \Gamma_{KL}} . \quad (3.7)$$

We note that if $_d T_{KL}$, $_d M_{KL}$, and Q_K are continuous functions of $\dot{C}_{KL}$, $\dot{\Gamma}_{KL}$, and $T,_K$, then from (3.5) it follows that

$$\dot{C}_{KL} = \dot{\Gamma}_{KL} = T,_K = 0 \text{ implies } _d T_{KL} = _d M_{KL} = Q_K = 0 . \quad (3.8)$$

Thus we have proved:

Theorem: The general form of constitutive equations (3.1), of smectic liquid crystals is thermodynamically admissible if and only if (3.4) to (3.7) are satisfied.

4.   CONSTITUTIVELY LINEAR THEORY

In this section our main objective is to construct suitable constitutive equations for the smectic substance. To this end we employ two of the most important principles of constitutive theory, namely, the principle of objectivity and the principle of material symmetry [33] (since the smectic compounds that we are interested in possess uniaxial symmetry as well as a center of inversion and other possible symmetries). As a result, we find that our constitutive relations, when subjected to these principles undergo considerable simplification.

In order to obtain a set of linear constitutive equations, we expand ψ as follows:

$$\rho_0 \psi = \Sigma_0 + A_{KL}(C_{KL} - \delta_{KL}) + B_{KL} \Gamma_{KL} + \frac{1}{2} A_{KLMN} \{(C_{KL} - \delta_{KL})$$

$$\cdot (C_{MN} - \delta_{MN})\} + \frac{1}{2} B_{KLMN} \Gamma_{KL} \Gamma_{MN} + \{C_{KLMN}$$

$$\cdot (C_{KL} - \delta_{KL}) \Gamma_{MN}\} , \qquad (4.1)$$

where Σ_0, A_{KL}, B_{KL}, A_{KLMN} and C_{KLMN} are material moduli which are functions of temperature only. The elastic parts of T_{KL} and M_{KL} are then obtained as

$$_e T_{KL} = \frac{\partial(\rho_0 \psi)}{\partial C_{KL}} = A_{KL} + A_{KLMN} (C_{MN} - \delta_{MN}) + C_{KLMN} \Gamma_{MN} ,$$

$$(.4.2)$$

$$_e M_{KL} = \frac{\partial(\rho_0 \psi)}{\partial \Gamma_{KL}} = B_{KL} + B_{KLMN} \Gamma_{MN} + C_{MNKL} (C_{MN} - \delta_{MN}) .$$

Linear expressions for the dissipative parts $_d T_{KL}$, $_d M_{KL}$, and Q_K can now be written in terms of their arguments C_{KL}, $\dot{C}_{KL}$, $\dot{\Gamma}_{KL}$ and $T,_K$ as follows:

$$\frac{\rho}{\rho_0} _d T_{KL} = a_{KLMN} \dot{C}_{MN} + \alpha_{KLMN} \dot{\Gamma}_{MN} , \qquad (4.3)$$

$$\frac{\rho}{\rho_0} _d M_{KL} = b_{KLMN} \dot{\Gamma}_{MN} + \beta_{KLMN} \dot{C}_{MN} + \gamma_{KLMN} G_{MN} ,$$

$$\frac{\rho}{\rho_0} Q_{KL} = D_{KLMN} \dot{C}_{MN} + d_{KLMN} \dot{\Gamma}_{MN} + \delta_{KLMN} G_{MN} ,$$

where the material moduli $\underset{\sim}{a}$, $\underset{\sim}{b}$, $\underset{\sim}{D}$, $\underset{\sim}{\alpha}$, β, $\underset{\sim}{d}$, γ and $\underset{\sim}{\delta}$ are functions of temperature only and where

$$G_{KL} = \frac{1}{2} T^{-1} \varepsilon_{KLM} T,_M ,$$

$$(4.4)$$

$$Q_{KL} = \varepsilon_{KLM} Q_M .$$

We note that in the natural state, i.e. when $x_{k,K} = \delta_{kK} = \chi_{kK}$, the stress and the couple stress tensors must vanish and the temperature field remains unchanged.

Material Symmetry. A smectic liquid crystal possesses uni-axial symmetry about its axis of orientation and a center of in-version (for simplicity we only consider those substances with center of inversion). With this in mind we examine the results of performing rigid rotations and inversion on equations (4.2) to (4.4) and requiring that these equations remain invariant under these rotations about the smectic liquid crystal's axis of orien-tation and after inversions through its center of inversion. The orthogonal group which characterizes this symmetry has the follow-ing forms [30]:

$$(S_{KL}) = \begin{bmatrix} \cos\phi & -\sin\phi & 0 \\ \sin\phi & \cos\phi & 0 \\ 0 & 0 & 1 \end{bmatrix} ; \begin{bmatrix} \pm 1 & 0 & 0 \\ 0 & \pm 1 & 0 \\ 0 & 0 & \pm 1 \end{bmatrix} \quad (4.5)$$

where ϕ is the orientation, which is given by the angle made by the molecular axis of orientation with the axis of orientation of the liquid crystal. The principle of material invariance [30] states that under the group of transformations of the material frame of reference X_K,

$$X_K^* = S_{KL} X_L . \quad (4.6)$$

Equations (4.2) to (4.4) should remain invariant. Hence, we have the following restrictions on the material moduli:

(i) $C_{KLMN} = 0 = \alpha_{KLMN} = \beta_{KLMN} = D_{KLMN} .$ (4.7)

(ii)   The non-vanishing components of $A_{KLMN} (B_{KLMN})$ are

(4.8)

$A_{1111} = A_{2222}, A_{1122} = A_{2211}, A_{1133} = A_{2233} = A_{3311} =$

$A_{3322}, A_{3333}, A_{1313} = A_{2323}, A_{3131} = A_{3232}, A_{1331} =$

$A_{3113} = A_{2332} = A_{3223}, A_{1212} = A_{2121}, A_{1221} = A_{2112} =$

$A_{1111} - A_{1122} - A_{1212} .$

(iii) The non-vanishing components of $a_{KLMN} (b_{KLMN}, d_{KLMN},$

δ_{KLMN})   are

$$a_{1111} = a_{2222}, \; a_{1122} = a_{2211}, \; a_{1133} = a_{2233}, \; a_{3311} =$$

$$a_{3322}, \; a_{3333}, \; a_{1313} = a_{2323}, \; a_{3131} = a_{3232}, \; a_{1331} =$$

$$a_{2332}, \; a_{3113} = a_{3223}, \; a_{1212} = a_{2121}, \; a_{1221} = \qquad (4.9)$$

$$a_{2112} = a_{1111} - a_{1122} - a_{1212}.$$

Using equations (4.7) to (4.9) and making use of the fact that the displacements and microrotations involved are assumed small, we obtain the linear constitutive equations for heat-conducting smectic liquid crystals with center of inversion.

$$t_{k\ell} = A_{k\ell mn} (u_{n,m} - \varepsilon_{mnp}\phi_p) + a_{k\ell mn}(v_{n,m} - \varepsilon_{mnp}v_p) \; ,$$

$$m_{k\ell} = B_{\ell kmn}\phi_{m,n} + b_{\ell kmn} \, v_{m,n} + \gamma_{\ell kmn} \, g_{mn} \; , \qquad (4.10)$$

$$q_b = (d_{\ell kmn} \, v_{m,n} + \delta_{\ell kmn} \, g_{mn}) \tfrac{1}{2} \, \varepsilon_{\ell kb} \; ,$$

where u_k and ϕ_k are, respectively, components of displacement and microrotation,

$$g_{mn} = \frac{1}{2} \, T^{-1} \, \varepsilon_{mnq} \, T_{,q} \; , \qquad (4.11)$$

and

$$A_{k\ell mn} = A_{KLMN}\delta_{kK}\delta_{\ell L}\delta_{mM}\delta_{nN} \; . \qquad (4.12)$$

Similar relations hold for $a_{k\ell mn}$, $B_{\ell kmn}$, $b_{\ell kmn}$, etc.

We note here (Eringen [33], Narasimhan [53]) that the reference state of smectic liquid crystals is any state that leaves density, axis of orientation, and layered structure unchanged. For heat-conducting smectic liquid crystals we need only add that the reference state can be taken as any state in which the temperature field remains unchanged, besides the other field properties mentioned above.

5.   THERMODYNAMIC RESTRICTIONS

In this section we formulate the thermodynamic restrictions

governing the material coefficients both in the reversible and irreversible part of the constitutive equations.

To obtain the thermodynamic restrictions for the material coefficients present in the reversible parts of the stress and couple stress we substitute the appropriate terms from the right-hand sides of equations $(4.10)_1$ and $(4.10)_2$, namely

$$_e t_{k\ell} = A_{k\ell mn}(u_{n,m} - \varepsilon_{mnp}\phi_p) \tag{5.1}$$

and

$$_e m_{k\ell} = B_{\ell kmn}\phi_{m,n} \tag{5.2}$$

into the energy equation (2.5) with $q_{k,k} = 0$ and $h = 0$. Thus we obtain for the non-dissipative part of the energy production

$$\rho_e\dot{\varepsilon} = A_{k\ell mn}(u_{n,m} - \varepsilon_{mnp}\phi_p)(v_{\ell,k} - \varepsilon_{k\ell r}v_r) + B_{\ell kmn}\phi_{m,n}v_{\ell,k}$$

$$= A_{k\ell mn}\{e_{mn} + \varepsilon_{mnp}(r_p - \phi_p)\}\{d_{k\ell} + \varepsilon_{k\ell r}(\omega_r - v_r)\} +$$

$$+ B_{\ell kmn}\phi_{m,n}v_{\ell,k}, \tag{5.3}$$

where

$$e_{k\ell} = \frac{1}{2}(u_{k,\ell} + u_{\ell,k}), \tag{5.4}$$

$$\dot{e}_{k\ell} = d_{k\ell} \equiv \frac{1}{2}(v_{k,\ell} + v_{\ell,k}), \tag{5.5}$$

$$r_{k\ell} = \frac{1}{2}(u_{k,\ell} - u_{\ell,k}), \tag{5.6}$$

$$r_k = \frac{1}{2}\varepsilon_{k\ell m}u_{m,\ell}, \tag{5.7}$$

$$\dot{r}_{k\ell} = w_{k\ell} \equiv \frac{1}{2}(v_{k,\ell} - v_{\ell,k}), \tag{5.8}$$

$$\dot{r}_k = \omega_k \equiv \frac{1}{2}\varepsilon_{k\ell m}v_{m,\ell}, \tag{5.9}$$

where (5.7) and (5.9) follow from the skew-symmetry of (5.6) and (5.8) respectively. We require for a non-dissipative medium that equation (5.3) be integrable, so we have, upon integration of (5.3),

$$\rho_e \varepsilon - \rho_e \varepsilon_0 = \frac{1}{2} A_{k\ell mn} \{ e_{mn} + \varepsilon_{mnp}(r_p - \phi_p) \} \{ e_{k\ell} + \varepsilon_{k\ell q}(r_q - \phi_q) \}$$
$$+ \frac{1}{2} B_{\ell kmn} \phi_{m,n} \phi_{\ell,k} \tag{5.10}$$

where $_e\varepsilon_0$ is the energy density function for a given reference configuration. We note that without loss of generality, we may take $_e\varepsilon_0 = 0$ in view of $_e\varepsilon_0$ being referred to a fixed (but arbitrary) reference configuration, and thus (5.10) becomes

$$\rho_e \varepsilon = \frac{1}{2} A_{k\ell mn} \{ e_{mn} + \varepsilon_{mnp}(r_p - \phi_p) \} \{ e_{k\ell} + \varepsilon_{k\ell q}(r_q - \phi_q) \}$$
$$+ \frac{1}{2} B_{\ell kmn} \phi_{m,n} \phi_{\ell,k} . \tag{5.11}$$

We require that $\rho_e \varepsilon$ be nonnegative for all possible variations of the independent variables e_{mn}, $r_p - \phi_p$, and $\phi_{m,n}$. Thus if we take $r_p - \phi_p = 0$ and $\phi_{m,n} = 0$ we are led to the following restrictions on the nonvanishing components of $A_{k\ell mn}$:

$$A_{1111} \geq 0, \; A_{1111} \geq |A_{1122}|, \; A_{3333} \geq 0, \; A_{1111} + A_{3333} \geq \tag{5.12}$$
$$|A_{1133}|, \; A_{1313} + A_{3131} \geq -2A_{1331}, \; A_{1212} + A_{1221} \geq 0.$$

Further restrictions on the $A_{k\ell mn}$ are obtained by letting $e_{k\ell} = 0$ and $\phi_{m,n} = 0$ and letting $e_{23} \neq 0$, $e_{13} \neq 0$, $r_1 - \phi_1 \neq 0$, $r_2 - \phi_2 \neq 0$, all other $e_{k\ell}$, $r_3 - \phi_3$, and $\phi_{m,n}$ vanish. Then

$$A_{1313} + A_{3131} \geq 2A_{1331}, \; A_{1212} \geq A_{1221}, \; A_{1313} \geq 0. \tag{5.13}$$

We note here that the first two inequalities in (5.13) and the last two in (5.12) combine to yield

$$A_{1313} + A_{3131} \geq |2A_{1331}|, \; A_{1212} \geq |A_{1221}| . \tag{5.14}$$

Finally, if we set $e_{k\ell} = 0$ and $r_i - \phi_i = 0$ we obtain the following set of restrictions for the $B_{\ell kmn}$:

$$B_{1111} \geq 0, \; B_{1212} \geq 0, \; B_{1313} \geq 0, \; B_{1212} \geq B_{1221}, \; B_{1111} \geq B_{1122},$$
$$B_{3131} B_{1313} \geq (B_{1331})^2, \; B_{3333}(B_{1111} + B_{1122}) \geq (B_{1133})^2 . \tag{5.15}$$

The results in (5.15) were obtained by investigating the conditions under which the following quadratic form would be non-
negative.

$$\pi_{ij} \, \xi_i \, \xi_j \geq 0, \qquad\qquad i,j = 1,2,\cdots, 9 , \qquad (5.16)$$

where

$$\xi_1 = \phi_{1,1}, \, \xi_2 = \phi_{1,2}, \, \xi_3 = \phi_{1,3}, \, \xi_4 = \phi_{2,1}, \, \xi_5 = \phi_{2,2}$$

$$\xi_6 = \phi_{2,3}, \, \xi_7 = \phi_{3,1}, \, \xi_8 = \phi_{3,2}, \, \xi_9 = \phi_{3,3} \; ;$$

$$\pi_{11} = \pi_{55} = \frac{1}{2} B_{1111}, \, \pi_{15}=\pi_{51} = B_{1122}, \, \pi_{19}=\pi_{91}=\pi_{59}=\pi_{95} = \frac{1}{2}B_{1133}$$

$$\pi_{22} = \pi_{44} = \frac{1}{2} B_{1212}, \, \pi_{24}=\pi_{42} = \frac{1}{2}B_{1221}, \, \pi_{33}=\pi_{66} = \frac{1}{2}B_{1313},$$
$$\hspace{9cm} (5.17)$$

$$\pi_{37}=\pi_{73}=\pi_{68}=\pi_{86} = \frac{1}{2}B_{1331}, \, \pi_{77}=\pi_{88} = \frac{1}{2}B_{3131}, \, \pi_{99} = \frac{1}{2}B_{3333},$$

All other π_{ij} vanish.

The relations (5.12) to (5.15) represent the thermodynamic
restrictions on the nondissipative material moduli of heat-
conducting smectic liquid crystals.

The $A_{k\ell mn}$ and $B_{k\ell mn}$ appearing in (5.12) to (5.15) are
given by equations of the type (4.12). They depend on the choice
of spatial coordinates x_k with respect to the material coordinates X_K in the natural state and hence are not material coefficients, whereas A_{KLMN}, B_{KLMN} are indeed the material coefficients. Thus for the relations (5.12) to (5.15) to represent
the thermodynamic restrictions on the nondissipative material moduli we would have to choose the x_k coincident with the X_K for
all $k,K = 1,2,3$ and thus $A_{k\ell mn}$, $B_{k\ell mn}$ become the actual material coefficients.

We now turn our attention to obtaining restrictions on the
dissipative material moduli. To this end we substitute equations
(4.3) into the inequality (3.5) and use the restrictions on the
material moduli due to symmetry properties of the smectic compound
to obtain the following inequality:

$$a_{KLMN} \dot{C}_{MN} \dot{C}_{KL} + b_{KLMN} \dot{\Gamma}_{MN} \dot{\Gamma}_{KL} + 2v_{KLMN} G_{MN} \dot{\Gamma}_{KL}$$

$$+ \delta_{KLMN} G_{MN} G_{KL} \geq 0 . \tag{5.18}$$

where we have defined

$$v_{KLMN} = \frac{1}{2} (\gamma_{KLMN} + d_{MNKL}) .$$

This inequality must be satisfied for all possible processes involving $\dot{C}$, $\dot{\Gamma}$, and G. In particular, for $G = 0$ we have

$$\dot{C}_{KL} \, \dot{C}_{MN} \, a_{KLMN} + \dot{\Gamma}_{KL} \, \dot{\Gamma}_{MN} \, b_{KLMN} \geq 0 . \tag{5.19}$$

Since $\dot{C}$ and $\dot{\Gamma}$ are independent variables, we can reduce separately each term on the left-hand side of (5.19) into a canonical form.

Since

$$\dot{C}_{KL} \, \dot{C}_{MN} \, a_{KLMN} = \frac{1}{2} \, \dot{C}_{KL} [\dot{C}_{MN} \, a_{KLMN} + \dot{C}_{MN} \, \dot{C}_{KL} \, a_{MNKL}]$$

$$= \dot{C}_{KL} \, \dot{C}_{MN} \, \frac{1}{2} \, [a_{KLMN} + a_{MNKL}] \tag{5.20}$$

we let

$$S_{KLMN} = \frac{1}{2} [a_{KLMN} + a_{MNKL}] = S_{MNKL} \tag{5.21}$$

and set

$$\dot{C}_{11} = \xi_1, \; \dot{C}_{12} = \xi_2, \; \dot{C}_{13} = \xi_3, \; \dot{C}_{21} = \xi_4, \; \dot{C}_{22} = \xi_5, \; \dot{C}_{23} = \xi_6,$$

$$\dot{C}_{31} = \xi_7, \; \dot{C}_{32} = \xi_8, \; \dot{C}_{33} = \xi_9 \, ;$$

and

$$\pi_{11} = S_{1111} = \pi_{55}, \; \pi_{22} = S_{1212} = \pi_{44}, \; \pi_{33} = S_{1313} = \pi_{66},$$

$$\pi_{37} = S_{1313} = \pi_{68}, \; \pi_{91} = S_{3311} = \pi_{95}, \; \pi_{99} = S_{3333},$$

$$\pi_{19} = S_{1133} = \pi_{59}, \; \pi_{15} = S_{1133} = \pi_{51}, \; \pi_{24} = S_{1221} = \pi_{42},$$

$$\pi_{73} = S_{3113} = \pi_{86}, \; \pi_{77} = S_{3131} = \pi_{88} . \tag{5.22}$$

Thus when $\dot{\underset{\sim}{\Gamma}} = \underset{\sim}{0}$, we can write equation (5.19) as

$$\dot{C}_{KL} \dot{C}_{MN} a_{KLMN} = \pi_{ij}\xi_i\xi_j \geq 0 \quad \text{with} \quad \pi_{ij} = \pi_{ji} \, . \qquad (5.23)$$

We now seek the necessary and sufficient conditions for the quadratic form (5.23) to be positive definite. This is accomplished by examining the determinant of the matrix (π_{ij}) and requiring that the determinants $\Delta_1, \Delta_2, \cdots, \Delta_9$, where $\Delta_k = \det(\pi_{ij})$, $i, j = 1, \cdots, k$, be positive. Adjoining to these conditions the necessary and sufficient conditions required to satisfy $\pi_{ij}\,\xi_i\,\xi_j = 0$ for all ξ we have

$$a_{1111} \geq 0, \; a_{1212} \geq 0, \; a_{1313} \geq 0, \; a_{1212} \geq a_{1221}, \; a_{1111} \geq a_{1122},$$

$$a_{3131}a_{1313} \geq \frac{1}{4}\,(a_{1331}+a_{3113})^2, \text{ and } a_{3333}(a_{1111}+a_{1122}) \geq$$

$$\frac{1}{2}\,(a_{1133} + a_{3311})^2. \qquad (5.24)$$

Next by taking $\dot{C}_{KL} = 0$ in (5.19) we can obtain similar results for the coefficients b_{KLMN} by an identical process. So we have

$$b_{1111} \geq 0, \; b_{1212} \geq 0, \; b_{1313} \geq 0, \; b_{1212} \geq b_{1221}, \; b_{1111} \geq b_{1122},$$

$$b_{3131}b_{1313} \geq \frac{1}{4}(b_{1331}+b_{3113})^2, \text{ and } b_{3333}(b_{1111}+b_{1122}) \geq$$

$$\frac{1}{2}\,(b_{1133} + b_{3311})^2 \, . \qquad (5.25)$$

Now consider $\dot{C}_{KL} = 0$ in (5.18); then we have

$$\dot{\Gamma}_{MN}\dot{\Gamma}_{KL}b_{KLMN} + 2G_{MN}\dot{\Gamma}_{KL}v_{KLMN} + G_{MN}G_{KL}\delta_{KLMN} \geq 0 \, . \qquad (5.26)$$

We wish to write this as a quadratic inequality of the form

$$A_{ij}\mu_i\mu_j \geq 0, \; i,j, = 1,2, \cdots 18; \quad A_{ij} = A_{ji} \, . \qquad (5.27)$$

To this end we make the following identifications:

$$\mu_1 = \dot{\Gamma}_{11}, \quad \mu_2 = \dot{\Gamma}_{12}, \quad \mu_3 = \dot{\Gamma}_{13} \, ,$$

$$\mu_4 = \dot{\Gamma}_{21}, \quad \mu_5 = \dot{\Gamma}_{22}, \quad \mu_6 = \dot{\Gamma}_{23},$$

$$\mu_7 = \dot{\Gamma}_{31}, \quad \mu_8 = \dot{\Gamma}_{32}, \quad \mu_9 = \dot{\Gamma}_{33},$$

$$\mu_{10} = G_{11}, \quad \mu_{11} = G_{12}, \quad \mu_{12} = G_{13}, \qquad\qquad (5.28)$$

$$\mu_{13} = G_{21}, \quad \mu_{14} = G_{22}, \quad \mu_{15} = G_{23},$$

$$\mu_{16} = G_{31}, \quad \mu_{17} = G_{32}, \quad \mu_{18} = G_{33};$$

$$t_{KLMN} = \frac{1}{2}\,(b_{KLMN} + b_{MNKL}) = t_{MNKL},$$
$$\qquad\qquad (5.29)$$
$$u_{KLMN} = \frac{1}{2}\,(\delta_{KLMN} + \delta_{MNKL}) = u_{MNKL};$$

$$A_{11} = A_{55} = t_{1111}, \quad A_{15} = A_{51} = t_{1122}, \quad A_{19} = A_{91} = A_{59} = A_{95} = t_{1133},$$

$$A_{1,10} = A_{10,1} = A_{5,14} = A_{14,5} = v_{1111},$$

$$A_{1,14} = A_{14,1} = A_{10,5} = A_{5,10} = v_{1122},$$

$$A_{1,18} = A_{18,1} = A_{18,5} = A_{5,18} = v_{1133},$$

$$A_{22} = A_{44} = t_{1212}, \quad A_{24} = A_{42} = t_{1221},$$

$$A_{2,11} = A_{11,2} = A_{13,4} = A_{4,13} = v_{1212},$$

$$A_{2,13} = A_{13,2} = A_{4,11} = A_{11,4} = v_{1221},$$

$$A_{33} = A_{66} = t_{1313}, \quad A_{3,12} = A_{12,3} = A_{6,15} = A_{15,6} = v_{1313},$$

$$A_{3,16} = A_{16,3} = A_{6,17} = A_{17,6} = v_{1331}, \quad A_{68} = A_{86} = A_{37} = A_{73} = t_{1331},$$

$$A_{77} = A_{88} = t_{3131}, \quad A_{7,12} = A_{12,7} = A_{8,15} = A_{15,8} = v_{3113},$$

$$A_{7,16} = A_{16,7} = A_{8,17} = A_{17,8} = v_{3131}, \qquad\qquad (5.30)$$

$$A_{99} = t_{3333}, \quad A_{9,10} = A_{10,9} = A_{9,14} = A_{14,9} = v_{3311},$$

$$A_{9,18} = A_{18,9} = v_{3333}, \quad A_{10,10} = A_{14,14} = u_{1111},$$

$$A_{10,18} = A_{18,10} = A_{14,18} = A_{18,14} = u_{1133} \text{ ,}$$

$$A_{11,11} = A_{13,13} = u_{1212}, \ A_{11,13} = A_{13,11} = u_{1221} \text{ ,}$$

$$A_{12,12} = A_{15,15} = u_{1313}, \ A_{12,16} = A_{16,12} = A_{15,17} = u_{1331} \text{ ,}$$

$$A_{16,16} = A_{17,17} = u_{3131}, \ A_{18,18} = u_{3333} \text{ ,}$$

and all other $A_{ij} = 0$.

By using equations (5.28) to (5.30) we can obtain the appropriate canonical form (5.27) for the inequality in (5.26). We then employ the exchange step method [54] in order to obtain the necessary and sufficient conditions for (5.27) to be positive definite and adjoin these to the necessary and sufficient conditions for

$$A_{ij} \, \xi_i \, \xi_j = 0, \quad i, \ j = 1, \ \cdots, \ 18 \text{ ,} \tag{5.31}$$

and obtain finally the following restrictions on the material moduli:

$$(\delta_{1212} b_{1212} - v_{1212}^2)(b_{1212}^2 - b_{1221}^2) \geq (v_{1221} b_{1212} - b_{1221} v_{1212})^2 \text{ ,}$$

$$(\delta_{1313} b_{1313} - v_{1313}^2)[4 b_{1313} b_{3131} - (b_{1331} + b_{3113})^2] \geq \tag{5.32}$$

$$[2 v_{1313} b_{1313} - v_{1313}(b_{1331} + b_{3113})]^2 \text{ ,}$$

$$[s_{16,16}^{(1)} s_{7,7}^{(1)} - (s_{7,16}^{(1)})^2][s_{12,12}^{(1)} s_{7,7}^{(1)} - (s_{7,12}^{(1)})^2] \geq$$

$$\times [s_{12,16}^{(1)} s_{7,7}^{(1)} - s_{7,12}^{(1)} s_{7,16}^{(1)}]^2, \quad s_{13,13}^{(2)} s_{11,11}^{(2)} \geq (s_{11,13}^{(2)})^2 \text{ ,}$$

$$s_{10,10}^{(3)} s_{9,9}^{(3)} \geq (s_{9,10}^{(3)})^2, \quad \{[s_{14,14}^{(3)} s_{9,9}^{(3)} - (s_{9,14}^{(3)})^2]$$

$$\times [s_{10,10}^{(3)} s_{9,9}^{(3)} - (s_{9,10}^{(3)})^2]\} \geq [s_{10,14}^{(3)} s_{9,9}^{(3)} - s_{9,10}^{(3)} s_{9,14}^{(3)}]^2 \text{ ,}$$

$$[s_{17,17}^{(4)} s_{8,8}^{(4)} - (s_{8,17}^{(4)})^2][s_{15,15}^{(4)} s_{8,8}^{(4)} - (s_{8,15}^{(4)})^2] \geq$$

$$[s_{15,17}^{(4)} s_{8,8}^{(4)} - s_{8,15}^{(4)} s_{8,17}^{(4)}]^2, \quad \{[s_{18,18}^{(5)} s_{10,10}^{(5)} - (s_{10,18}^{(5)})^2] \times$$

$$\times \; [S_{14,14}^{(5)}S_{10,10}^{(5)} - (S_{10,14}^{(5)})^2]\} \geq [S_{14,18}^{(5)}S_{10,10}^{(5)} - S_{10,14}^{(5)}S_{10,18}^{(5)}]^2$$

where

$$S_{p,q}^{(1)} = A_{p,q}A_{3,3} - A_{3,p}A_{3,q} \quad \text{for} \quad p,q = 7, \; 12, \; 16 \; ;$$

$$S_{p,q}^{(2)} = M(A_{p,q}A_{2,2} - A_{2,p}A_{2,q}) - (A_{4,p}A_{2,2} - A_{2,4}A_{2,p})$$

$$\times \; (A_{4,q}A_{2,2} - A_{2,4}A_{2,q}) \quad \text{for} \quad p,q = 11, \; 13 \; ;$$

$$S_{p,q}^{(3)} = N(A_{p,q}A_{11} - A_{1,p}A_{1,q}) - (A_{s,p}A_{1,1} - A_{1,s}A_{1,p})$$

$$\times \; (A_{s,q}A_{1,1} - A_{1,5}A_{1,q}) \quad \text{for} \quad p,q = 9, \; 10, \; 14, \; 18 \; ;$$

$$S_{p,q}^{(4)} = A_{p,q}A_{6,6} - A_{6,p}A_{6,q} \quad \text{for} \quad p,q = 8, \; 15, \; 17 \; ; \quad (5.33)$$

$$S_{p,q}^{(5)} = [S_{p,q}^{(3)}S_{9,9}^{(3)} - S_{9,p}^{(3)}S_{p,q}^{(3)}]/N^2A_{11}^2 \quad \text{for} \quad p,q = 10, \; 14, \; 18 \; ;$$

$$M = b_{1212}^2 - b_{1221}^2 \quad \text{and} \quad N = b_{1111}^2 - b_{1122}^2 \; .$$

Thus we have proved the following theorem:

Theorem. The constitutive equations (4.10) are thermodynamically admissible if and only if the restrictions (5.12) to (5.15)
and (5.24) to (5.33) hold for the elastic and dissipative material
moduli, respectively.

We note here that the thermodynamic restrictions for the dissipative material coefficients are similar to those obtained for
heat-conducting nematic liquid crystals [51]. This is physically reasonable since both the smectics and nematics are made up
of rod-like or cigar-shaped molecules with an overall parallel
alignment. This similarity manifests itself in the expression for
the dissipative stress, dissipative couple stress, and the temperature gradient bivector; all of which appear in the Clausius-
Duhem inequality from which the thermodynamic restrictions on dissipative moduli are obtained.

The difference between the nematics and smectics, that is the
layered structure which is present in smectics but not in nematics,
is featured in the elastic parts of the stress and couple stress.

Here we have determined the thermodynamic restrictions on these
material moduli characterizing the elastic properties of the smectics.

6. DISCUSSION OF RESULTS

It follows from the constitutive equations $(4.10)_3$, that the
heat-conduction equation contains the term $\nu_{m,n}$ which represents
the contribution from the microgyration effects. The second term
in $(4.10)_3$ is the contribution from externally imposed temperature
gradient. Hence even if the temperature gradient is absent, heat-
conduction could still occur in the medium, in general due to the
micropolar nature of the flow. This suggests the occurence of
non-Fourier type heat-conduction which is derived directly from
the constitutive equations rather than assuming a heat-conduction
law, a priori.

One of the main contributions of the present theory of the
smectic mesophase is the obtaining of explicit forms for thermodynamic restrictions governing the material coefficients of the
heat-conducting smectic liquid crystal, which is an invaluable information while dealing with practical applications. The present
work also shows that material symmetry conditions should be taken
into consideration in order to reduce the number of material coefficients to an absolute minimum consisting of only distinct non-
vanishing coefficients so that theoretical as well as experimental
workers may know precisely to what extent comparisons of the theoretical predictions with available experimental data would be possible.

Thus, the present constitutive theory, at least, in the linear
case should prove to be of great value in practical applications,
which we have considered in a subsequent paper demonstrating that
the theory developed in this work would actually lead to prediction of results which allow for direct and gratifying comparisons
with available experimental results.

REFERENCES

1. G. Friedel, Ann. Physique 19, 273 (1922).
2. G. H. Brown, Ind. Res. 53, May (1966).
3. J. L. Fergason, Sci. Am. 211, No. 2, 77 (1964).
4. Groupe d'Etudes des Cristaux Liquides (Orsay), J. Chem. Phys.
   51, 816 (1969).
5. G. W. Gray, "Molecular Structure and the Properties of Liquid
   Crystals, Acad. Press, New York (1962).

6. J. L. Fergason and G. H. Brown, J. Am. Oil Chem. Soc. 45, 120 (1968).

7. R. S. Porter and J. F. Johnson, eds., "Ordered Fluids and Liquid Crystals", Adv. in Chem. Series, No. 63, Am. Chem. Soc. (1967).

8. R. S. Porter, E. M. Barrall II, and J. F. Johnson, J. Chem., Phys. 45, 1452 (1966).

9. R. S. Porter and J. F. Johnson, J. Appl. Phys. 34, 51(1963).

10. R. S. Porter and J. F. Johnson, J. Phys. Chem., 66, 1826 (1962).

11. E. M. Barrall II, R. S. Porter, and J. F. Johnson, J. Phys. Chem., 68, 2810 (1964).

12. R.S. Porter and J. F. Johnson, J. Appl. Phys. 34, 55(1963).

13. R. S. Porter and J. F. Johnson, "The Rheology of Liquid Crystals", Ch. 5, in "Rheology", 4, ed., F. R. Eirich, 317 (1967).

14. S. Chandrasekhar, Mol. Cryst. Liquid Cryst. 2, 71 (1966).

15. W. Maier and A. Saupe, Z. Naturforsch. 13A, 564 (1958); 14A, 882 (1959); 15A, 287 (1960).

16. K. K. Kobayashi, Phys. Letters 31A, 125 (1970); J. Phys.Soc. Japan 29, 101 (1970).

17. A. Saupe, Angew. Chem. International Edit. 7, No. 2, 97(1968).

18. A. Saupe, Mol. Cryst. Liquid Cryst. 7, 59 (1969).

19. I. G. Chistyakov, Soviet Physics USPEKHI, 9, No.4, 551 (1967).

20. C. W. Oseen, Trans. Faraday Soc. 29, 883 (1933).

21. H. Zocher, Trans. Faraday Soc. 29, 945 (1933).

22. A. Anzelius, Uppsala Univ. Arsskr. Mat. Och Naturvet, 1 (1931).

23. F. C. Frank, Discussions Faraday Soc. 25, 19 (1958).

24. P. C. Martin, P. S. Pershan, and J. Swift, Phys. Rev. Lett., 25, No. 13, 844 (1970).

25. M. Stephen, Phys. Rev. A2, 1558 (1970).

26. T. C. Lubensky, Phys. Rev. A2, 2497 (1970).

27. D. Forster, et al, Phys. Rev. Lett., 26, 1016 (1971).

28. H. Huang, Phys. Rev. Lett., 26, 1525 (1971).

29. H. Schmidt and J. Jahnig, Annals of Physics, 71, 129 (1972).

30. J. D. Lee and A. C. Eringen, J. Chem. Phys. 54, 5027 (1971).

31. J. D. Lee and A. C. Eringen, J. Chem. Phys. 55, 4504 (1971).

32. J. D. Lee and A. C. Eringen, J. Chem. Phys. 55, 4509 (1971).

33. J. D. Lee and A. C. Eringen, J. Chem. Phys. 58, 4203 (1973).

34. A. C. Eringen and J. D. Lee, "Liquid Crystals and Ordered Fluids", Am. Chem. Soc. Symposium 2, 383 (1973).

35. J. L. Ericksen, Arch. Rat. Mech. Anal. 4, 231 (1960).

36. J. L. Ericksen, Trans. Soc. Rheol. 5, 23 (1961).

37. J. L. Ericksen, Arch. Rat. Mech. Anal. 9, 371 (1962).

38. J. L. Ericksen, Arch. Rat. Mech. Anal. 23, 266 (1966).

39. J. L. Ericksen, Phys. of Fluids, 9, 1205 (1966).

40. J. L. Ericksen, Quart. J. Mech. Appl. Math., 19, 455 (1966).

41. J. L. Ericksen, Appl. Mech. Rev., 20, 1029 (1967).

42. J. L. Ericksen, J. Fluid Mech., 27, 59 (1967).

43. J. L. Ericksen, Mol. Cryst. Liquid Cryst. 7, 153 (1969).

44. J. L. Ericksen, "Liquid Crystals and Ordered Fluids", Am.Chem.
     Soc. Symposium, $\underline{1}$, (1970), ed. R. S. Porter and J. F. Johnson.
45. F. M. Leslie, Quart. J. Mech. Appl. Math., $\underline{19}$, 357 (1966).
46. F. M. Leslie, Arch. Rat. Mech. Anal. $\underline{28}$, 265 (1968).
47. F. M. Leslie, Proc. Roy. Soc. London $\underline{A\ 207}$, 359 (1968).
48. F. M. Leslie, Mol. Cryst. Liquid Cryst. $\underline{7}$, 407 (1969).
49. A. C. Eringen, Proc. Eleventh International Congr. Appl. Mech.,
     ed., H. Görtler (Springer, Berlin), 131 (1966).
50. A. C. Eringen, J. Math. Mech. $\underline{15}$, 909 (1966).
51. M.N.L. Narasimhan and A. C. Eringen, Int. J. Eng. Sci., $\underline{13}$,
     233 (1975).
52. M.N.L. Narasimhan and A. C. Eringen, Mol. Cryst. Liquid Cryst.,
     $\underline{29}$, 57 (1974).
53. M.N.L. Narasimhan, "Continuum Theory of Heat-Conducting Smec-
     tic Liquid Crystals", in Proc. Twelfth Annual Meeting, Soc.
     Eng. Sci., ed., Morris Stern, 61 (1975).
54. W. Nef, "Linear Algebra", (Translated from the German by J.C.
     Ault), McGraw Hill, New York, 212 (1967).

ORIENTATIONAL EFFECTS IN HEAT-CONDUCTING

SMECTIC LIQUID CRYSTALS

M.N.L. Narasimhan and T.E. Kelley

Oregon State University, Department of Mathematics

Corvallis, Oregon 97331

ABSTRACT

The mutually competing orienting influences on heat-conduct-ing smectic liquid crystals of temperature gradients, solid bound-aries and shear flows are investigated. The case of Couette flow of smectic liquid crystals between concentric rotating cylinders is treated by solving the governing field equations based on the micropolar continuum theory of A. C. Eringen. Included in the so-lution of this problem is the derivation of explicit expressions for apparent viscosity coefficient, the orientation and micro-rotation fields, as well as heat-conduction. These expressions lead to a direct and more satisfying comparison with experimental results. The behavior of apparent viscosity under various shear-rates and temperatures as well as its dependence on the gap-width between the cylinders is investigated reaching good agreement with experimental results.

1. INTRODUCTION

Liquid crystal research has experienced a phenomenal growth during the last five years. This is not at all surprising since liquid crystals are found to be present both in animate and inani-mate systems with which we are involved in every day life as well as in the research laboratory. Liquid crystals possess unique structural characteristics which are responsible for their high sensitivity to external stimuli such as heat, light, chemical vapor, electric and magnetic fields, shear, and solid boundaries, a behavior virtually unsurpassed even by the solid crystalline matter. For instance, as J. L. Fergason [1] has stated, certain

319

types of quartz rotate the plane of polarization of a light beam
passing normal to its bounding surface about 20 degrees per millimeter of depth and are considered optically active. But an optically active cholesteric liquid crystal, on the other hand, rotates the plane of polarization of light through an angle of as
much as 18,000 degrees or 50 rotations per millimeter. This enor-
mous optical activity of liquid crystals is due to the unique pack-
ings of their molecules which are readily orientable by external
fields. Orientational changes and the resulting optical phenomena
such as double refraction, circular dichroism and an iridescent
color display under the influence of even minute fluctuations of
temperature or uneven conduction of heat are familiar to all re-
searchers in the field. Exploitation of such high sensitivity of
liquid crystals both in industry and biomedical fields for various purposes should naturally be expected. The need for nematic
liquid crystals, for instance, for especially display technology
has provided the impetus to search for new as well as synthetic
materials that will function over a wide temperature range and
even well below room temperature. The twisted nematics, or the
cholesteric liquid crystals are utilized for a variety of devices
such as detection of tumor, or to chart out blood flow in newly
grafted skin in the case of burn victims, and many other potential
applications in living systems. It is considered that the brain
itself consists of liquid crystalline matter and so does much of
the human body; and on the industrial side, detection of leakage
of current in integrated circuits and electro-optic systems is a
much exploited application of liquid crystals for nondestructive
testing of materials. Liquid crystals are finding important uses
in the research laboratory in such areas as chromatography and
chemical kinetics. Furthermore, the potential application of
liquid crystals of all the three basic types - the nematic, the
smectic and the cholesteric - to high speed flight and space ex-
ploration has been recently discussed in detail by R. A. Champa
[2]. The sophistication of instrumentation needed on board such
high speed vehicles include weight reduction of the equipments
even to the extent of miniaturization of those devices, high sensitivity, remote sensing of temperatures, reliability of performance and power efficiency. Liquid crystals of all the three
basic types and even, perhaps, their mixtures, are ideally suited
because of their high sensitivity and less consumption of power
for their operation as diagnostic devices which will be an invaluable aid in extending and complementing man's senses and actions. Other potential applications include precise detection of
new energy sources consistent with economic feasibility with the
help of sophisticated sensors consisting of liquid crystal media
and also laser beam profile analysis used for a variety of strate-
gic as well as biomedical research. Success of the above stated
applications depends largely on encapsulation of liquid crystals
from the standpoint of sensitivity, temperature range of operation

and simplicity of application.

In order to understand the multifaceted nature of liquid crystals, a full knowledge of their molecular order, alignment and molecular mobility is essential. For example, structure and mobility of aggregates, or ordered domains of liquid crystals are involved in polymerization processes which are of great importance in industry such as textiles as well as in the field of medicine. It is well known that monomers which cannot be polymerized in the crystalline state (Blumstein et al [3]) can easily polymerize in a mesophase because of molecular mobility. Molecular mobility implies deformation of aggregates or ordered domains of liquid crystals. Thus a thorough rheological study of mesophases becomes very essential for a proper understanding of many mechanical processes involving liquid crystals. To this end, besides experimental studies, two fundamental theoretical approaches exist, as is natural to expect. One is the molecular theoretical approach and the other is the continuum or more specifically the microcontinuum approach. Presently, it is not the question as to which of these attempts is preferable to follow, but rather, how best we can understand and benefit from both of the approaches, and bridge the gap between them, while holding the final outcome from both to be ultimately tested against experimental results.

In a few recent papers, Narasimhan and Eringen [4,5], Narasimhan [6], and Narasimhan and Kelley [7] have constructed a microcontinuum theory of heat-conducting nematics and smectics, based on the micropolar theory of Eringen [8]. We are encouraged by the fact that Narasimhan and Eringen's theory [4,5] has successfully predicted orientational patterns and shear-rate dependence of viscosity, formation of adsorption layer in capillary flows, in the case of nematics reaching gratifying agreement with very carefully performed experiments of Porter and Johnson [9], Peter and Peters [10], Becherer and Kast [11] and Miesowicz [12].

The present paper deals with the application of the recent Narasimhan and Kelley's theory [7] to the analytical investigation of orientational effects in heat-conducting smectic liquid crystals with reference to shear-rate dependence as well as channel gap-width dependence of the apparent coefficient of viscosity. Despite the handicap of nonavailability of adequate experimental data to compare all the rheological aspects of the smectic phase in our study, we have availed ourselves of a few existing experimental works, for instance, those of Tamamushi and Matsumoto [13] for comparing our theoretical results. Since most rheological studies are carried out in viscometric flows of liquid crystals, we investigate mathematically the application of the microcontinuum theory [4-7] to the flow of a smectic phase in a rotational, coaxial cylindrical annulus based on the micropolar theory of Eringen [8]. In this investigation, we obtain solutions for the velocity and pressure fields, stress and couple stress, heat flux distributions, microrotation and orientation fields. An expression for the apparent coefficient of viscosity is obtained and its dependence on shear-rate and the gap-width of the annulus are studied. It is found that our theoretical predictions are in good agreement with those of the available experiments thus building confidence in the application of the micropolar theory [8] and its subsequent ramifications into microcontinuum theory of liquid crystals [4-7].

2. BASIC EQUATIONS

The equations of motion of a general micropolar continuum [8] and constitutive equations for heat-conducting smectic liquid crystals [7] reduce in the linear incompressible case to the following:

Equation of Continuity:

$$v_{k,k} = 0 , \tag{2.1}$$

Microinertia:

$$\frac{\partial i_{k\ell}}{\partial t} + i_{k\ell,m} v_m + e_{\ell mr} v_r i_{km} + e_{kmr} v_r i_{m\ell} = 0 , \tag{2.2}$$

Linear Momentum:

$$t_{ji,j} + \rho (f_i - \dot{v}_i) = 0 , \tag{2.3}$$

Moment of Momentum:

$$m_{ji,j} + e_{ik\ell} t_{k\ell} + \rho (\ell_i - \dot{\sigma}_i) = 0 , \tag{2.4}$$

Conservation of Energy:

$$\rho \dot{\varepsilon} = t_{k\ell} (v_{\ell,k} - e_{k\ell m} v_m) + m_{k\ell} v_{\ell,k} + q_{k,k} + \rho h , \tag{2.5}$$

Stress Constitutive Equation:

$$t_{k\ell} = A_{k\ell mn} (u_{n,m} - e_{mnp} \phi_p) + a_{k\ell mn} (v_{n,m} - e_{mnp} v_p) , \tag{2.6}$$

Couple Stress Equation:

$$m_{k\ell} = B_{\ell kmn} \, \phi_{m,n} + b_{\ell kmn} \, v_{m,n} + \gamma_{k\ell mn} \, g_{mn} \, , \tag{2.7}$$

Heat-Flux Vector:

$$q_b = (d_{\ell kmn} \, v_{m,n} + \delta_{\ell kmn} \, g_{mn}) \, \frac{1}{2} \, e_{\ell kb} \, , \tag{2.8}$$

where t_{ij}, m_{ij}, and q_b are respectively, the stress tensor, the couple stress tensor and the heat-flux vector, f_i is the body force, v_i is the velocity vector, u_n is the displacement vector, ε is the internal energy density per unit mass, h is the heat supply per unit mass, ℓ_i is the body couple, ρ is the mass density, ϕ_m is the orientation vector, $v_r = \dot{\phi}_r$ is the microgyration vector and g_{mn} given by

$$g_{mn} = \frac{1}{2} T^{-1} e_{mnq} T_{,q} \tag{2.9}$$

is the temperature gradient bivector, T is the temperature, e_{mnq} is the alternating tensor and the spin-inertia $\dot{\sigma}_i$ is given by

$$\dot{\sigma}_i = (i_{mm} \delta_{ij} - i_{ij}) \dot{v}_j - e_{kij} \, i_{km} \, v_m \, v_j \tag{2.10}$$

in which $i_{k\ell} = i_{\ell k}$ is the microinertia tensor satisfying (2.2) and δ_{ij} is the Kronecker delta. The generalized Clausius-Duhem inequality has been derived and used in [7] to obtain the necessary thermodynamic restrictions governing the constitutive equations.

In the above equations (2.1) to (2.10) we have employed spatial and material rectangular coordinate systems $x_i (i = 1,2,3)$ and $X_I (I = 1,2,3)$. The coefficients $a_{k\ell mn}$ in the spatial system are related to the corresponding a_{KLMN}, in the material coordinate system by the relation

$$a_{k\ell mn} = a_{KLMN} \, \delta_{kK} \, \delta_{\ell L} \, \delta_{mM} \, \delta_{nN} \tag{2.11}$$

and are functions of temperature only, δ_{kK} being the direction cosines between the spatial and material coordinates. Similar statments as (2.11) hold for the coefficients $A_{k\ell mn}$, $b_{k\ell mn}$, $B_{k\ell mn}$, $\gamma_{k\ell mn}$, $d_{k\ell mn}$ and $\delta_{k\ell mn}$. Thermodynamic and material symmetry restrictions on these material moduli have been derived previously [7]. The motion of a material point X_K is given by $x_k = x_k(X_K, t)$ and χ_{kK} denotes the micromotion.

3. FORMULATION OF THE PROBLEM

We consider a steady Couette flow of heat-conducting smectic liquid crystals between two long concentric circular cylinders of radii r_1 and r_2 $(r_1 < r_2)$ which are rotating with constant angular velocities Ω_1 and Ω_2 respectively. We choose cylindrical polar coordinates (r, θ, z) such that the z-axis coincides with the long axis of the cylinders. Let (x,y,z) and (X,Y,Z) represent the spatial and material coordinates. We assume that the material and spatial frames are coincident and hence in view of (2.11) the coefficients $a_{k\ell mn}$, etc., become the actual material coefficients. In addition the material frame is chosen in such a way that the X-axis coincides with the axis of orientation of the liquid crystal. The velocity field and the microgyration velocity field are given by

$$v_r = 0 , \quad v_\theta = r w (r,\theta) , \quad v_z = 0 ; \tag{3.1}$$

$$v_r = 0 , \quad v_\theta = 0 \qquad , \quad v_z = \nu(r,\theta) . \tag{3.2}$$

In the velocity field, the z-dependence may be excluded due to infinite channel length and noting that the equation of continuity in cylindrical coordinates has the form

$$\frac{\partial v_r}{\partial r} + \frac{\partial v_\theta}{r\partial \theta} + \frac{v_r}{r} = 0 ,$$

which reduces to $0 = \partial w/\partial\theta$ by use of (3.1). Thus we have w as a function of r only. The temperature field is $T = T(r)$ and we allow arbitrary spatial dependence for the microgyration velocity ν to depend on both r and θ.

The micromotion is given by

$$\chi_{kK} = \begin{bmatrix} \cos \phi & -\sin \phi & 0 \\ \sin \phi & \cos \phi & 0 \\ 0 & 0 & 1 \end{bmatrix} \tag{3.3}$$

where

$$\underset{\sim}{\phi} = (0,\ 0,\ \phi\ (r,\theta)) \qquad\qquad (3.4)$$

is an axial vector normal to the plane of motion (i.e. normal to the $r\theta$-plane). Finally, the displacement field u is given by

$$u_r = 0\ ,\quad u_\theta = u(r,\theta)\ ,\quad u_z = 0\ . \qquad (3.5)$$

In cartesian coordinates (3.1), (3.2), and (3.5) become

$$v_x = rw(r)\sin\ \theta\ ,\quad v_y = rw(r)\cos\ \theta\ ,\quad v_z = 0\ ; \qquad (3.6)$$

$$\nu_x = 0 \qquad\quad ,\quad \nu_y = 0 \qquad\quad ,\quad \nu_z = \dot{\phi} = \nu(r,\theta)\ ; \quad (3.7)$$

$$u_x = -u\ \sin\ \theta\ \ ,\quad u_y = u\ \cos\ \theta\quad\ ,\quad u_z = 0\ . \qquad (3.8)$$

We shall now substitute (3.4), (3.6), (3.7), and (3.8) into equations (2.6) - (2.8) and make use of the material symmetry conditions [7] on the $A_{k\ell mn}$, $B_{k\ell mn}$, $a_{k\ell mn}$, $b_{k\ell mn}$, $\gamma_{k\ell mn}$, $d_{k\ell mn}$, and $\delta_{k\ell mn}$, in order to obtain the following results:

The stress components are:

$$t_{xx} = t_{11} = (A_{1111} - A_{1122})\ \sin\ \theta\ \cos\ \theta\ [\frac{u}{r} - \frac{\partial u}{\partial r}] + (A_{1111}\sin^2\theta +$$

$$+ A_{1122}\cos^2\theta)\ \frac{\partial u}{r\partial\theta} - \frac{1}{2}\ (a_{1111} - a_{1122})rw'\ \sin\ 2\theta\ ,$$

$$t_{yy} = t_{22} = -(A_{1111} - A_{1122})\sin\ \theta\ \cos\ \theta\ [\frac{u}{r} - \frac{\partial u}{\partial r}] + (A_{1111}\cos^2\theta +$$

$$+ A_{1122}\sin^2\theta)\ \frac{\partial u}{r\partial\theta} + \frac{1}{2}\ (a_{1111} - a_{1122})rw'\ \sin\ 2\ \theta\ ,$$

$$t_{xy} = t_{12} = (A_{1212}\ \cos^2\ \theta - A_{1221}\sin^2\ \theta)\ \frac{\partial u}{\partial r} + (A_{1212}\sin^2\ \theta -$$

$$- A_{1221}\ \cos^2\ \theta)\ \frac{u}{r} - (A_{1212} + A_{1221})\cos\ \theta\ \sin\ \theta\ (\frac{\partial u}{r\partial\theta}) -$$

$$- (A_{1212} - A_{1221})\ \phi + rw'\ [\frac{1}{2}(a_{1212} - a_{1221}) +$$

$$+ \frac{1}{2}(a_{1111} - a_{1122})\cos\ 2\ \theta] + (a_{1212} - a_{1221})(w-\nu)\ ,$$

$$t_{yx} = t_{21} = (A_{1221}\cos^2\theta - A_{1212}\sin^2\theta)\frac{\partial u}{\partial r} + (A_{1221}\sin^2\theta -$$

$$- A_{1212}\cos^2\theta)\frac{u}{r} - (A_{1212}+A_{1221})\cos\theta\,\sin\theta\,(\frac{\partial u}{r\partial\theta}) +$$

$$+ (A_{1212} - A_{1221})\phi + rw'[\frac{1}{2}(a_{1111}-a_{1122})\cos 2\theta -$$

$$- \frac{1}{2}(a_{1212} - a_{1221})] - (a_{1212}-a_{1221})(w - v) ,$$

$$t_{zz} = t_{33} = A_{3311}(\frac{\partial u}{r\partial\theta}) ,$$

$$t_{xz} = t_{13} = 0 = t_{zx} = t_{yz} = t_{zy} . \qquad (3.9)$$

The couple stress components are:

$$m_{xx} = m_{11} = 0,\ m_{yy} = m_{22} = 0,\ m_{zz} = m_{33} = 0,\ m_{xy} = m_{12} = 0 ,$$

$$m_{yx} = m_{21} = 0 , \qquad (3.10)$$

$$m_{zx} = m_{31} = B_{3131}\,\phi_{,x} + b_{3131}v_{,x} + \frac{1}{2}T^{-1}(\gamma_{1331}-\gamma_{1313})T_{,y} ,$$

$$m_{xz} = m_{13} = B_{3131}\,\phi_{,x} + b_{3131}v_{,x} + \frac{1}{2}T^{-1}(\gamma_{3131}-\gamma_{3113})T_{,y} ,$$

$$m_{zy} = m_{32} = B_{2332}\phi_{,y} + b_{2332}v_{,y} + \frac{1}{2}T^{-1}(\gamma_{2323}-\gamma_{2332})T_{,x} ,$$

$$m_{yz} = m_{23} = B_{3232}\phi_{,y} + b_{3232}v_{,y} + \frac{1}{2}T^{-1}(\gamma_{3223} - \gamma_{3232})T_{,x} .$$

The heat-flux vector components are:

$$q_x = q_1 = \frac{1}{2}(d_{2332} - d_{3232})v_{,y} + \frac{1}{4}T^{-1}(\delta_{3232}+\delta_{2323}-\delta_{3223} -$$

$$- \delta_{2332})\,T_{,x} ,$$

$$q_y = q_2 = -\frac{1}{2}(d_{2332} - d_{3232})v_{,x} + \frac{1}{4}T^{-1}(\delta_{3232} + \delta_{2323}-\delta_{3223} -$$

$$- \delta_{2332})\,T_{,y} , \qquad (3.11)$$

$$q_z = q_3 = 0 .$$

We now define the following parameters involving the material coefficients and temperature in order to facilitate their use later
in the paper.

$$k_1 = \frac{1}{2}(a_{1111}-a_{1122}) = k_4, \quad k_2 = a_{3311}, \quad k_3 = \frac{1}{2}(a_{1212}-a_{1221}) \,,$$

$$h_1 = A_{1111}, \quad h_2 = A_{1122}, \quad h_3 = A_{1133}, \quad h_4 = A_{3333}, \quad h_5 = A_{1212} \,,$$

$$h_6 = A_{1221}, \quad h_7 = \frac{1}{2}(A_{1111}-A_{1122}) = \frac{1}{2}(A_{1212}+A_{1221}), \quad h_8 = A_{1313},$$

$$h_9 = A_{3113} = A_{1331}, \quad h_{10} = (A_{1212} - A_{1221}) \,, \qquad (3.12)$$

$$k_5 = B_{1331}, \quad k_6 = b_{1331}, \quad k_7 = \frac{1}{2}T^{-1}(\gamma_{1331} - \gamma_{3113}),$$

$$k_{11} = \frac{1}{2}(d_{2332}-d_{3232}), \quad k_{12} = \frac{1}{4}T^{-1}(\delta_{3232}+\delta_{2323}-\delta_{3223}-\delta_{2332}) \,.$$

On using (3.12) in (3.9) to (3.11) the stress, the couple
stress and the heat-flux components reduce to the following simplified forms:

$$t_{xx} = t_{11} = 2h_7\sin\theta\cos\theta\left(\frac{u}{r} - \frac{\partial u}{\partial r}\right)+(h_1\sin^2\theta + h_2\cos^2\theta)\frac{\partial u}{r\partial\theta}$$

$$- k_1rw'\sin 2\theta \,,$$

$$t_{yy} = t_{22} = -2h_7\sin\theta\cos\theta\left(\frac{u}{r} - \frac{\partial u}{\partial r}\right)+(h_1\cos^2\theta + h_2\sin^2\theta)\frac{\partial u}{r\partial\theta}$$

$$+ k_1 rw'\sin 2\theta \,,$$

$$t_{xy} = t_{12} = (h_5\cos^2\theta - h_6\sin^2\theta)\frac{\partial u}{\partial r} +(h_5\sin^2\theta - h_6\cos^2\theta)\frac{u}{r}$$

$$- 2h_7\sin\theta\cos\theta\left(\frac{\partial u}{r\partial\theta}\right) - h_{10}\phi + rw'(k_3+k_1\cos 2\theta)$$

$$+ 2k_3(w - v) \,, \qquad (3.13)$$

$$t_{yx} = t_{21} = (h_6\cos^2\theta - h_5\sin^2\theta)\frac{\partial u}{\partial r} + (h_6\sin^2\theta - h_5\cos^2\theta)\frac{u}{r}$$

$$- 2h_7\sin\theta\cos\theta\left(\frac{\partial u}{r\partial\theta}\right)+h_{10}\phi + rw'(k_1\cos 2\theta -k_3)$$

$$- 2k_3(w - v) \,,$$

$$t_{xz} = t_{13} = 0 = t_{zx} = t_{yz} = t_{zy} \ ,$$

$$t_{zz} = t_{33} = h_3 \frac{\partial u}{r \partial \theta} \ .$$

$$m_{xx} = 0 = m_{yy} = m_{zz} = m_{xy} = m_{yx} \ ,$$

$$m_{zx} = m_{31} = k_5 \ \phi,_x + k_6 \nu,_x + k_7 T,_y \ ,$$

$$m_{xz} = m_{13} = k_8 \phi,_x + k_9 \nu,_x + k_{10} T,_y \ , \tag{3.14}$$

$$m_{zy} = m_{32} = k_5 \phi,_y + k_6 \nu,_y - k_7 T,_x \ ,$$

$$m_{yz} = m_{23} = k_8 \phi,_y + k_9 \nu,_y - k_{10} T,_x \ .$$

$$q_x = q_1 = k_{11} \nu,_y + k_{12} T,_x \ ,$$

$$q_y = q_2 = -k_{11} \nu,_x + k_{12} T,_y \ , \tag{3.15}$$

$$q_z = q_3 = 0 \ .$$

Equations (3.13) - (3.15) are now substituted into the balance equations of motion and energy (2.3) - (2.5) to obtain the field equations governing $w(r)$, $T(r)$, $\nu(r,\theta)$, and $u(r,\theta)$ (in the absence of external body forces, body couples, and heat sources):

$$h_5 \left(\frac{u}{r^2} - \frac{\partial u}{r \partial r} - \frac{\partial^2 u}{\partial r^2}\right) - h_1 \frac{\partial^2 u}{r^2 \partial \theta^2} - (h_1 + h_5) \cot \theta \ \left(\frac{\partial u}{r^2 \partial \theta}\right)$$

$$+ (h_1 - h_5)\left(\frac{\partial^2 u}{r \partial \theta \partial r}\right) \cot \theta - (k_1 + k_3)(rw'' + 3w')$$

$$+ 2 k_3 \left(\frac{\partial \nu}{\partial r} + \frac{\cot \theta}{r} \frac{\partial \nu}{\partial \theta}\right) = 0 \ , \tag{3.16}$$

$$- h_5 \left(\frac{u}{r^2} - \frac{\partial u}{r \partial r} - \frac{\partial^2 u}{\partial r^2}\right) + h_1 \frac{\partial^2 u}{r^2 \partial \theta^2} - (h_1 + h_5) \tan \theta \ \left(\frac{\partial u}{r^2 \partial \theta}\right)$$

$$+ (h_1 - h_5) \tan \theta \ \left(\frac{\partial^2 u}{r \partial \theta \partial r}\right) + (k_1 + k_3)(rw'' - 3w') - 2k_3 \left(\frac{\partial \nu}{\partial r} - \frac{\tan \theta}{r} \frac{\partial \nu}{\partial \theta}\right) = 0$$

$$\tag{3.17}$$

$$B_{3131} \nabla^2 \phi + b_{3131} \nabla^2 \nu + 2k_3 (rw' + 2w) - 4k_3 \nu - \rho j \dot{\nu}$$

$$- 4h_7 \phi + h_{10} \left(\frac{\partial u}{\partial r} + \frac{u}{r} \right) = 0 , \qquad (3.18)$$

$$\nabla^2 T = 0 , \qquad (3.19)$$

where

$$\nabla^2 = \frac{\partial^2}{\partial r^2} + \frac{1}{r} \frac{\partial}{\partial r} + \frac{1}{r^2} \frac{\partial^2}{\partial \theta^2} , \qquad j = i_{11} - i_{22} . \qquad (3.20)$$

In order for a solution set to exist for equations (3.16) to (3.19) taking into account the micromotions of local substructures under appropriate boundary conditions it will be necessary to have

$$k_3 = \frac{1}{2} (a_{1212} - a_{1221}) \neq 0 \quad \text{and} \quad a_{1212} \neq 0 \qquad (3.21)$$

which is permitted by the thermodynamic restrictions on a_{1212} and a_{1221} obtained as demonstrated in [7], that is,

$$a_{1212} \geq 0, \quad a_{1212} - a_{1221} \geq 0 . \qquad (3.22)$$

Therefore we require $a_{1212} - a_{1221} > 0$ and $a_{1212} > 0$.

The coefficient B_{3131} is subject to the thermodynamic restrictions [7], namely, it is nonnegative, but we note that its vanishing would eliminate part of the explicit expression for microrotation ϕ in equation (3.18). In view of this we should require that $B_{3131} = 0$ and $h_7 = \frac{1}{2} (A_{1111} - A_{1122}) = \frac{1}{2} (A_{1212} + A_{1221}) = 0$ not occur simultaneously. Thus we must consider three cases: (i) $B_{3131} \neq 0$ and $A_{1212} \neq A_{1221}$, (ii) $B_{3131} \neq 0$ and $A_{1212} = -A_{1221}$, and (iii) $B_{3131} = 0$ and $A_{1212} \neq -A_{1221}$. We note also that the coefficient b_{3131} appearing in equation (3.18) is required to be nonnegative by the thermodynamic restrictions [7].

As for equation (3.19) governing the temperature field we now show that it can be obtained in the following manner from the energy equation (2.5). The term $\rho \dot{\varepsilon}$ on the left-hand side of equation (2.5) has the form

$$\rho\dot{\varepsilon} = \rho(\dot{\psi}+\dot{\eta} \ T + \eta \ \dot{T}) \ \underset{\sim}{} \ \rho\dot{\psi}_0(\rho^{-1}, T) + \rho\dot{\eta}T = 0 \qquad (3.23)$$

in the linear micropolar theory which we are using. Here we have
used the fact that in the steady state case owing to the geometry
of the channel the material time derivative denoted by D/Dt or by
a superposed dot is given by

$$(3.24)$$

$$\frac{D}{Dt} = w(r) \ \frac{\partial}{\partial\theta} = w\frac{\partial}{\partial\theta}, \quad \frac{D\phi(r,\theta)}{Dt} = \dot{\phi} = w \ \frac{\partial\phi}{\partial\theta}, \quad \dot{\nu}(r,\theta) = w \ \frac{\partial\nu}{\partial\theta} \ \underset{\sim}{} \ 0 \ .$$

The last equation in (3.24) is due to the fact that in the linear
theory w and ν are taken to be first-order, small quantities.
Also, from the linear theory we have

$$\dot{\psi} = \dot{\psi}_0(\rho^{-1},T) + \text{Higher order terms} \ \underset{\sim}{} \ \dot{\psi}(\rho^{-1},T)$$

$$(3.25)$$

$$= w \ \frac{\partial}{\partial\theta} \ [\psi_0(\rho^{-1},T(r))] = 0$$

and

$$\dot{\eta} = \frac{D}{Dt} \ (- \frac{\partial\psi}{\partial t}) \ \underset{\sim}{} \ w \ \frac{\partial}{\partial\theta}[\psi_0(\rho^{-1},T(r))] + \text{Higher order terms} = 0,$$

$$(3.26)$$

where we make use of the fact that ρ^{-1} is constant due to the
incompressibility condition and $T = T(r)$. The right-hand side of
equation (2.5) contains only one term of the first order, $q_{k,k}$,
the other terms may be neglected in comparison with w,ν, and
∇T in the linear theory. Use of equations (3.15), and (3.23)
to (3.26) in the energy equation will then yield (3.19).

4.   SOLUTION OF THE FIELD EQUATIONS

From equations (3.16) and (3.17) we obtain the following
two equations

$$2k_3 \ \frac{\partial\nu}{\partial\theta} = (h_1 + h_5) \ \frac{\partial u}{r\partial\theta} + (h_5 - h_1) \ \frac{\partial^2 u}{\partial r\partial\theta} \qquad (4.1)$$

and

$$2k_3 \ \frac{\partial\nu}{\partial r} = a_{1212}(rw'' + 3w')+h_5 \ \frac{\partial}{\partial r}(\frac{u}{r} + \frac{\partial u}{\partial r})+h_1 \ \frac{\partial^2 u}{r^2\partial\theta^2} \ . \qquad (4.2)$$

Differentiation of both sides of equation (4.1) with respect to
r and of equation (4.2) with respect to θ and the use of the
facts that $w = w(r)$ and the interchange of partial derivatives

is allowed lead to the following differential equation for the displacement u.

$$h_1 \left[\frac{\partial^2 \tau}{\partial r^2} - \frac{\partial}{\partial r} \left(\frac{\tau}{r} \right) + \frac{\partial^2 \tau}{r^2 \partial \theta^2} \right] = 0 \qquad (4.3)$$

where

$$\tau = \frac{\partial u}{\partial \theta} \quad . \qquad (4.4)$$

Equation (4.3) requires that $h_1 = A_{1111} \neq 0$ since we require a nontrivial solution to our problem and that the following differential equation be satisfied by τ.

$$\frac{\partial^2 \tau}{\partial r^2} - \frac{\partial}{\partial r} \left(\frac{\tau}{r} \right) + \frac{\partial^2 \tau}{r^2 \partial \theta^2} = 0 \qquad (4.5)$$

Equation (4.5) may be solved using the technique of separation of variables and the requirement that τ must be real-valued leads to

$$\tau = Lr \qquad (4.6)$$

where L is a constant. We now make use of the following relation between displacement $u(r,\theta)$ and the tangential velocity $v_\theta(r) = rw(r)$, namely

$$\frac{Du}{Dt} = v_\theta \quad . \qquad (4.7)$$

However, since we are using the constitutively linear theory we have $D/Dt = w \frac{\partial}{\partial \theta}$ and hence (4.7) may be rewritten

$$w \frac{\partial u}{\partial \theta} = w\tau = v_\theta = rw \quad . \qquad (4.8)$$

Thus we must have $L = 1$ and τ from (4.4), (4.6) and (4.8) becomes:

$$\tau = \frac{\partial u}{\partial \theta} = r \quad . \qquad (4.9)$$

Integration of (4.9) with respect to θ now yields

$$u(r,\theta) = r\theta + U(r), \qquad \theta \in [-\pi, \pi) , \qquad (4.10)$$

where $U(r)$ is an arbitrary function of r.

Without loss of generality we may choose $U(r) = 0$ since the fluid particles are in continuous rotation about the common axis of the cylinders describing circles with centers on the cylinders' common axis (z - axis). We now define the displacement field which satisfies (4.10) in the interval $[-\pi,\pi]$ as well as any extended interval for all θ. Since the circular cylinders in our problem are performing continuous rotations about their common axis the displacement field $u(r,\theta)$ must be a periodic function of θ with periodicity 2π. Thus we define the displacement field to be bounded and have a finite number of maxima and minima in $[-\pi,\pi)$.

We have

$$u(r,\theta) = u(r,\theta_0), \quad \theta_0 = \theta \bmod 2\pi, \quad \theta_0 \in [-\pi,\pi) \qquad (4.11)$$

and then by Dirichlet's theorem, the Fourier series for $u(r,\theta)$ converges to $\frac{1}{2}[u(r,\theta^+) + u(r,\theta^-)]$ at every value of θ. Here the notations $u(r,\theta^+)$ and $u(r,\theta^-)$ are defined by $u(r,\theta^\pm) = \lim_{\varepsilon\to 0} u(r,\theta\pm\varepsilon)$ where ε is an arbitrary positive number.

Equations (4.1) and (4.2) may now be rewritten as

$$\frac{\partial v}{\partial r} = \frac{1}{\alpha}\left[\frac{d}{dr}(rw' + 2w)\right], \qquad \frac{\partial v}{\partial \theta} = \gamma_1 , \qquad (4.12)$$

where

$$\alpha = \frac{a_{1212} - a_{1221}}{a_{1212}} > 0 \quad \text{and} \quad \gamma_1 = \frac{2A_{1212}}{a_{1212} - a_{1221}} . \qquad (4.13)$$

In addition, we use the expression for u in equation (3.18) and differentiate both sides of the resulting equation with respect to θ to obtain:

$$B_{3131}\nabla^2\left(\frac{\partial\phi}{\partial\theta}\right) - 2(A_{1212}+A_{1221})\frac{\partial\phi}{\partial\theta} = 2(A_{1212}+A_{1221}) . \qquad (4.14)$$

It is clear from equation (4.14) that we must consider three different cases depending on which of the coefficients in (4.14) are taken to vanish. Thus we must consider:

Case i. $B_{3131} = 0$ and $A_{1212} \neq -A_{1221}$ $\qquad (4.15)$

Case ii. $B_{3131} \neq 0$ and $A_{1212} = -A_{1221}$ (4.16)

Case iii. $B_{3131} \neq 0$ and $A_{1212} \neq -A_{1221}$ (4.17)

We now seek the expression for the angular velocity $w = w(r)$ for each of the three cases (4.15)-(4.17). For case (i) equation (4.14) reduces to

$$\partial\phi/\partial\theta = -1$$ (4.18)

and we immediately obtain

$$\nu = \frac{D\phi}{Dt} = w\,\frac{\partial\phi}{\partial\theta} = -w \ .$$ (4.19)

Since $w = w(r)$ equations $(4.12)_2$ and $(4.13)_2$ require that $A_{1212} = 0$.

The condition $A_{1212} = 0$ together with the thermodynamic restriction [7], namely, $A_{1212} \geq |A_{1221}|$ requires that

$$A_{1212} = A_{1221} = 0 \ .$$ (4.20)

In view of (4.20) we must conclude that case (i) cannot occur for smectic liquid crystals in this type of viscometric flow.

In order to solve for w in case (ii) we must first solve

$$\nabla^2 \left(\frac{\partial\phi}{\partial\theta}\right) = 0$$ (4.21)

for $\partial\phi/\partial\theta$. Equation (4.21) has as solution

$$\frac{\partial\phi}{\partial\theta} = C_0 + C_1\, \ell nr + \theta(C_3 + C_4\, \ell nr)$$ (4.22)

where C_0, C_1, C_3 and C_4 are constants to be determined.

We now employ the relation $\nu = w\partial\phi/\partial\theta$ and equation $(4.12)_2$ to obtain the following expression for w:

$$w = \gamma_1/[C_3 + C_4\, \ell n\ r] \ .$$ (4.23)

The constants C_3 and C_4 in (4.23) may now be determined by use of the boundary conditions on w at the inner and outer cylinder, respectively.   Thus we have

$$w(r) = \Omega_2 \ln R / [\overline{\Omega}\ln R + (1 - \overline{\Omega}) \ln \overline{r}] \qquad (4.24)$$

where

$$R = \frac{r_2}{r_1} , \quad \overline{r} = \frac{r}{r_1} , \quad \overline{\Omega} = \frac{\Omega_2}{\Omega_1} \qquad (4.25)$$

Now equation (4.24) is immediately suspect since we note that if (4.24) holds then the only way for w to vanish is for $w(r_2) = \Omega_2 = 0$ and then $w = 0$ for all $r \in [r_1, r_2]$; a situation which is not borne out experimentally.   Likewise we expect that when the two cylinders are rotating in opposite senses that there be an $r_0 \in (r_1, r_2)$ such that $w(r_0) = 0$.   Our misgivings about equation (4.24) are justified mathematically by substituting for w from the latter and checking the expression for w in equation $(4.12)_1$ at the endpoints $r = r_1$ and $r = r_2$.   We obtain

$$C_0 + C_1 \left(\frac{\ln r_1 - \overline{\Omega}\ln r_2}{1 - \overline{\Omega}}\right) = \frac{2}{\alpha} \left(1 - \frac{1 - \overline{\Omega}}{\overline{\Omega}\ln R}\right) \qquad (4.26)$$

and

$$C_0 + C_1 \frac{\ln r_1 - \overline{\Omega}\ln r_2}{1 - \overline{\Omega}} = \frac{2}{\alpha} \left(1 - \frac{1 - \overline{\Omega}}{\ln R}\right) \qquad (4.27)$$

which when subtracted yield

$$\overline{\Omega} = 1, \quad \text{that is} \quad \Omega_1 = \Omega_2 . \qquad (4.28)$$

But we have required that $\Omega_1 \neq \Omega_2$ since for $\Omega_1 = \Omega_2$ we would have simply a rigid rotation.   Thus we conclude that $C_3 = C_4 = 0$ and $A_{1212} = 0$, in view of $\nu = w\partial\phi/\partial\theta$ as well as (4.22) and $(4.12)_2$ which leads to $A_{1221} = 0$ since $A_{1212} = -A_{1221}$.   The expression for $\partial\phi/\partial\theta$ is now given by

$$\frac{\partial\phi}{\partial\theta} = a + b \ln \overline{r} \qquad (4.29)$$

where we have defined

$$a = \frac{\partial\phi}{\partial\theta}(r_1); \quad b = \frac{\hat{b} - a}{\ln R} ; \quad \hat{b} = \frac{\partial\phi}{\partial\theta}(r_2), \quad \overline{r} = \frac{r}{r_1} . \qquad (4.30)$$

We note that the vanishing of C_3 and C_4 means that $\partial\phi/\partial\theta$ and hence ν is a function of r only. Using this fact, equation $(4.12)_1$ is integrated with respect to r on both sides to obtain

$$w(a + b \ln \bar{r}) = \frac{1}{\alpha} (rw' + 2w) + C \qquad (4.31)$$

where C is the constant of integration. Now equation (4.31) has the solution

$$w(r) = r^\lambda \{\Delta_0 \, \text{erf}(\eta) + \Delta_1\} \qquad (4.32)$$

where

$$\lambda = \lambda(r) = A+B \ln r, \quad \lambda_1 = \lambda(r_1), \quad \lambda_2 = \lambda(r_2),$$

$$A = -2-\alpha(a-b \ln r_1), \quad B = -\frac{1}{2}\alpha b, \quad \eta = \eta(r) = B^{\frac{1}{2}}\ln r + \frac{1}{2}AB^{-\frac{1}{2}},$$

$$\eta_1 = \eta(r_1) = -(2+\alpha a)[-\ln r/2\alpha(b-a)]^{\frac{1}{2}},$$

$$\eta_2 = \eta(r_2) = -(2+\alpha b)[-\ln R/2\alpha(b-a)]^{\frac{1}{2}},$$

$$\text{erf}(\eta) = \sqrt{\frac{2}{\pi}} \int_0^{\sqrt{2}\eta} \exp(-x^2/2)dx, \quad E = \text{erf}(\eta_2)-\text{erf}(\eta_1),$$

$$\Delta_0 = \frac{1}{E}[\Omega_2 r_2^{-\lambda_2} - \Omega_1 r_1^{-\lambda_1}] = \frac{1}{E}\Omega_2 r_1^{-\lambda_1} [R^s - \Omega_0],$$

$$\Delta_1 = \frac{1}{E} [\Omega_1 r_1^{-\lambda_1} \text{erf}(\eta_2) - \Omega_2 r_2^{-\lambda_2} \text{erf}(\eta_1)]$$

$$= \frac{1}{E} \Omega_2 r_1^{\lambda_1} [\Omega_0 \text{erf}(\eta_2) - R^s \text{erf}(\eta_1)],$$

$$\Omega_0 = \Omega_2/\Omega_1, \quad s = 2 + \alpha(\frac{a+b}{2}). \qquad (4.33)$$

We remark here that the sixth equation in (4.33) containing $B^{\frac{1}{2}}$ requires $B > 0$. Since $B = -\frac{1}{2}\alpha b$ and $\alpha > 0$ we must have $b < 0$ which, in view of (4.29) requires that

$$\hat{b} = \frac{\partial \phi}{\partial \theta}(r_2) < \frac{\partial \phi}{\partial \theta}(r_1) = a . \qquad (4.34)$$

If it should happen that $b = 0$, i.e. $\hat{b} = a$ then we have from (4.29) $\frac{\partial \phi}{\partial \theta} = a$, and so equation (4.31) would have the solution

$$w(r) = \frac{\Omega_1 r_1^k - \Omega_2 r_2^k}{r_1^k - r_2^k} + \frac{r_1^k r_2^k (\Omega_2 - \Omega_1)}{r_1^k - r_2^k} \cdot \frac{1}{r^k} , \qquad (4.35)$$

where $k = 2 - \alpha a$.

Note that if $a = b = 0$ then $k = 2$ and we have the usual classical solution for the angular velocity w due to Couette flow.

We now consider case (iii) where we have $B_{3131} \neq 0$, $A_{1212} \neq -A_{1221}$ so we must solve

$$\nabla^2 \left(\frac{\partial \phi}{\partial \theta}\right) - p^2 \left(\frac{\partial \phi}{\partial \theta}\right) - p^2 = 0 \qquad (4.36)$$

where we have defined

$$p^2 = \frac{2(A_{1212} + A_{1221})}{B_{3131}} . \qquad (4.37)$$

This equation has solution

$$\frac{\partial \phi}{\partial \theta} = \bar{a} I_0(pr) + \bar{b} K_0(pr) - 1 + \theta\{a_1 I_0(pr) + b_1 K_0(pr)\} \quad (4.38)$$

where I_0 and K_0 are zero-order modified Bessel functions of the first and second kind, respectively, and $\bar{a}$, $\bar{b}$, a_1, b_1 are constants to be determined by the boundary conditions. If the same procedure applied in case (ii) is employed we find that the last term on the right-hand side of (4.38) must vanish, i.e. $a_1 = b_1 = 0$ and we also must conclude that $A_{1212} = 0$. Once again the thermodynamic restriction [7] leads to $A_{1221} = 0$ and hence case (iii) does not occur for smectic liquid crystals in this viscometric flow.

The temperature field $T(r)$ is obtained by solving equation (3.19) subject to the boundary conditions

$$T(r_1) = T_1 \quad \text{and} \quad T(r_2) = T_2 . \qquad (4.39)$$

Thus we obtain for the temperature field

$$\overline{T} = \frac{T(r)}{T_1} = \frac{\ln R + (\kappa-1)\ln \overline{r}}{\ln R}$$

where

$$R = r_2/r_1, \qquad \kappa = T_2/T_1 . \qquad (4.40)$$

Heat-conduction can be obtained from the equations (3.11) by using the appropriate expressions for $v(r,\theta)$ and $T(r)$. The components of the heat-flux vector are

$$q_r = \frac{1}{2} d_0 \frac{\partial v}{r\partial\theta} + \frac{1}{4} T^{-1} \delta_0 \frac{\partial T}{\partial r} , \qquad q_\theta = -\frac{1}{2} d_0 \frac{\partial v}{\partial r} \qquad (4.41)$$

where

$$d_0 = d_{2332} - d_{3232}, \quad \delta_0 = \delta_{3232} + \delta_{2323} - \delta_{3223} - \delta_{2332} . \qquad (4.42)$$

Thus, the solution set for the governing field equation can be expressed in non-dimensional form where $\Omega_1 \neq 0$, $T_1 \neq 0$, $R \neq 1$ and $a > 0$:

Angular Velocity Field

For $a \neq \hat{b}$:

$$\overline{w} = \frac{w(r)}{\Omega_1} = \overline{r} \frac{-[2+\alpha(a+\frac{\hat{b}-a}{2\ln R} \ln \overline{r})]}{E} \{\overline{\Delta}_0 \text{erf}(\eta) + \overline{\Delta}_1\}, \qquad (4.43)$$

where

$$\overline{\Delta}_0 = \overline{\Omega} R^{[2+\frac{\alpha}{2}(a+\hat{b})]} -1 , \quad \overline{\Delta}_1 = \text{erf}(\eta_2)-\overline{\Omega}\text{erf}(\eta_1)R^{[2+\frac{\alpha}{2}(a+\hat{b})]}$$

$$(4.44)$$

For $a = \hat{b}$:

$$\bar{w} = w(r)/\Omega_1 = (1-\bar{\Omega}R^k)/(1-R^k) + [(\bar{\Omega}-1)R^k]/\bar{r}^{-k}(1-R^k), \quad (4.45)$$

where, as before $k = 2 - \alpha a$.

Microgyration Field

For $a \neq \hat{b}$:

$$\bar{\nu} = \nu/\Omega_1 = \frac{1}{E}(a+b \ln \bar{r})[\bar{\Delta}_0 erf(\eta) + \bar{\Delta}_1]\bar{r}^{\hat{s}} , \qquad (4.46)$$

where $b = (\hat{b}-a)/\ln R$, $\hat{s} = -[2+\alpha(a+\frac{b}{2}\ln \bar{r})]$.

For $a = \hat{b}$:

$$\bar{\nu} = \nu/\Omega_1 = a[1-\bar{\Omega}R^k + \frac{(\bar{\Omega}-1)R^k}{\bar{r}^{-k}}]/(1-R^k) . \qquad (4.47)$$

Macrodisplacement Field

$$\bar{u} = u(r,\theta)/r_1 = \bar{r}\theta , \qquad \theta \in [-\pi,\pi) . \qquad (4.48)$$

Temperature Field

$$\bar{T} = T(r)/T_1 = [(\kappa-1)\ln \bar{r} + \ln R]/\ln R . \qquad (4.49)$$

Heat-conduction

$$\bar{q}_r = q_r/q_1 = [\delta T^{-1}(\kappa-1)]/4\bar{r} \ln R) \qquad (4.50)$$

where $q_1 = 1/\delta_0 r_1$.

For $a \neq \hat{b}$:

$$\bar{q}_\theta = q_\theta/q_0 = \frac{\bar{w}}{\bar{r}} [b-2\bar{s}-\alpha(\bar{s})^2] + \frac{\bar{s} \bar{\Delta}_0}{\bar{r} E}\sqrt{\frac{-2\alpha b}{\pi}} R^{s_1} \qquad (4.51)$$

where $q_0 = -\Omega_1 d_0/2r_1$, $\bar{s} = a+b \ln \bar{r}$, $s_1 = \frac{2[1+\alpha a+(\alpha a)^2]}{\alpha(\hat{b}-a)}$.

For $a = \hat{b}$:

$$\bar{q}_\theta = q_\theta/q_0 = [ak R^k(\bar{\Omega}-1)]/\bar{r}^{k+1}(1-R^k) . \qquad (4.52)$$

5. ORIENTATION FIELD

In section 4 we have found an expression for $\partial\phi/\partial\theta$ with, as yet, undetermined boundary conditions at the inner and outer cylinder walls. We can expect a variety of boundary conditions here which are physically meaningful since smectic liquid crystals orient themselves at the boundary in a way that depends on the prior treatment of the boundary material. Hence, we have left the expressions for boundary conditions flexible, preferring to denote

$$\frac{\partial\phi}{\partial\theta}(r_1) = a \quad \text{and} \quad \frac{\partial\phi}{\partial\theta}(r_2) = \hat{b} \, . \tag{5.1}$$

We now proceed to develop an analytical expression for the orientation field $\phi(r,\theta)$.

Integration of equation (4.29) with respect to θ yields

$$\phi(\bar{r},\theta) = \theta[a + b \ln \bar{r}] + H(\bar{r}) \tag{5.2}$$

where $H(\bar{r})$ is an arbitrary function of integration and $\bar{r} = r/r_1$. In order to determine $H(\bar{r})$ we shall use the expression for the torque on the cylinders. The torque per unit length about the z-axis on the cylinder of radius r is denoted by τ_0 and besides the usual contribution to the torque from the shear stress $t_{r\theta}$ we will also have a contribution from the couple stress, namely m_{rz}.

Thus we have

$$\tau_0 = \int_0^{2\pi} (r^2 t_{r\theta} + rm_{rz})d\theta \, , \tag{5.3}$$

where

$$t_{r\theta} = a_{1212} \, r \, w' + 2 \, k_3 \, (w-\nu) \, ,$$
$$m_{rz} = B_{3131} \frac{\partial\phi}{\partial r} + b_{3131} \frac{\partial\nu}{\partial r} \, . \tag{5.4}$$

Upon carrying out the θ-integration in (5.3), using the known expressions for w, ν, and ϕ, and making use of the law of balance of torques which requires that the torque remain constant for any cylinder of radius r, $r_1 \leq r \leq r_2$, we obtain the following differential equation in the dimensional form for the unknown function H(r):

$$H'(r) = \frac{1}{r} [\frac{\tau_0}{2\pi \, B_{3131}} -2 \, b\pi] - \frac{b_{3131}}{B_{3131}} [\frac{1}{\alpha}(rw'' + 3w')] +$$

$$+ (a_{1212} + a_{1221}) \, rw - \frac{2a_{1212}\Delta_0(\sqrt{B/\pi})r}{B_{3131}\exp(A^2/4B)}, \qquad (5.5)$$

$$H'(r) = \frac{1}{r} [\frac{\tau_0}{2\pi B_{3131}}] - \frac{b_{3131}}{B_{3131}} [\frac{1}{\alpha}(rw'' + 3w')] + ra_{1221} \frac{\Omega_2 r_2^k - \Omega_1 r_1^k}{r_2^k - r_1^k}$$

$$+ \frac{1}{r^{k-1}} [(a_{1212} + a_{1221}) \frac{(\Omega_1-\Omega_2)r_1^k r_2^k}{r_2^k - r_1^k}], \qquad (5.6)$$

where the equation in (5.5) is for $a \neq \hat{b}$ and the equation in (5.6) is for $a = \hat{b}$. Integration of (5.5) and (5.6) with respect to r yields

$$H(r) = (\ln r)[\frac{\tau_0 - 4b\pi^2 B_{3131}}{2\pi B_{3131}}] - \frac{b_{3131}}{B_{3131}}[\frac{1}{\alpha}(rw'+2w)] - r^2\{\exp(-A^2/4B)$$

$$\times a_{1212} \, \Delta_0 \, \sqrt{B/\pi}\} + (a_{1212}+a_{1221}) \int^r tw(t)dt + C_A . \qquad (5.7)$$

$$H(r) = (\ln r)[\frac{\tau_0}{2\pi B_{3131}}] - \frac{b_{3131}}{B_{3131}} [\frac{1}{\alpha}(rw' + 2w)] +\{\frac{a_{1221}r^2}{2}$$

$$\times \frac{\Omega_2 r_2^k - \Omega_1 r_1^k}{r_2^k - r_1^k}\} - \frac{(a_{1212}+a_{1221})}{\alpha a r^{\alpha a}} \cdot \frac{(\Omega_1-\Omega_2)r_1^k r_2^k}{r_2^k - r_1^k} + C_B ,$$

$$(5.8)$$

where C_A and C_B are arbitrary constants of integration for the cases $a \neq b$ and $a = b$, respectively. However, the presence of C_A and C_B in the expressions for the orientation field derived from putting (5.7) and (5.8) into (5.2) does not affect the behavior of the orientation and its dependence on the spatial coordinates r and θ, the radius ratio R and the angular velocity ratio of the cylinders and neither does C_A or C_B appear in any of the solutions given by equations (4.43) to (4.52). In view of these facts we may as well omit the constants from the expressions for the orientation field for the two cases. The torque τ_0 appearing in (5.7) and (5.8) can be physically measured in a given experiment and hence is considered to be known. The classical expression for τ_0 in the case of Couette flow between two concentric rotating cylinders is given by

$$\tau_0 = \frac{4\pi r_1^2 (\Omega_2 - \Omega_1)\mu}{1 - R^2} \qquad (5.9)$$

where μ is the viscosity of the Newtonian fluid between the cylinders. Finally we remark that the integral in (5.7) may be integrated numerically for a given experiment and hence the expressions for the orientation fields are completely known for each case.

6.   APPARENT VISCOSITY

In order to gain some idea of the non-Newtonian behavior of the smectic liquid crystal we define an apparent viscosity. The apparent viscosity is defined such that it is equal to the constant coefficient of viscosity for the flow of an incompressible Newtonian fluid. Therefore, we choose:

$$\eta_{app} = [\frac{\text{Shear stress}}{\text{Shear-rate on the outer cylinder}}]_{\text{at } \theta = 0, \ r = r_2} \qquad (6.1)$$

which becomes, on using the known solutions for w and v, and also the expressions for shear stress and shear-rate:

$$\eta_{app} = a_{1212} + \frac{1}{\gamma} [2k_3\Omega_2(1-\hat{b})] , \qquad (6.2)$$

where

$$\gamma = \frac{2\Delta_0\sqrt{B/\pi}}{e^{A^2/4B}} - \Omega_2(2+\alpha b), \qquad a \neq \hat{b} , \qquad (6.3)$$

$$\gamma = \frac{(2+\alpha a)(\Omega_2-\Omega_1)}{R^k - 1} , \qquad a = \hat{b} , \qquad (6.4)$$

$$2k_3 = a_{1212} - a_{1221} , \tag{6.5}$$

where according to [7], a_{1212} and a_{1221} are functions of temperature only. We recall that a and $\hat{b}$ in equations (6.3) and (6.4) involve the boundary conditions on the orientation field at the cylinder walls.

Determination of the Material Coefficients

We shall now proceed to demonstrate a method by which the material coefficients a_{1212} and a_{1221} which are functions of temperature only [7], can be calculated for a given smectic substance. In order to do this we utilize the experimental work of Tamamushi and Matsumoto [13], wherein the authors investigated the non-Newtonian behavior of the viscosity (as a function of shear-rate and temperature) of ammonium laurate which exhibits the smectic mesophase in the temperature range from 106° to 112°C. The apparatus used for determination of the viscosity at given constant temperature was a viscometer of Couette type which worked in the range of shear-rate up to 5000 dyn. cm^{-2}. The actual shear-rates used in the experiments ranged from 41 sec^{-1} to 325 sec^{-1} while the corresponding shear force ranged from 0 to 1500 dyn. cm^{-2}.

To determine a_{1212} we note that according to equation (6.2) the main contribution to η_{app} at high shear-rates comes from the first term on the right-hand side. Indeed, the experimental results indicate a leveling off of the viscosity values for shear-rates above 325 sec^{-1} and an examination of ratios of apparent viscosities at successive experimental shear-rates leads us to make the following estimate for a_{1212}:

$$a_{1212} = 0.9 \, [\eta_{app} \quad \text{for} \quad 325 \, sec^{-1}] . \tag{6.6}$$

The determination of a_{1221} proceeds in the following manner. Since only the shear-rate γ and η_{app} are given in Tamamushi and Matsumoto's experiments we make the following choices for Ω_2 and $1 - \hat{b}$ based on experimental evidence and the thermodynamic restrictions on a_{1212} and a_{1221} obtained in [7]. Peter and Peters [10] have conducted Couette viscometer experiments with nematic liquid crystals with an outer cylinder radius of 4.00 cm and an inner cylinder radius of 3.96 cm. The outer cylinder was given an angular velocity of 11.84 rad./sec. in their experiments. Thus a reasonable choice for Ω_2 in (6.2) would be 19 rad./sec.

Our choice of a value for $1 - \hat{b}$ must be guided by the thermodynamic restrictions on a_{1212} and a_{1221} since little experimental data concerning the boundary conditions on the orientation field are available. The thermodynamic restrictions on a_{1212} and a_{1221} are

$$a_{1212} \geq 0 \ , \quad a_{1212} - a_{1221} > 0 \ . \tag{6.7}$$

We rewrite equation (6.2) using our value chosen above for Ω_2 and obtain

$$\frac{1}{19} \ (\eta_{app} - a_{1212}) = \frac{1 - \hat{b}}{\gamma} \ (a_{1212} - a_{1221}) \ . \tag{6.8}$$

Now the left-hand side of equation (6.8) is positive and attains its largest values for low shear-rates. Using the values for a_{1212} and η_{app} at $\gamma = 41 \ \text{sec}^{-1}$, and the restrictions (6.7) we obtain the following value for $1 - \hat{b}$ which satisfies equation (6.8) as well as the restrictions in (6.7):

$$1 - \hat{b} = 10 \ . \tag{6.9}$$

The value for $1 - \hat{b}$ reflects the change in the orientation field at the outer cylinder wall and can be maintained by the use of external means such as the imposition of a magnetic field of suitable strength or a thermal gradient.

Now the experimental data for $\gamma = 41 \ \text{sec}^{-1}$ and the values for a_{1212}, Ω_2 and $1 - \hat{b}$ as given above, can be used in the following equation obtained from (6.2) wherein $(1-\hat{b})\Omega_2/\gamma \approx 4.63$,

$$\eta_{app} (\text{at } 41 \ \text{sec}^{-1}) = a_{1212} + 4.63(a_{1212} - a_{1221}) \ , \tag{6.10}$$

to determine a_{1221} for various temperatures in the smectic range. As mentioned before, a_{1212} and a_{1221} are functions of temperature only [7].

Thus we obtain the following table of values for a_{1212} and a_{1221} .

Table 1.   Values for a_{1212} and a_{1221}.

	106°	108°	110°	112°	114°	116°	°C
a_{1212}	4.68	5.22	4.59	2.25	.27	.18	Poise
a_{1221}	2.15	2.98	1.06	.45	.07	.01	Poise

These values for a_{1212} and a_{1221} given in the above table can now be used in equation (6.8) in order to obtain our analytical formula for η_{app} in the case of the smectic compound, ammonium laurate.   The results of our analytical computation of η_{app} versus the shear-rate for different smectic temperatures are shown in Figure 1.   The solid line denotes our theoretical results

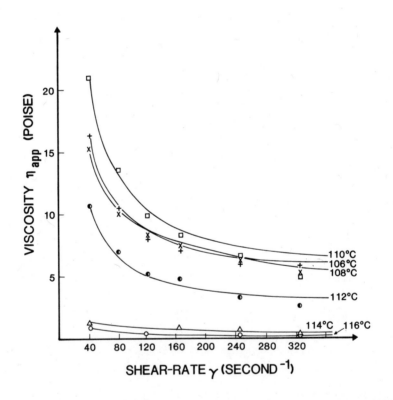

Fig. 1.   Behavior of apparent viscosity with shear-rate.   Solid
         line denotes our theoretical results; the symbols,
         + 106°C; × 108°C; □ 110°C; ● 112°C; Δ 114°C; O 116°C,
         denote experimental results of Tamamushi and Matsumoto.

and the symbols denote experimental results obtained from the
graphs of Tamamushi and Matsumoto [13]. It is clear from the
graph that our theoretical results are found to be in close agreement with the above experimental results.

7. DISCUSSION

The main results obtained in this paper consist of explicit
analytical derivation of expressions for the velocity, pressure,
temperature, heat-conduction, microgyration velocity, and orientation fields for the flow of a smectic mesophase in a rotational
viscometer. The expression for the apparent viscosity coefficient
(6.2) is found to be a function of the shear-rate, gap-width of
the concentric annulus and consists of some material coefficients
that are functions of temperature only and governed by appropriate
thermodynamic restrictions obtained by us in a companion paper
[7]. In the present study, we have demonstrated the method of determining the above referred material moduli by utilizing the experimental data of Tamamushi and Matsumoto [13] for ammonium
laurate which exhibits smectic mesophase in the temperature range
106°C to 112°C. The procedure we have followed is analogous to
the one used in [5] and is as follows. By using our theoretical
formula for apparent viscosity, equation (6.2), and matching it
with the experimental data [13] for just one value of the shear-
rate, say, $41 \sec^{-1}$., the values of the material coefficients
a_{1212} and a_{1221} (which are purely temperature dependent functions according to our constitutive theory developed in [7]), have
been obtained for various smectic temperatures for ammonium laurate.
Thus, in the expression,

$$\eta = \eta(\gamma, R, a_{1212}, a_{1221}) , \qquad (7.1)$$

the explicit form of which is given by equation (6.2), a_{1212} and
a_{1221} are determined once and for all for the ammonium laurate,
for its various smectic temperatures. This leads to the complete
analytical characterization of the apparent viscosity coefficient
of ammonium laurate (smectic) as a function of shear-rate and gap-
width as is clear from (6.2) and (7.1). This resulting formula
for the apparent viscosity is then used to predict its behavior
for a wide range of shear-rate values for the rotational viscometric flow of ammonium laurate. We have recorded our analytical
findings, in Figure 1, which are depicted by the solid line. For
purposes of comparison of our theoretical results with those of
the experiments, we have shown in the same figure, the experimental values of η for ammonium laurate computed from the graph
of Tamamushi and Matsumoto [13]. As Figure 1 shows, there is a

gratifying agreement between our theoretical results and the
experimental results of [13], thus building confidence in our theoretical analysis based on the microcontinuum theory of Eringen [8].
We also mention in passing that our present study is similar to
the previous investigations of Narasimhan and Eringen [5] in the
case of heat-conducting nematics (p-azoxyanisole), in which the
theoretical mesophase rheology results were found to be in remarkable agreement with those of the experiments of Porter and
Johnson [9] and others. Thus we are encouraged to feel confident
that this work leads to further development and applications of
our analytical study in close verification and agreement with
experimental results involving the smectic mesophase.

REFERENCES

1. J. L. Fergason, Sci. Am. 211, No. 2, 77 (1964).
2. R. A. Champa,"Liquid Cryst. and Ordered Fluids", 2, 507 (1974),
    ed., J. F. Johnson and R. S. Porter, Plenum Press, N.Y.
3. A. Blumstein, et al, "Liquid Cryst. and Ordered Fluids", 2,
    277 (1974), ed., J. F. Johnson and R. S. Porter, Plenum Press,
    N.Y.
4. M. N. L. Narasimhan and A. C. Eringen, Int. J. Eng. Sci., 13,
    233 (1975).
5. M. N. L. Narasimhan and A. C. Eringen, Mol. Cryst. Liquid
    Cryst., 29, 57 (1974).
6. M. N. L. Narasimhan, "Continuum Theory of Heat-Conducting
    Smectic Liquid Crystals", in Proc. Twelfth Annual meeting,
    Soc. Eng. Sci., ed. Morris Stern, 61 (1975).
7. M. N. L. Narasimhan and T. E. Kelley, "Liquid Cryst. and
    Ordered Fluids", 3 (1977), ed., J. F. Johnson and R. S. Porter,
    Plenum Press, N.Y.
8. A. C. Eringen, Proc. Eleventh International Cong. Appl. Mech.,
    ed., H. Görtler (Springer, Berlin), 131 (1966).
9. R. S. Porter and J. F. Johnson, "Orientation of Nematic Meso-
    phases", J. Phys. Chem. 66, 1826 (1962).

10. S. Peter and H. Peters, Zeit. für Physik. Chemie., 3, S103
    (1955).
11. G. Becherer and W. Kast, Ann. Phys. 41, 355 (1942).
12. M. Miesowicz, Nature, 158, 27 (1946).
13. B. Tamamushi and M. Matsumoto, "Liquid Cryst. and Ordered
    Fluids", 2, 711 (1974), ed., J. F. Johnson and R. S. Porter,
    Plenum Press, N.Y.

FLUCTUATIONS AND ORDERING IN NEMATIC LIQUID CRYSTALS

Roger Chang

Science Center, Rockwell International

Thousand Oaks, California 91360

ABSTRACT

The paper discusses the odd-even fluctuation of the various thermodynamic properties of two homologous series of nematic liquid crystals near the nematic/isotropic transition temperature. A new alignment phenomenon not previously discussed in the literature (associated with horizontally aligned films between glass substrates) is reported. The degree of alignment near the nematic/isotropic transition temperature increases, while the order parameter decreases, reversibly with increasing temperature. Development of a rigorous theory of anisotropic intermolecular interactions to interpret quantitatively these experimental observations is suggested.

Although fluctuations and ordering in nematic liquid crystals have been frequently discussed in the literature, basic understanding of these phenomena is still not very satisfactory. This is partly due to the fact that good quantitative experimental data are hard to come by. A liquid crystal sample, presumably in thin film form, must be held within some physical constraints during experimental measurements and is unavoidably subjected to boundary effects. A convenient way to simplify the boundary effects, aside from the use of magnetic or electric fields, is to rub or treat the confinement surfaces so the liquid crystal molecules are unidirectionally aligned either perpendicular or parallel to the confinement substrate. We assume in the following discussions that the nematic liquid crystal molecules are unidirectionally aligned parallel to the confinement surfaces. Such a film has two principal physical parameters which can be measured experimentally. First is the nematic order parameter S characterisitc of the liquid crystal and independent of the nature of boundary constraints. Second is an

347

average alignment parameter Φ which is zero for randomly oriented
and unity for perfectly aligned liquid crystal directors averaged
over all the molecules in the film plane. Birefringence and light
scattering measurements can be made to yield the two parameters
quantitatively as functions of temperature. Figure 1A is a typical
Cary-14 spectrophotometer trace where, at a given wave length λ,
the order parameter S is related directly to $\Delta\lambda$ and the alignment
parameter Φ to ΔI. Calculations of the order parameter S from $\Delta\lambda$
as functions of the wave length of light have been discussed by
the author in previous publications.[1,2] The relationship between
the alignment parameter Φ and ΔI can be visualized in the following
way. When a uniaxial film normal to the Z-axis of a Cartesian
coordinate system is placed between a polarizer and an analyzer
such that the extraordinary refractive index n_y in the Y-direction
and the ordinary refractive index n_x in the X-direction are fixed,
the transmitted intensities and phase shifts will depend on the
orientations of the polarizer and analyzer with respect to the film
in the XY plane. Following Born and Wolf,[3] the intensity of light
transmitted by the analyzer is,

$$I = I_o[\cos^2\psi - \sin 2\phi \; \sin 2(\phi-\psi)\sin^2(\delta/2)] \qquad (1)$$

$$\delta = [2\pi(n_y - n_x)d]/\lambda \qquad (2)$$

where I_o is the intensity of the incident beam, ϕ is the angle that
the polarizer makes with the X-axis and ψ is the angle between the
polarizer and analyzer. The modulation of the transmitted intensity
as functions of wave length λ depends on the phase shift δ which is
directly proportional to $(n_y - n_x)$ according to equation (2). For a
randomly oriented film (in the film plane), $n_y = n_x$, hence $(n_y - n_x) = 0$
and $\Delta I = 0$ (zero modulation). For a perfectly aligned film $(n_y - n_x) =$
$\Delta n = n_{||} - n_{\perp}$ (where $n_{||}$ and $n_{\perp}$ are, respectively, the refractive
indices parallel and perpendicular to the liquid crystal directors)
and ΔI is a maximum.

Only a few investigations of the birefringence of nematic liquid
crystals exist in the literature and even fewer studies cover homologous series of nematic liquid crystals.[4-7] The odd-even fluctuations of the thermodynamic properties of two homologous series of
nematic liquid crystals (the azoxybenzenes and benzylideneanilines)
as functions of end chain length are summarized in Tables I and II.
The following thermodynamic properties are listed: temperature of
nematic/isotropic transition (T_{NI}), enthalpy of nematic/isotropic
transition (ΔH), nematic order parameter near $T_{NI}(S_c)$, intermolecular
interaction energy (U) and normalized temperature coefficient of U
near T_{NI} ($\partial \ln U/\partial\varepsilon$, where $\varepsilon = T/T_{NI}$, T being the absolute temperature
of measurement). The U and $\partial \ln U/\partial\varepsilon$ parameters are calculated according to the mean field theory[8] as follows,

$$U = 4.55 \ k_B T_{NI} \qquad (3)$$

$$\frac{\partial \ln U}{\partial \varepsilon} = 1 - \frac{2\Delta H}{4.55 \ S_c^2 k_B T_{NI}} \qquad (4)$$

where k_B is the Boltzmann constant. The positive values of $\partial \ln U / \partial \varepsilon$ and odd-even fluctuation of U and other thermodynamic properties near T_{NI} suggest that more refined treatments of the anisotropic inter-molecular interactions than that proposed by Maier and Saupe[9] are needed. An approximate treatment by Marcelja[10] provides a quali-tative picture of the odd-even effect. Intuitively, for two nematic liquid crystals of the same homologous series having different T_{NI} and order parameter at some reduced temperature below T_{NI}, the con-dition giving rise to odd-even fluctuation near T_{NI} would be for one liquid crystal to have smaller U but larger $\partial \ln U / \partial \varepsilon$ and the other liquid crystal to have larger U but smaller $\partial \ln U / \partial \varepsilon$. The condition is observed experimentally for both homologous series of nematic liquid crystals shown in Table I and II.

The pronounced increase in the alignment parameter Φ (shown by the increase in ΔI) as the temperature of the liquid crystal approaches T_{NI} is amply demonstrated in Fig. 1 for MBBA and Fig. 2 for the biphenyl liquid crystal mixture (BDH E5). The absorption traces shown in Figs. 1 and 2 are reproducible with respect to both increases and decreases of the measuring temperature. The reversible behavior is not expected if the alignment of the liquid crystal molecules in contact with the substrate improves with increasing temperature. It is believed rather that the orientation of these liquid crystal molecules in immediate contact with the substrate is fixed and that the long range intermolecular interaction among the molecules increases as their temperature approaches T_{NI}. Some support of the hypothesis comes from the computer simulation work on a lattice model of nematic liquid crystal by Meirovitch[12] who found that the mean field param-eter expressing the average effect of the longer range intermolecular interaction of all the molecules increases as their temperature approaches T_{NI}.

It is likely that all the experimental observations can be rigorously interpreted theoretically according to some lattice model where both the short and long range intermolecular interactions are properly taken care of in future investigations.

Table I.   Fluctuation of Thermodynamic Properties
Near T_{NI} for 4,4'-di-n-alkoxyazoxybenzenes
$C_nH_{2n+1}O$-⟨O⟩-N=N-⟨O⟩-$OH_{2n+1}C_n$
$\downarrow$
O

n	T_{NI},°K	ΔH,cal/mol	S_c	U,cal/mol	$\partial \ell n U / \partial \varepsilon$
1	408.5	137	.43	3680	.602
2	440.7	327	.46	3970	.227
3	396.8	161	.36	3580	.315
4	409.9	247	.43	3700	.285
5	396.4	173	.37	3580	.302
6	402.3	250	.42	3640	.225

T_{NI} and ΔH data from "Liquid Crystals and Plastic Crystal" edited
by G. W. Gray and P. A. Winsor, Vol. 2, p. 261 (1974), Chapter X
by E. M. Barrall and J. F. Johnson.
S_c data from A. Pines, D. J. Rubin and S. Allison, Phys. Rev.
Letters 33, 1002 (1974).

Table II.   Fluctuation of Thermodynamic Properties Near
T_{NI} for 4,4'-n-alkoxybenzylidenebutylanilines
$C_nH_{2n+1}O$-⟨O⟩-CH=N-⟨O⟩-C_4H_9

n	T_{NI},°K	ΔH,cal/mol	S_c	U,cal/mol	$\partial \ell n U / \partial \varepsilon$
1	319.1	100	.300	2870	.235
2	352.1	140	.324	3190	.168
3	330.1	110	.297	2980	.170
4	349.2	140	.322	3150	.150

Data from R. Chang, F. B. Jones, Jr., and J. J. Ratto, Mol. Cryst.
and Liq. Cryst. 33, 13 (1976).

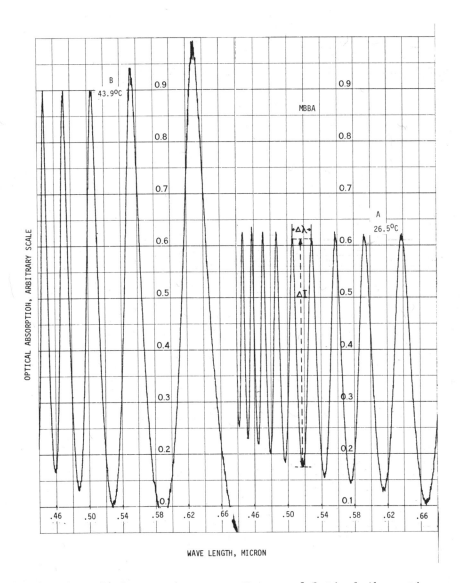

Fig. 1. Cary-14 Spectrophotometer Traces of Optical Absorption
Versus Wave Length, MBBA, Film Thickness ~ 30 Microns,
Horizontally Aligned, Polarizer Parallel to Analyzer,
A--26.5°C, B--43.9°C.

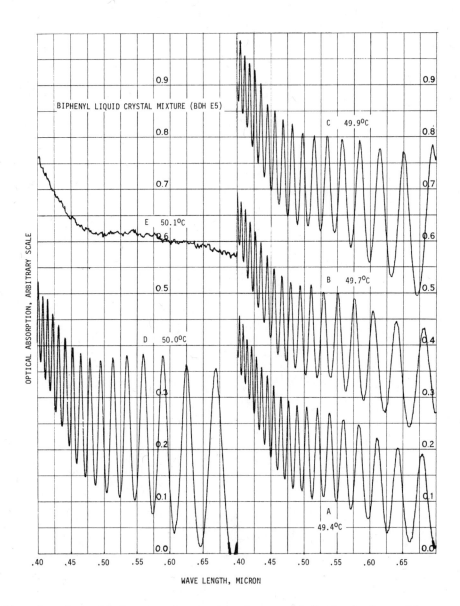

Fig. 2. Cary-14 Spectrophotometer Traces of Optical Absorption
         Versus Wave Length, Biphenyl Liquid Crystal Mixture
         (BDH E5, Atomergic Chemetals Co.), Film Thickness ~ 78
         Microns, Horizontally Aligned, Polarizer Parallel to
         Analyzer, A--49.4°C, B--49.7°C, C--49.9°C, D--50.0°C,
         E--50.1°C (just passes T_{NI}).

REFERENCES

(1) R. Chang, Mol. Cryst. and Liq. Cryst. 30, 155 (1975).

(2) R. Chang, F. B. Jones, Jr., and J. J. Ratto, Mol. Cryst. and Liq. Cryst. 33, 13 (1976).

(3) M. Born and E. Wolf, "Principles of Optics", MacMillan, New York, 1964, p. 965.

(4) G. Pelzl and H. Sackmann, Sym. Faraday Soc. No. 5, 68 (1971).

(5) See Reference (2).

(6) E. G. Hanson and Y. R. Shen, Mol. Cryst. and Liq. Cryst. 36, 143 (1976).

(7) G. R. van Hecke, B. D. Santarsiero, and L. J. Theodore, Mol. Cryst. and Liq. Cryst. (in press).

(8) P. G. de Gennes, "The Physics of Liquid Crystals", Clarendon Press, Oxford, 1974.

(9) W. Maier and A. Saupe, Z. Naturforsch., 13a, 564 (1958); 14a, 882 (1959); 15a, 287 (1960); 16a, 816 (1961).

(10) S. Marcelja, J. Chem. Phys. 60, 3599 (1974).

(11) A. Pines, D. J. Ruben, and S. Allison, Phys. Rev. Letters 33, 1002 (1974).

(12) H. Meirovitch, Chem. Physics 21, 251 (1977).

CORRELATION FUNCTION APPROACH TO THE STUDY OF RAMAN BAND SHAPES OF LIQUID CRYSTALS

Kenneth Brezinsky* and Bernard J. Bulkin**

Hunter College, CUNY, New York, N.Y. 10021, and

Polytechnic Institute of N.Y., Brooklyn,N.Y.11201

*CUNY; **PINY, address correspondence to
this author.

ABSTRACT

Correlation functions computed from vibrational
spectroscopic band shapes can yield information about
liquid crystals, but they must be applied with care.
This paper outlines the approach to be used, and shows
results for one case 4,cyano, 4'octyloxybiphenyl.

Introduction

For the past decade, it has been fruitful to apply
Gordon's formulation (1) of the connection between
infrared and Raman band shapes and correlation functions
to study rotational motion in liquids (2,3). These
approaches have proved much more powerful than the more
limited study of the half widths of the vibrational
bands (4) or even the direct study of the functional
forms of the band shape (5).

To apply such an approach to liquid crystals
requires considerable caution, however. The work of
Gordon, Rothschild and others has been done with the
type of orientational averaging present only in
isotropic phases. One must examine the possible con-

sequences of anisotropy for such a formulation. In
addition, most prior work confirming the Gordon
approach has been carried out on small molecules,
where rotational relaxation provides the most efficient
mechanism of energy dissipation on the time scale
involved (0-4ps). For large molecules (all mesogens
now known are large molecules) this may not be the
case. Rothschild (6) has already shown that for a few
large molecules vibrational relaxation dominates the
correlation function.

As rotational molecular motion is of great
interest to the theory of liquid crystals, it is not
surprising that there have already been some attempts
to study it using vibrational spectroscopy.

Evans (7) obtained a correlation function from the
far infrared spectrum of N- 4-methoxybenzylidene-4'-n-
butylaniline (MBBA) for use in his comparison of various
models of molecular motion; however, he did not draw
any detailed conclusions about the molecular motion of
MBBA. Lugomer (8) calculated a correlation function
from the infrared spectrum of bis-4,4', heptyloxyazoxy-
benzene and related the results to benzene ring rotations
of the molecules. He used a formulation that is strictly
applicable only to molecules in an isotropic environment.

Some work has been done to determine rotational
motion in liquid crystals with the more limited approach
of monitoring peak half widths. Kirov and Simova (9)
related the half width from infrared spectra of PAA to
the barrier to rotation around the long molecular axis
and around axes perpendicular to it. Fontana and
Bini (10) relate half widths of low frequency Raman peaks
of TBBA to rotational motion.

In this paper, we first reintroduce a formulation
due to Callender and Pershan (11) which we believe could
be applicable to liquid crystals if rotational relaxation
processes were important on the 0-4 psec time scale.
We then show that for one case the correlation function
on this time scale is dominated by vibrational relaxation.
Our experiments have been carried out on well aligned
materials, because this greatly facilitates the use of
the Callender and Pershan approach, originally formulated
for crystals.

Formalism

Correlation functions can be straightforwardly calculated from the Raman spectra of liquids (12). However, for aligned liquid crystals none of the averaging over all molecular orientations that is permissible with a liquid can be done. Therefore to use the correlation approach with liquid crystals a formulation of correlation functions that is not restricted to an isotropic environment, but rather explicitly considers the geometry of the Raman scattering experiment, is necessary. Such a formulation is provided by Callender and Pershan in their work on crystals (11). The normalized correlation function obtainable from a Raman spectrum and of use in our work is

$$\langle \beta_{yz}^{\nu}(t)\,\beta_{yz}^{\nu}(o)\rangle\langle Q^{\nu}(t)\,Q^{\nu}(o)\rangle = \frac{\displaystyle\int_{-\infty}^{\infty} I_{vh}(\omega)\exp\{+i(\omega+\omega_{\nu})t\}d\omega}{\displaystyle\int_{-\infty}^{\infty} I_{vh}(\omega)\,d\omega}$$

eq.1

where ω is the circular frequency shift of the scattered light and ω_{ν} is the circular frequency of the ν th vibrational mode. The β terms are the anisotropic elements of the polarizability tensor for the ν th mode; the Q's are the time dependent normal coordinates for the ν th mode in the Heisenberg representation. The VH subscript of the frequency dependent intensity ($I(\omega)$) refers to the scattering geometry shown in Fig. 1, in which the scattered radiation is collected along x only with polarization perpendicular to the polarization direction of the (linearly polarized) incident beam. The angle brackets represent an ensemble average. The correlation function containing the anisotropic tensor elements, $\langle \beta_{yz}^{\nu}(t)\,\beta_{yz}^{\nu}(0)\rangle$, indicates the time evolution of a tumbling system. The correlation function $\langle Q^{\nu}(t)\,Q^{\nu}(0)\rangle$ will indicate the types of vibrational relaxation processes.

For liquids a similar expression is obtained (9) for the normalized correlation function

$$\langle \mathrm{Tr}\overline{\beta}^{\nu}(t)\,\overline{\beta}^{\nu}(o)\rangle\langle\overline{Q}^{\nu}(t)\,\overline{Q}^{\nu}(o)\rangle = \frac{\displaystyle\int_{-\infty}^{\infty} I_{vh}(\omega)\exp\{+i\,(\omega+\omega_{\nu})\,t\}d\omega}{\displaystyle\int_{-\infty}^{\infty} I_{vh}(\omega)\,d\omega}$$

eq.2

where Tr denotes the trace of the β matrix elements. In this case, however, the vibrational correlation function can be written explicitly as a direct consequence of the orientational averaging that is possible for liquids:

$$\langle Q^{\nu}(t)Q^{\nu}(o)\rangle = \int_{-\infty}^{\infty} \hat{I}_{vib}(\omega)\exp\{+i\,(\omega+\omega_{\nu})\,t\}d\omega$$

eq.3

where

$$\hat{I}_{vib}(\omega) \equiv \frac{I_{vv}(\omega) - 4/3\,I_{vh}(\omega)}{\displaystyle\int_{-\infty}^{\infty} \{I_{vv}(\omega) - 4/3\,I_{vh}(\omega)\}d\omega}$$

Note that these equations also should apply to a mesogen in its isotropic phase.

Correlation functions, as the ones developed above, have been used in statistical mechanics (13) (14) and nuclear magnetic resonance spectroscopy (15) to examine fluctuating systems. For a property V(t) of the fluctuating system, correlation functions indicate the average degree to which the value of V(t) is related to the value of V(0). If the property V(t) oscillates around a mean value of zero because of the system

fluctuations, then for large enough values of t, the
positive and negative oscillations will cause the
correlation function <V(t)V(0)> to go to zero. Different fluctuating systems can be distinguished by the
rate of decay to zero. In addition, the form of the
correlation function curve as it decays will differ
in systems with different underlying causes of fluctuation. Note that all systems will have the maximum
correlation between V(t) and V(0) at t=0, i.e.

$$<V(0)V(0)> = \text{maximum value.}$$

Thus when normalized to the maximum value, the correlation function approaches unity as t approaches zero :

$$\frac{<V(t)V(0)>}{<V(0)V(0)>} \xrightarrow[t\to o]{} 1$$

This normalization is conventionally carried out for
vibrational band correlation functions.

In aligned liquid crystal systems, tumbling motion
will cause the molecular long axis to fluctuate around
the director. Such movement could affect the way $\beta(t)$
is related to $\beta(0)$ and $Q(t)$ is related to $Q(0)$. Through
an examination of the decay rate and the form of the
correlation functions involving these terms, an understanding of tumbling motion might be possible.

Results

To study the applicability of the formalism, we
have examined the CN stretching mode at 2225 cm^{-1} in
the Raman spectrum of 4 cyano 4' octyloxybiphenyl
in its smectic, nematic and isotropic phases. For
the liquid crystalline phases, both homeotropic and
two homogeneous alignments have been studied. In all
cases, the correlation functions have been computed
from the I_{vh} spectrum, as described above. The alignments (shown in Fig. 1) studied thus represent all
three possible orientations of the molecular long
axis with respect to the incident and scattered
polarizations.

To illustrate the spectroscopic results the 2225
cm^{-1} CN stretching bands (I_{vh}) for the homeotropically
aligned sample in the smectic (62°C), nematic (74°C),
and isotropic phases (96°C) are shown in Fig. 2. Each
of the peaks shown has a full width at half height of

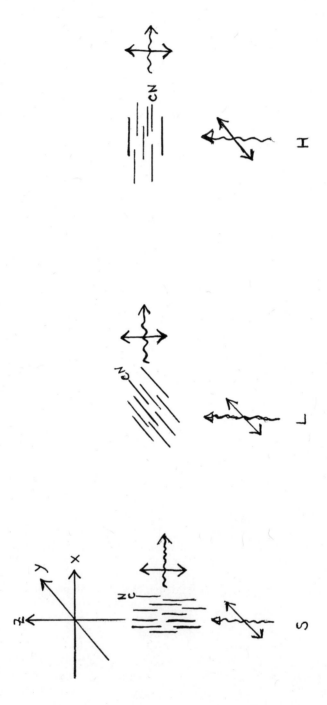

Figure 1: Scattering geometries for I_{VH} spectra. Light is incident along z, scattered light observed along x. The $\vec{E}$ vectors isolate the yz component. Three orientations of the director are shown.

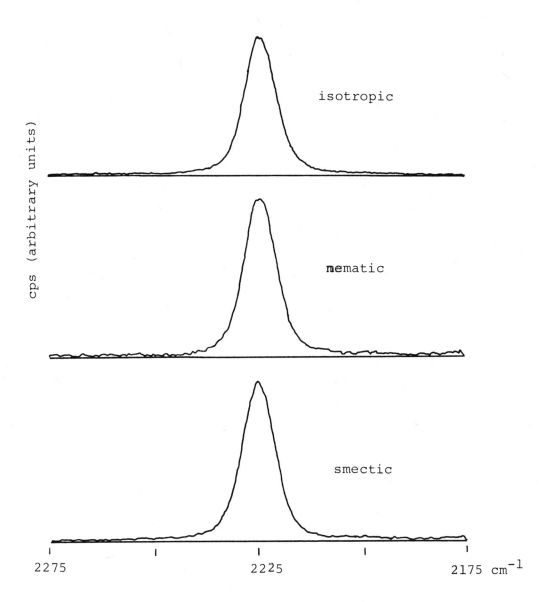

Figure 2: Cyano stretching peak of homeotropically
aligned smectic and nematic phase as well as isotropic
phase.

9 cm^{-1}. There seems to be no difference between these
results. This observation is borne out by the correlation functions calculated from these bands which are
shown in Fig. 3. The correlation function curves are
identical to within experimental error. Note that the
t=0 values are all unity as expected for normalized
correlation functions, and that the functions are all
decaying toward zero. There is no apparent difference
in the rate or form of decay. The small oscillations
in the curves arise from the computation.   On this
semilogarithmic scale the curves are not linear as
would be expected (3) if random, diffusional rotational
motion were dominating the correlation function; however,
the accumulation of errors, which become significant at
longer times, make it difficult to ascertain if the
correlation functions are starting to become linear at
longer times. Very similar results were obtained from
the Raman spectra of the homogeneously aligned samples.

The correlation functions obtained from the aligned
liquid crystal samples indicate that there are no large
differences in rotational tumbling motion between the
smectic, nematic and isotropic phases. To examine the
degree to which the nitrile stretch band shape is
affected by rotation, the vibrational correlation
function was calculated from the isotropic results
using equation 3. The results of this calculation and
the correlation function calculated from the isotropic$_1$
results are compared in Fig. 4. Clearly, the 2225 cm^{-1}
peak in this compound is dominated by vibrational
relaxation in the isotropic phase. This is presumably
the case in the liquid crystalline phases as well.

In an attempt to further determine the significance
of vibrational relaxation to the band shape, Raman
spectra of solutions of the liquid crystal in non-polar
solvents were taken. The correlation functions calculated from the benzene solution results are shown in
Fig. 5. These curves clearly show differences in the
rate of decay with increasing dilution. Note that in
our experimental arrangement, accuracy is improved in
solutions. The differences seen in these correlation
functions are significant. The correlation functions
calculated from the more dilute solutions show a
tendency toward linear decay.

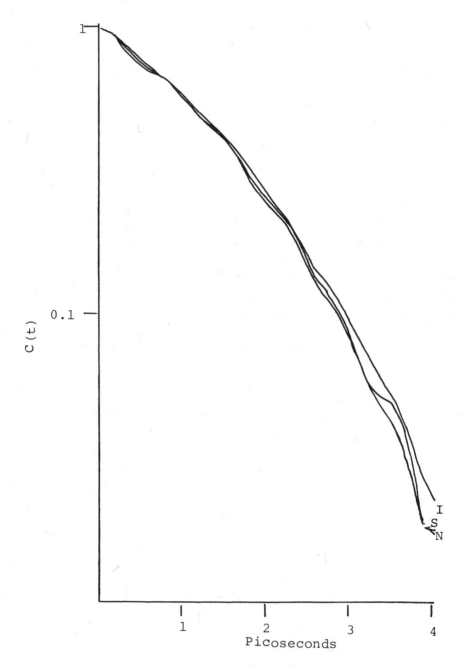

Figure 3: Correlation functions of homeotropically aligned smectic (S) and nematic (N) phase spectra and isotropic (I) phase spectrum.

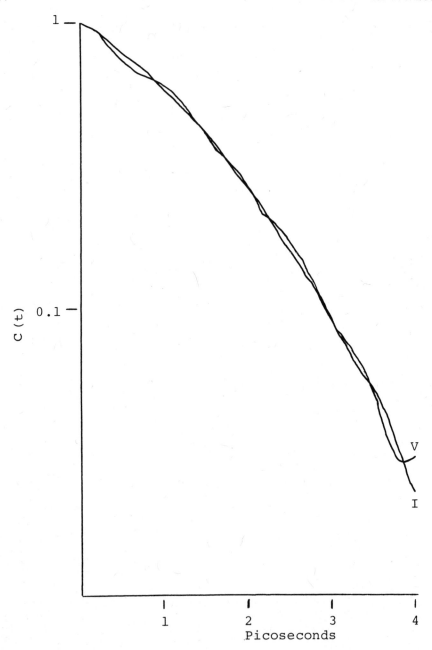

Figure 4:   V=correlation function from $I_{vv}-4/3\ I_{vh}$

I=correlation function from isotropic phase I_{vh} spectra

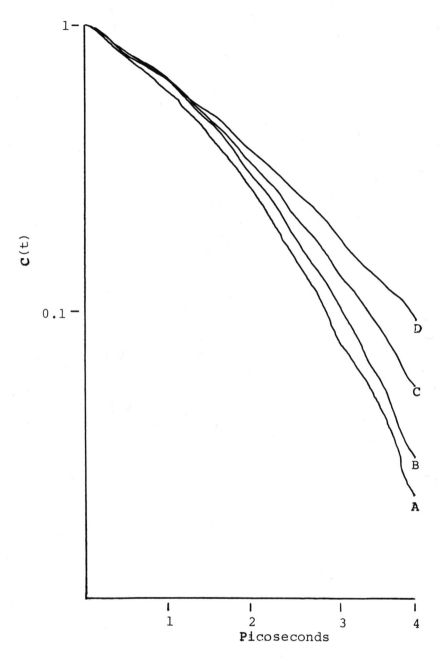

Figure 5: Correlation functions from 4-Octyloxy-4'
cyanobiphenyl in benzene. Mole fractions: A-1.00
(isotropic phase), B=0.51, C=0.14, D=0.04.

Discussion

If, as the data suggest, the nitrile stretching
band shape is determined by vibrational relaxation
processes, then what processes are the important ones?
Three processes significant in liquids are (16): 1.)
dipole - dipole induced transitions; 2.) resonance
vibrational energy transfer; 3.) vibrational dephasing.
The dipole - dipole relaxation mechanism is a dissipative mechanism which shortens the vibrational lifetime
and transfers the vibrational energy into rotational
and translational motions of the liquid "lattice".
Resonance vibrational energy transfer is also a lifetime
shortening mechanism and involves the transfer of
vibrational energy into the same vibrational mode on
a nearby molecule. Both of these relaxation mechanisms
will cause broadening of the vibrational bandshape as
they shorten the excited state lifetime (17). Vibrational
dephasing refers to the effect the time dependent surrounding medium has on a vibrating molecule (16). The
medium acts through a stochastic potential to perturb
the vibrational energy levels and cause a distribution
of values around a mean. This mechanism, though a
band broadening one, does not involve a shortening of
the vibrational lifetime.

Rothschild has shown (18) that dipole - dipole
transitions and resonance vibrational energy transfer
do not contribute as significantly as dephasing to
vibrational band broadening. In addition, results on
the large, slowly rotating molecule quinoline are
quite similar to ours (6) and can be explained well
in terms of vibrational dephasing (16), (6).

That our correlation functions show a slower rate
of decay in solution is consistent with the fact that
the peak shape is dominated by vibrational relaxation
processes. All of the processes mentioned above will
lead to slower decay in solution (3,6) as the vibrational
relaxation becomes less efficient. Slower rates of decay
of the correlation function can be identified with
narrower band widths. If, by contrast, orientational
motion were a large factor in determing the bandshape,
broader rather than narrower peaks would be expected in
solution as a consequence of the greater rotational
freedom.

We conclude that, molecular tumbling motion involving the CN group on the four picosecond time scale is not significant for the molecule studied. This implies that differences in tumbling motion in the smectic, nematic and isotropic phases, if they exist, are not apparent on this time scale. We also conclude that the reason the correlation functions are so similar in the three phases examined is that the same degree of vibrational dephasing exists in each phase; the distribution of local environments persists despite the lessening of lateral forces as the temperature is raised. This is consistent with other results indicating that short range order exists in mesogens in the isotropic phase (19).

That the correlation functions are not linear indicates that the local environments are slowly varying with respect to the time of vibration (16). The benzene solution results indicate that the local structures which depend on the intermolecular forces of the liquid crystal molecules can be broken up by the non-polar solvent molecules. This leads to a narrower lineshape, and slower vibrational relaxation. The linearity of the dilute solution correlation functions is consistent with a model of rapidly varying environments which tend to average to a uniform localized structure (16).

Further work on this and other systems is continuing.

REFERENCES

(1)   R. G. Gordon, Advan. Magn. Resonance 3, (1968).

(2)   W. G. Rothschild, J. Chem. Phys. 57, 991 (1972).

(3)   H. S. Goldberg and P. S. Pershan, J. Chem. Phys. 58, 3816 (1973).

(4)   F. J. Bartoli and T. A. Litovitz, J. Chem. Phys. 56, 404 & 413 (1972).

(5)   K. S. Seshadri and R. N. Jones, Spectrochim. Acta, 19, 1013 (1963).

(6)   W. G. Rothschild, in "Molecular Motions in Liquids" (J. Lascombe, ed.), p. 247, Reidel, Dordrecht, 1974.

(7)   M. Evans, J. Chem. Soc., Faraday Trans. II 71, 2051 (1975).

(8)   S. Lugomer, Mol. Cryst. Liquid Cryst. 29, 141 (1974).

(9)   P. Simova and N. Kirov, Spectrochim. Acta A 29, 55 (1973).

(10) M. P. Fontana and S. Bini, Phys. Rev. A, 14, 1555 (1976).

(11) R. Callender and P. S. Pershan, Phys. Rev. A, 2, 672 (1970).

(12) L. A. Nafie and W. L. Peticolas, J. Chem. Phys. 57, 3145 (1972).

(13) N. Davidson, "Statistical Mechanics", McGraw-Hill, New York, 1962.

(14) T. M. Reed and K. E. Gubbins, "Applied Statistical Mechanics", McGraw-Hill, New York, 1973

(15) A. Abragam, "The Principles of Nuclear Magnetism", Clarendon, Oxford, 1961.

(16) W. G. Rothschild, J. Chem. Phys. 65, 455 (1976).

(17) K. A. Valiev, Opt. Spectry. 11, 253 (1961).

(18)   W. G. Rothschild, G. J. Rosasco, and
       R. C. Livingston, J. Chem. Phys. 62, 1253 (1975).

(19)   H. Gruler, Z. Naturfi, 28a, 474 (1972)

LATTICE MODEL TREATMENTS OF THE FORMATION OF ORDERED PHASES IN STIFF CHAIN POLYMER-DILUENT SYSTEMS

Elizabeth L. Wee and Wilmer G. Miller

Department of Chemistry, University of Minnesota

Minneapolis, Minnesota   55455

The Flory-Huggins lattice treatment has been the cornerstone for many years for understanding the solution thermodynamics of flexible polymers. The effect of chain stiffness was investigated by Flory, again using a lattice approach,[1,2] where the stability of both ordered and disordered phases was considered. The phase equilibria for a rigid rod-diluent system, plotted as the volume fraction v_i, are shown in Figure 1 for a rod of 150 axial ratio, where χ is the van Laar polymer-solvent heat of mixing parameter. Alternatively, χ may be looked upon as proportional to the excess free energy of mixing. The phase equilibria are determined by equating chemical potentials μ_i for the ordered phase

$$(\mu_1-\mu_1^0)/RT = \ln(1-v_2) + [(y-1)/x]v_2 + 2/y + \chi v_2^2 \tag{1}$$

$$(\mu_2-\mu_2^0)/RT = \ln(v_2/x) + (y-1)v_2 + 2 - \ln y^2 + \chi x(1-v_2)^2 \tag{2}$$

with the corresponding ones for the disordered phase.

$$(\mu_1-\mu_1^0)/RT = \ln(1-v_2) + (1-1/x)v_2 + \chi \, v_2^2 \tag{3}$$

$$(\mu_2-\mu_2^0)/RT = \ln(v_2/x) + (x-1)v_2 - \ln x^2 + \chi \, x(1-v_2)^2 \tag{4}$$

371

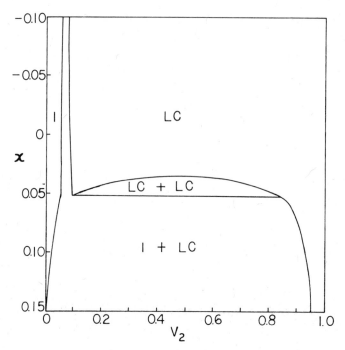

<u>Figure 1.</u>   Lattice model phase diagram for rigid, impenetrable
             rods of axial ratio 150.   Phases are isotropic (I)
             or liquid crystalline (LC)

In the Flory treatment for the ordered phase a rod of axial ratio
x is placed on the lattice as y submolecules, as shown in Figure 2.

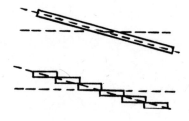

Figure 2

The equilibrium value of y $(1 \leqslant y \leqslant x/2)$ is a measure of disorder in the ordered phase, with its mean value a function of rod concentration. The expressions for the chemical potentials given in equations (1) and (2) are restricted to the equilibrium disorder, as given by equation (5).

$$v_2 = [x/(x-y)][1 - \exp(-2/y)] \qquad (5)$$

We have previously[3-6] determined the temperature-composition phase diagram for the rod-like polymer poly-γ-benzyl-α,L-glutamate (PBLG) in dimethylformamide, shown in Figure 3. The similarities between the theoretical and experimental phase diagrams are striking, although there are significant differences. Two experimental findings are of importance to us here. One is the observation that a region of two ordered phases of different composition is stable, with a rather flat critical region. The other is the observation that anytime a solution is taken across the phase boundary into the wide biphasic region, it gels. This occurs irrespective of whether we are starting from an isotropic region, or an ordered region. We have tentatively attributed this to the kinetics of phase formation when passing into the wide biphasic region.[7] In particular, bicontinuous phases are postulated to form by the Cahn-Hilliard spinodal decomposition mechanism,[8] where phase separation is occurring in a thermodynamically unstable region ($\partial^2 G/\partial x^2 < 0$). The flat ordered-ordered critical region should insure that the spinodal and binodal lie close to each other, making it relatively easy to pass through the metastable into the unstable region over a wide range in concentration. It, thus, becomes of interest to determine the spinodal, which may be accomplished by equating the second derivative of the free energy, or the first derivative of the chemical potential, with respect to composition equal to zero. This is shown in Figure 4 for the ordered phase of axial ratio 100, together with the corresponding binodal. The binodal, depending on the simultaneous solution of two equations, has no solutions when y approaches $x/4$. The spinodal becomes anomalous when y exceeds $x/4$. Before this is reached the ordered-ordered biphasic solution becomes unstable with respect to the ordered-disordered biphasic solution. The spinodal must be calculated from a continuous function, hence one cannot merely change from the ordered to the disordered free energy expression at $y = x/4$. The spinodal may be extended to lower concentrations by holding y fixed at some arbitrary value, ignoring equation 5, as shown in Figure 4. At high concentration, where $y = 1$, the spinodal may be calculated also by ignoring equation 5, as done when calculating the binodal at high concentration.

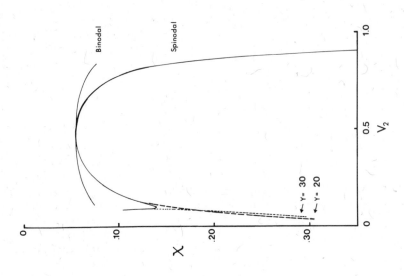

Figure 4. The binodal and spinodal for the ordered phase for rods of axial ratio 100. The dashed lines correspond to the spinodal calculated with fixed y values.

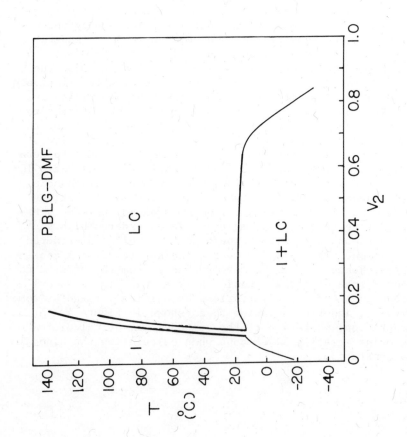

<u>Figure 3.</u>   Temperature-composition phase diagram for PBLG (310,000) in DMF.

The low concentration phase boundary in Figure 3, from 1 to 7 volume percent polymer, has a linear relationship between $\log v_2$ and $1/T$. The theoretical phase boundary in the same concentration range has a linear relationship between $\log v_2$ and χ. Through these relationships a χ-T relationship was established. Using this scale the theoretical temperature-composition phase diagram is shown in Figure 5, together with the spinodal. The unstable region is seen to be quite accessible, experimentally, over a considerable concentration range. It is also our observation that gelation occurs over a wide range in rod lengths. It thus becomes relevant to examine the rod length dependence of the Flory model, considering its success in predicting the main features of the phase diagram for rod-like polypeptides.

Phase equilibria for rod axial ratios 100, 50 and 40 are shown in Figures 6-8, respectively. When the axial ratio is greater than about 50, a region always exists where the stable state is two ordered phases. Below x = 50, two phase regions are always composed of one ordered and one disordered phase. No spinodals can be computed for the low (X<50) axial ratio cases.

We turn our attention next to modifications of the Flory lattice model, especially those relevant to the wide biphasic region. Vapor sorption studies clearly indicate that the dissolution of the rod-like polypeptides is dominated by the mixing of solvent with flexible polymer side chains.[9-11] The rod backbone is effectively inaccessible to the solvent. The statistical thermodynamics can be viewed in two parts; one involving the configurations of the rods on the lattice with no rod-solvent interaction, the other involving the configurational statistics of placing flexible side chains on the lattice, and their interaction with the solvent.

Two models have been developed to approximately account for the flexible side chains, while still using the basic lattice approach of Flory. One of these we will call the cut-off side chain (COSC) model. This model views the system as effectively a three component system - n_1 molecules of solvent, n_2 'hard core' rods each of axial ratio x, and mxn_2 flexible side chains each occupying c lattice positions. If the total number of lattice sites is n_o, then

$$n_o = n_1 + n_2 x(1+mc) \qquad (6)$$

In filling the lattice, the probability that each succeeding site is vacant in a row is given by the mole fraction of vacant sites, i.e., a random distribution of submolecules, side-chains and vacant sites, is assumed. In deriving the partition function for the

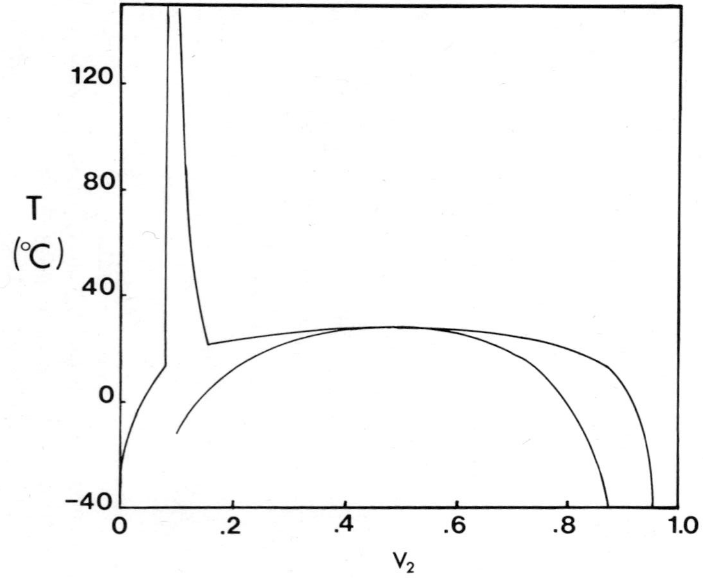

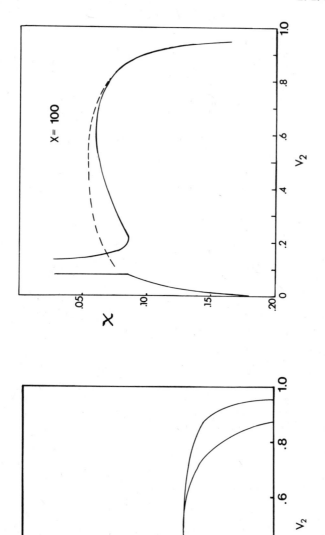

Figure 5. Temperature-composition phase
diagram and spinodal for axial ratio 100,
using a χ–1/T relationship appropriate to
polybenzylglutamatrin DMF.

Figure 6. Order-order (———) and order-disorder phase equilibria for axial ratio 100.

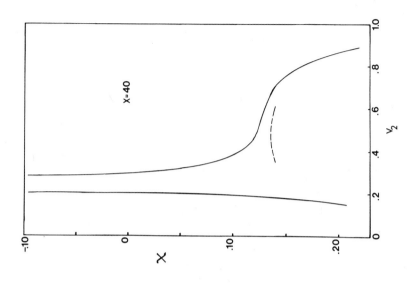

Figure 8. Order-order (---) and order-disorder phase equilibria for axial ratio 40.

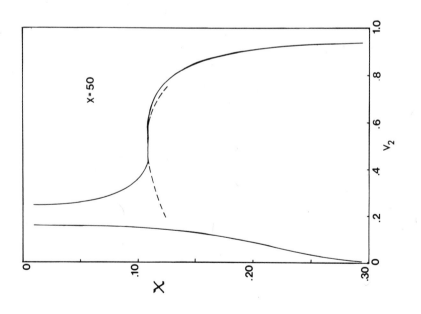

Figure 7. Order-order (---) and order-disorder phase equilibria for axial ratio 50.

ordered phase the 'hard core' rods are replaced by y submolecules, as in the Flory model (Figure 2). After adding j solute molecules ('hard core' plus side chain) the $(j+1)$ st 'hard core' is added, followed by adding mx cut-off side chains. The number of 'hard core' configurations and cut-off side chain configurations are estimated following the assumptions of Flory for rigid rods and of Flory-Huggins for random coils, respectively. With v_{2hc} taken as the volume fraction of 'hard core' (n_2x/n_o) and v_{2sc} the volume fraction of side-chain (n_2mxc/n_o), the free energy of mixing ordered, 'hard core', side chains and solvent becomes

$$\Delta G_M/RT = n_1 \ln(1-v_2) + n_2 \ln v_2 - n_2(\ln x y^2 - y + 1)$$
$$- (n_1 + n_2 xmc + n_2 y) \ln[1-v_{2hc}(1-\tfrac{y}{x})]$$
$$+ n_2 mx \ln v_{2sc} + n_2 mx[(c-1)-\ln c]$$
$$- n_2 mx(c-1)\ln(z-1) - n_2 \ln(1+mc) + \chi n_2 xmc(1-v_2) \qquad (7)$$

where the heat of mixing side chains with solvent is incorporated in a van Laar form as a χ parameter. The disorientation index is assumed to adopt the value which minimizes ΔG_M, which leads to the relationship

$$v_2 = [x(1+mc)/(x-y)][1 - \exp(-2/y)] \qquad (8)$$

For equilibrium degree of disorder, equation 7 becomes

$$\frac{\Delta G_M}{RT} = n_1 \ln(1-v_2) + n_2 \ln v_2 - n_2[\ln x y^2 - y - 1] + \frac{2(n_1 + n_2 xmc)}{y}$$
$$+ n_2 xm \ln v_{2sc} + n_2 xm[(c-1)-\ln c] - n_2 \ln(1+mc) \qquad (9)$$
$$- n_2 mx(c-1)\ln(z-1) + \chi n_2 mxc(1-v_2)$$

The chemical potentials for solute and solvent are obtained by taking the derivative of the free energy with respect to n_2 and n_1, respectively.

$$\frac{\mu_1 - \mu_1^o}{RT} = \frac{v_2 m(c-1)}{(1+mc)} + \frac{v_2}{(1+mc)}\frac{(y-1)}{x} + \chi v_2^2 \frac{mc}{1+mc}$$
$$+ \ln(1-v_2) + \frac{2}{y} \qquad (10)$$

$$\frac{\mu_2 - \mu_2^o}{RT} = mx(1-c)\ell n(z-1) - mx(1-c)v_2 + (y-1)v_2 + \frac{2mxc}{y}$$

$$+ 2 + mx\ell n \frac{v_2^m}{1+mc} + \ell n \frac{v_2}{x(1+mc)} - \ell ny^2$$

$$+ \chi \ mxc(1-v_2)^2 \tag{11}$$

The chemical potentials for the isotropic phase are obtained by substituting y = x in equation 7 and taking derivatives

$$\frac{\mu_1 - \mu_1^o}{RT} = \ell n(1-v_2) + v_2[1 - \frac{1}{x(1+mc)}] - \frac{v_{2sc}}{c} + \chi v_2 v_{2sc} \tag{12}$$

$$\frac{\mu_2 - \mu_2^o}{RT} = \ell n \frac{v_2}{x(1+mc)} + v_2[x(1+mc)-mx] + mx\ell n \frac{v_{2sc}}{c}$$

$$+ \chi \ mcx(1-v_2)^2 - mx(c-1)\ell n(z-1) - \ell nx^2 \tag{13}$$

From equations 10-13 the various phase equilibrium may be determined. This model will clearly overemphasize the entropy of mixing side chains with solvent. However, this can be compensated somewhat by considering m and c to be variables, not necessarily directly related to the actual side chain density and length.

We turn to another approximate accounting of the side chain-solvent mixing problem using a model which we will call the at-tached side chain (ASC) model. The notation and approach is as with the COSC model, the difference being in the counting statistics of the side chains. In adding the (j+1)st solute molecule, the 'hard core' is added as in the COSC model. In placing the first segment of the first side chain of the (j+1)st solute molecule, it can be placed in only z-2 ways, instead of in the number of ways equal to the number of empty sites. The first segment of all other side chains of the (j+1)st solute molecule can be placed in only one way. Other than this difference the development is the same as with the COSC model.

The free energy of mixing solvent with ordered polymer is given by

$$\Delta G_M/RT = n_1 \ln(1-v_2) + n_2 \ln v_2 - (n_1 + n_2 y + n_2 xmc)$$

$$+ \ln[1 - v_{2hc}(1 - \frac{y}{x})] - n_2[\ln y^2 - y + 1]$$

$$- n_2[mx(c-1) + 1]\ln(z-1) + n_2 mxc$$

$$- n_2 \ln x(1+mc) + \chi\, n_2 mxc(1-v_2) \tag{14}$$

If the free energy is minimized with respect to y, equation 8 is again the constraint. The chemical potentials are obtained by differentiation, and are given by equations 15 and 16 for the ordered

$$\frac{\mu_1 - \mu_1^{\,o}}{RT} = \ln(1-v_2) + \frac{v_2}{(1+mc)x}(y-1) + \frac{2}{y} + \frac{v_2 mc}{1+mc}$$

$$+ \chi\, \frac{mc}{1+mc}\, v_2{}^2 \tag{15}$$

$$\frac{\mu_2 - \mu_2^{\,o}}{RT} = \ln\frac{v_2}{x(1+mc)} + v_2(y-1+mcx) + 2 - \ln y^2 + \frac{2mcx}{y}$$

$$+ [mx(1-c)-1]\ln(z-1) + \chi mcx(1-v_2)^2 \tag{16}$$

state, and by equations 17 and 18 for the disordered state.

$$\frac{\mu_1 - \mu_1^{\,o}}{RT} = \ln(1-v_2) + v_2(1 - \frac{1}{x(1+mc)}) + \chi\, \frac{mc}{1+mc}\, v_2{}^2 \tag{17}$$

$$\frac{\mu_2 - \mu_2^{\,o}}{RT} = \ln\frac{v_2}{(1+mc)x} + v_2[x(1+mc)-1] - \ln x^2$$

$$- [mx(c-1)+1]\ln(z-1) + \chi\, mxc(1-v_2)^2 \tag{18}$$

When $m \to o$ and $c \to o$, both models reduce to the Flory model, except for the heat of mixing term, which residues entirely in side chain related terms in the COSC and ASC models. Phase equilibria calculated with the ASC model depend only on mc, whereas in the COSC model both m and c must be specified. As the flexible side chains are reduced in length or density, and approach the Flory rigid, impenetrable rod model, the χ value necessary to enter the wide biphasic region becomes large for the ASC model and even larger for the COSC model because of the larger entropy of mixing.

In many of the flexible side chain , helical polypeptides the mass of the side chains constitute a significant fraction of the polymer volume, hence the lattice positions occupied by side chains may be more than that occupied by the helical hard core. However, if m and c are treated as empirical parameters in fitting the vapor sorption data to experimental data on polymer dissolution, the effective m and c are much smaller than would be expected from mass considerations.[11] This may be a deficiency in the model, or to steric conflicts among the side chains. In Figure 9 is shown the phase equilibria using the ASC model with side chain parameters appropriate to the polybenzylglutamate-DMF data.[11] The high concentration, wide biphasic equilibrium line is seen to be more in conformity with the experimental data (Figure 3) than in the Flory model (Figure 1), but the ordered-ordered two phase region is not stable relative to an ordered-disordered system. Thus a spinodal cannot be calculated. Decreasing the value of mc to 0.05, Figure 10, brings an ordered-ordered two phase region back into thermodynamic favor over a narrow region. It should also be noted that the critical composition is shifted towards much lower concentration. This is a result of a significant contribution of the flexible side chain to the thermodynamics.

Each of the models will give ordered-ordered biphasic regions under some conditions, but not under all conditions.

SUMMARY

The Flory model for rigid, impenetrable rods exhibits a stable ordered-ordered biphasic region at high rod axial ratios, but not at low axial ratios. The limits of the unstable region, the spinodal, can be calculated only under limited composition ranges without artificially forcing a solution. The region of instability for the disordered-ordered biphasic region cannot be calculated from the model, but may in fact be well approximated by the ordered-ordered spinodal due to the closeness of the binodals for the two types of biphasic regions at high χ.

The lattice model has been modified to include flexible side chains, which experimentally are known to contribute importantly to the solution thermodynamics of many rod-like molecules. The wide biphasic region is changed by this modification, as well as the relative stability of the ordered-ordered biphasic region. The calculation of the region of instability suffers from the same problems as the rigid, impenetrable rod model.

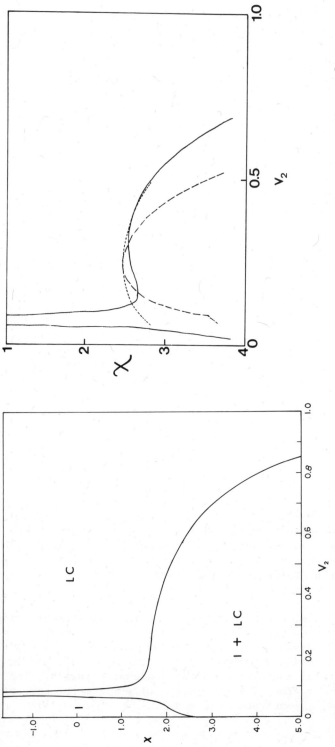

Figure 9. Lattice model phase diagram for rigid rods with flexible, permeable side chains (ASC model). Axial ratio is 150 and mc is 0.15.

Figure 10. Lattice model phase diagram for rigid rods with flexible, permeable side chains (ASC model). Axial ratio is 150 and mc is 0.05. The ordered-ordered biphasic binodal (–·–·–) as well as spinodal (————) are shown, together with the disordered-order binodal (————).

ACKNOWLEDGMENTS

We wish to thank L. Kou for calculating the spinodals, and to the USPHS (GM-16122, GM-16922) for financial support.

REFERENCES

1. P. J. Flory, Proc. Roy. Soc. (London) A234, 60 (1956).

2. P. J. Flory, Proc. Roy, Soc. (London) A234, 73 (1956).

3. E. L. Wee and W. G. Miller, J. Phys. Chem., 75, 1446 (1971).

4. W. G. Miller, J. H. Rai and E. L. Wee, in Ordered Fluids and Liquid Crystals, Vol. 2, R. Porter and J. Johnston, eds., Plenum Publ. Corp., 1974, p. 243.

5. W. G. Miller, C. C. Wu, E. L. Wee, G. L. Santee, J. H. Rai, and K. D. Goebel, Pure and Applied Chem., 38, 37 (1974).

6. C. C. Wu, Ph.D. thesis, U. of Minn., 1973.

7. W. G. Miller, L. Kou, K. Tohyama and V. Voltaggio, J. Polymer Sci., Symposium 58, in press.

8. J. W. Cahn, J. Chem. Phys., 42, 93 (1965).

9. P. J. Flory and W. J. Leonard, J. Amer. Chem. Soc., 87, 2102 (1965).

10. J. H. Rai and W. G. Miller, Macromolecules, 5, 45 (1972).

11. J. H. Rai and W. G. Miller, Macromolecules, 6, 257 (1973).

A LATTICE OF RIGID RODS WITH LONG RANGE ANISOTROPIC ATTRACTIONS

Michael S. Rapport and Andrew G. De Rocco

The Institute for Physical Science and Technology and
the Department of Physics and Astronomy, The University
of Maryland, College Park, Maryland 20742

INTRODUCTION

Until recently, two points of view concerning the forces
which shape the distinctive first order transition between the
nematic liquid crystalline phase and the normal isotropic liquid
prevailed. From one perspective, that advocated by Maier and
Saupe,[1] long ranged, anisotropic attractive interactions were
viewed as primarily responsible for forging the unique properties
of the transition. From the other perspective,[2] the first order
transition was considered to arise exclusively from the effects of
hard core steric repulsions between elongated rod-like molecules.
Neither position, however, has been without its shortcomings.

An assessment of the Maier and Saupe model, for instance, has
been made by a number of investigators.[3] Criticisms have also been
levied against the advocates of the "steric models"; foremost has
been the observation that these "steric" models make no provision
for attractive interactions which are, of course, known to operate
in real systems. Recently, this neglect of attractive interactions
has been remedied.

In particular, a number of investigators,[4] supported by an
accumulation of experimental evidence,[5] have suggested that both
short ranged steric forces and the longer ranged attractive interactions contribute, roughly comparably, to the first-order,
anisotropic-to-isotropic phase transition. The inclusion of
attractive interactions in "steric" rigid rod systems has proved
both a tractable and reasonably successful approach. One of the
earliest examples of such a model was the calculation of Cotter

385

and Martire[6] in which were included attractive forces between otherwise rigid rods on a lattice whose statistics were estimated by the
counting formulas of DiMarzio.[7] The blending of attractive and
repulsive energies was, in their model, complete, in the sense that
the transition (nematic-isotropic) arose when the orientational
alignment energy just balanced the configurational entropy characteristic of the isotropic phase. In more recent models, for
example Wulf,[3] the transition is compelled by the short range
repulsive forces yet assisted in some important details by the
overall attractive interactions. His work suggests that it may
not be the anisotropy of the attractive force which is important,
for the anisotropy may be insignificant compared to the overall
part played by the attractive forces through radial correlations
and the density change at the transition.

A second approach, one employing the tools of perturbation
theory, was presented by Brenner and McQuarrie,[8] who have extended
the earlier Zwanzig[9] model. More recently Cotter[10] has appended
Maier and Saupe-like forces to a rigid "scaled particle" system
and with considerable success. This calculation, as well as that
of Gelbart and Baron,[11] are properly referred to as "van der Waals
theories" of the phase transition (hard cores and a gentle mean
field attraction) and represent a realization of the viewpoint
adumbrated by one of us (AGDR).[12] Such a calculation, in spite of
its general achievements, gives a portrait of the nematic-isotropic
transition deficient in those aspects which doubtless depend on
short range correlations, both orientational and translation.

The present model--yet another hybrid--investigates the effect
of orientation-dependent, long ranged attractions on the first
order anisotropic to isotropic transition in a system of long
rigid rods. The present system is treated within the spirit of the
DiMarzio[7] and Wulf and De Rocco[13] restricted orientational lattice
approximation, and the attractive interactions are included within
the mean field approximation.

We demonstrate here that, relative to the purely steric model,
the inclusion of attractive interactions leads, superficially, to
less realistic values of the characteristic transition parameters.
In particular, the relative density discontinuity at the transition is considerably larger than in the steric model and, in some
cases, nearly two orders of magnitude larger than that observed in
real mesophase systems, where $\Delta\rho/\bar{\rho} \lesssim 0.005$, typically.

Despite this anticipated but somewhat disappointing outcome, it
is suggested that the inclusion of anisotropic attractions demands,
of thsoe molecules which would exhibit more realistic mesophase
behavior, a characteristically smaller axial ratio-- an observation
made previously by other investigators and one certainly in

agreement with observations on real mesomorphic molecules, where
$\ell \gtrsim 3\text{-}4$. Moreover, if we extend this observation to a system of
semiflexible molecules (at least of the type described previously
by Wulf and De Rocco) then mesophase-like behavior can be achieved,
with the inclusion of attractive anisotropic interactions, provided
either (i) the semiflexible rods, for a given bond flexibility, are
made "shorter" or (ii) for a specified length, the molecular flexibility is increased. These remarks remain tentative and will be
investigated more completely, and reported on at a later date.

THE MODEL

The point of departure for the present model is the DiMarzio
and Wulf and De Rocco (WDR) rigid rod, restricted orientational
lattice model which we now review briefly. With the WDR scheme,
N rigid rods of axial ratio ℓ, are packed, according to the
DiMarzio prescription, onto a cubic lattice of V unit sites, with
each molecule permitted to orient in any one of three mutually
orthogonal directions. Short ranged repulsive interactions,
exclusively, are present; these repulsions, moreover, are approximated by the hard core "steric" forces which prevent molecular
overlap.

The configuration of this rigid rod lattice is characterized
by the distribution of molecules aligned in each of the "allowed"
orientations, $\{N_i\}$, with N_i representing the number of molecules
oriented in the i-th direction, i = 1,2,3 and $N_1 + N_2 + N_3 = N$.
With the introduction of an orientational order parameter, s, such
that

$$N_1 = N_2 = sN \tag{1}$$

$$N_3 = (1-2s)N$$

where $0 \leqslant s \leqslant 1/3$, the orientational distribution exhibits the
unaxial symmetry (3-axis, preferred) characteristic of real nematic
mesophases. Those configurations with s = 1/3 represent a state
of an isotropic orientational distribution reminiscent of the normal isotropic liquid; s = 0 describes a state of complete orientational alignment; and, intermediate values of s correspond, of
course, to states of partial orientational alignment not unlike
the nematic mesophase.

A first order, isotropic (s = 1/3) to anisotropic (s < 1/3)
orientational phase transition arises, in this system, solely from
the packing of rigid rod-like molecules which are prevented from
overlapping by hard core repulsions. The statistical mechanics of
this system is contained in the configurational partition function,

$$Q_N = \sum_s G(\ell;\rho;s) \tag{2}$$

where $G(\ell;\rho;s)$ is the number of ways of packing N rods into the lattice with an orientational distribution specified in eq. (1); ρ, is the reduced density of the system, where

$$\rho = (V_0/V) \tag{3}$$

with V_0, the close packed density, $N\ell$; and, $0 \leqslant \rho \leqslant 1$. Moreover, $G(\ell;\rho;s)$ is, according to the DiMarzio recipe and in the notation of WDR,

$$G(\ell;\rho;s) = \prod_{\nu=1}^{3} \{V-(\ell-1)N_\nu\}! / \prod_{\nu=1}^{3} N_\nu! (N!)^2 (V-N\ell)!$$

$$N_1 = N_2 = sN, \quad N_3 = (1-2s)N \tag{4}$$

$$\rho = N\ell/V$$

The partition function is estimated by the well established maximum term approximation,

$$\frac{1}{N} \ln Q_N \simeq \frac{1}{N} \ln G(\ell;\rho;s^*) \tag{5}$$

where s^* is the value of the orientational parameter which corresponds to the minimum of the free energy of the system, namely

$$\left(\frac{\partial \ln G(\ell;\rho;s)}{\partial s} \right)_{s^*} = 0 \tag{6a}$$

or

$$g_\ell(s,\rho) = - \frac{1}{(\ell-1)} \ln \frac{s}{1-2s} - \ln \frac{1-(\ell-1/\ell)\rho s}{1-(\ell-1/\ell)\rho(1-2s)} \tag{6b}$$

Characteristic of steric models, two types of solution occur depending on the density. First, for all densities, $s = 1/3$ satisfies eq. (6b) and corresponds physically to an orientationally isotropic phase of the system. Moreover, for densities greater than a critical density, $\rho^* = 3/(\ell-1)$, a second solution also appears, $s^* < 1/3$ and corresponds to the possibility of a partially aligned orientational distribution associated with an anisotropic phase.

For $\rho > \rho^*$, the solution which corresponds to the stable configuration of the system is determined by comparing the respective free energies of the system for $s^* = 1/3$ and $s^* < 1/3$, at the given density. Accordingly, that value of s which yields the lower free energy, at the density of concern, corresponds to the stable solution. For low densities, the isotropic phase ($s^* = 1/3$) is the stable configuration of the system; for densities, $\rho > \rho^*$, anisotropy characterizes the orientational state of the system and as ρ moves beyond ρ^*, the degree of order increases, indicating a progressive increase in the extent of alignment as the system becomes more compact.

This transition from an isotropic to an anisotropic state is distinctively first order with a concomitant density change between the isotropic and anisotropic phases. The parameters of this transition--the mean transition density and the relative density discontinuity--are obtained by investigating the behavior of the equation of state

$$p\beta = - \frac{\partial}{\partial V}\left(\beta F(s^*;\ell;\rho)\right) \tag{7}$$

which in the vicinity of the "critical density" exhibits a van der Waals-like loop which may be interpreted as the signature of a first order phase transition. The loop, treated by the standard Maxwell construction, yields the appropriate transition parameters.

The present model extends the DiMarzio - WDR caricature by introducing into the system long ranged attractive interactions. These attractions are chosen to be pairwise additive, independent of the relative separation of the interacting pair and <u>aniso-tropic</u>. The form of these attractions is, accordingly,

$$V_{\alpha,\beta} = -\tilde{\rho}^n\{\lambda_{\parallel} (\vec{v}_\alpha \cdot \vec{v}_\beta)^2 + \lambda_\perp (1-(\vec{v}_\alpha \cdot \vec{v}_\beta)^2)\} \tag{8}$$

The density ($\tilde{\rho} = N/V$) dependence in the pair interaction is a consequence of performing an appropriate "average"[14] over all relative separations of the interacting pair; $\vec{v}_\alpha, \vec{v}_\beta$ are unit vectors specifying the orientations, respectively, of the α-th and β-th molecules (in this restricted orientational model, $\vec{v} = \vec{1},\vec{2},\vec{3}$). In addition, $\lambda_{\parallel}$ and $\lambda_\perp$ represent the interaction strength of molecules, respectively, parallel and orthogonally oriented with respect to one another. The choice $\lambda_{\parallel} >> \lambda_\perp$, moreover, insures that a configuration of two molecules in parallel alignment is preferred, energetically, to one in which the molecules are orthogonal to one another. The interaction, in addition, does not distinguish between those aligned configurations in which the molecules are laterally situated with respect to one another and those in which the molecules are terminally situated or

collinear. All parallel alignments are assigned equal interaction
strength, and we exclude, thus, the possibility of a smectic
phase.

In terms of an isotropy parameter, $\tau = \lambda_{\perp}/\lambda_{\parallel}$, $V_{\alpha\beta}$ becomes

$$V_{\alpha,\beta} = -\lambda_{\parallel}\tilde{\rho}^n\{(\vec{v}_\alpha \cdot \vec{v}_\beta)^2(1-\tau) + \tau\} \tag{9}$$

Interestingly, for $\tau = 1$, the interaction $V_{\alpha\beta}$ is orientation-
independent and van der Waals-like in form

$$V_{\alpha,\beta}^{(1)} = -\lambda_{\parallel}\tilde{\rho}^n \tag{10}$$

And the choice, $\tau = -\frac{1}{2}$, yields

$$V_{\alpha,\beta}^{(-\frac{1}{2})} = -\lambda_{\parallel}\tilde{\rho}^n\{\frac{1}{2}(3(\vec{v}_\alpha \cdot \vec{v}_\beta)^2 - 1)\} \tag{11}$$

which for $n = 2$, leads to precisely the results of the Maier and
Saupe mean field interaction.

We proceed to calculate the configurational energy Φ_N of N
rigid rods packed onto a lattice and characterized by the orientational distribution

$$f(\vec{v})_s = s\delta_{\vec{v},\vec{1}} + s\delta_{\vec{v},\vec{2}} + (1-2s)\delta_{\vec{v},\vec{3}} \tag{12}$$

where

$$\delta_{\vec{v},\vec{v}'} = \begin{cases} 1 \text{ for } \vec{v} = \vec{v}' \\ 0 \text{ for } \vec{v} \neq \vec{v}' \end{cases}$$

In the simplest approximation, the attractive interactions are
treated in the mean field approximation--each molecule interacting
with the remaining (N-1) molecules in the system. Within this
scheme, the configurational energy of the system is

$$\Phi_N = \frac{1}{2}Nz \sum_{\nu_\alpha=1}^{3} \sum_{\nu_\beta=1}^{3} f_s(\vec{v}_\alpha)f_s(\vec{v}_\beta)V_{\alpha\beta}(\vec{v}_\alpha,\vec{v}_\beta) \tag{13}$$

with z, the number of "interaction" neighbors; in this case,
z = (N-1) implying, clearly, weak two-body interactions. After
considerable simplification,

$$\Phi_N = -\tfrac{1}{2}N(N-1)\tilde{\rho}^n\lambda_{\|}\{(6s^2-4s)(1-\tau)+1\} \tag{14}$$

Further simplification is achieved by noting that as the system approaches close packed conditions $V \to N\ell$, a state of total orientational alignment is expected. An energy $-\Phi_{\|}$ is assigned to that configuration, where

$$\lim_{\substack{v \to v_0 \\ s \to 0}} \Phi_N = -\tfrac{1}{2}N(N-1)\lambda_{\|}\left(\frac{N}{V_0}\right)^n = -\Phi_{\|} \tag{15}$$

which is easily identified as the total configurational energy of N parallel molecules. Finally,

$$\Phi_N = -\Phi_{\|}\rho^n\{(6s^2-4s)(1-\tau)+1\} \tag{16}$$

where ρ is the reduced density.

For the present system, then, the configurational partition function is

$$Q_N = \sum_s G(\ell,\rho,s)\exp\{-\beta\Phi_N(s,\rho,\tau)\} \tag{17}$$

where $\beta = 1/kT$; moreover, $G(\ell,\rho,s)$ which represents the number of ways of packing, onto the lattice, N molecules, axial ratio ℓ, with the resulting orientational distribution $f_s(\vec{v})$, is precisely the result derived earlier by DiMarzio (see eq. 4).

The partition function for the present model is a natural extension of the corresponding partition function for the purely steric model, namely

$$Q_N^{(0)} = \sum_s G(\ell,\rho,s) \tag{18}$$

In the present system, the lattice is packed according to the DiMarzio prescription with each configuration assigned an appropriate configurational energy with proper weighting according to Boltzmann statistics.

Unfortunately, Q_N cannot be determined directly; instead it is evaluated by the maximum term approximation

$$\frac{1}{N} \ln Q_N \approx \frac{1}{N} \ln \{G(\ell,\rho,s) \ \exp(-\beta\Phi_N(\rho,s,\tau))\} \tag{19}$$

where s^* maximizes $G(\ell,\rho,s)e^{-\beta\Phi_N}$ and consequently satisfies the extremum condition

$$g_\ell(s^*,\rho) - \omega\rho^n(1-3s^*) = 0 \tag{20}$$

which has been simplified, in form at least, by defining a parameter

$$\omega = \frac{2\beta\Phi_{\shortparallel}(1-\tau)}{N(\ell-1)} \tag{21}$$

with $\Phi_{\shortparallel}/N$ interpreted as the average energy of a single molecule in a state of complete orientational alignment ($s = 0$). Moreover, the reduced energy parameter

$$\Lambda = \beta\left(\frac{\Phi_{\shortparallel}}{N}\right) \tag{22}$$

is introduced here and is regarded as one of the fundamental parameters of the model.

THE ISOTROPIC TO ANISOTROPIC TRANSITION

The present model, like the steric model of WDR, exhibits a transition between orientationally disordered ($s = 1/3$) and ordered ($s < 1/3$) phases. The transition is, moreover, a first order transition.

In the present system, as revealed in Table I, the critical density at which the isotropic to anisotropic transition occurs is observed to decrease as the energy parameter $\Lambda(1-\tau)$ increases. It is clear from this behavior that the inclusion of sufficiently strong anisotropic attractive interactions will cause the system to form an anisotropic phase at lower densities than that predicted for a system of rodlike molecules interacting solely via hard core repulsions.

The table below illustrates this effect for a particular choice of the interaction density dependence, namely n=2; the effect is also present for n = 1,3,4 and less pronounced for the latter two cases. The use of n = 3,4 has the disadvantage of a certain thermodynamic inconsistency,[15] although n=4 has been suggested on experimental grounds.[5,16]

TABLE I

ORIENTATIONAL PARAMETER, s^*, AS A FUNCTION OF REDUCED
DENSITY AND INTERACTION PARAMETER $\Lambda(1-\tau)$: $\ell = 10$

n=2 $\Lambda(1-\tau)$	ρ							
	.23	.24	.25	.26	.27	.28	.29	.30
0.5	*	*	*	*	*	*	*	*
1.0	*	*	*	*	*	*	*	.131
1.5	*	*	*	*	*	*	.143	.104
2.0	*	*	*	*	*	*	.112	.086
3.0	*	*	*	*	*	.125	.093	.073
4.0	*	*	*	*	.118	.087	.068	.055
4.5	*	*	*	.118	.087	.067	.053	.042
6.0	*	*	.149	.101	.075	.059	.047	.037

Note: $s^* = 1/3$: For the steric model, $\rho^* = 0.33$ for $\ell = 10$.

This result is not unexpected, since in addition to the propensity toward alignment resulting from packing effects, the presence of attractive anisotropic forces provides yet another mechanism favoring orientational alignment.

The properties of the first order phase transition are derived from examining the equation of state of this system in the vicinity of the "critical density". Again, the presence of a "loop" signals the first order transition; the effect of the additional contribution to the equation of state on the transition is illustrated (for a typical case) in the scaleless figure below. Note,

$$(p\beta\ell)^{ATTRACTIVE} = (p\beta\ell)^{STERIC} - n\Lambda\rho^{n+1}\{(1-\tau)(6s^{*2}-4s^*)+1\} \quad (23)$$

where s^* satisfies eq. (20).

In addition to the value of the order parameter, the transition is characterized by the mean transition density, $\bar{\rho}$,

$$\bar{\rho} = \tfrac{1}{2}(\rho_{aniso} + \rho_{iso})$$

and by the density discontinuity $\Delta\rho/\bar{\rho}$. The van der Waals loop when treated, conventionally, by the standard Maxwell construction yields values for $\bar{\rho}$ and $\Delta\rho$.

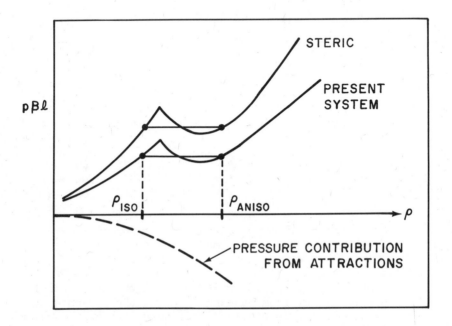

FIGURE 1. $p\beta\ell$ vs. reduced density ρ.

RESULTS

The present model system predicts both a stable low density anisotropic phase and, concomitantly, a first order transition between the anisotropic and isotropic phases. The primary effect of including long ranged attractive interactions in this lattice system has been to produce a significant increase in the magnitude of the density discontinuity at the transition. This not unexpected feature of the present model contrasts with that for real nematic systems, where the density discontinuity is typically of the order of 0.005, whereas, in the present system, $\Delta\rho/\bar{\rho} \gtrsim 0.27$, in all cases investigated; for some values of the interaction parameters, density discontinuities exceeding 50% (0.50) were

noted. In fact, the "best result" in this system is a density
discontinuity $\approx 19\%$, which is expected to occur for relatively
weak attractive interactions; in this case, of course, the system
begins to resemble the hard core steric model of WDR.

The density discontinuity data is displayed, partially, in
Table II for various combinations of the parameters n, Λ, τ. The
computations were restricted to a single value of the axial ratio,
namely $\ell = 10$. Also, in the present model, the critical density
at which the isotropic to anisotropic transition occurs is determined primarily by the hard core repulsions between elongated
molecules, at least for relatively weak attractive interactions
whose density dependence goes as ρ^n, with n = 3,4, roughly. For
a van der Waals-like interaction, dependent on the first power of
the density ($\Phi = -\lambda\rho$), the attractive interactions are observed
to play a significant role in shifting the critical density below
the critical density characteristic of the purely steric model.
The density discontinuity, in this case, is again rather large,
generally exceeding 33%.

The mean transition density, on the other hand, is affected
relatively less by the presence of attractive interactions, at
least of the type introduced in this system. In general, the
mean transition density is somewhat higher (approximately 5-6%, on
the average) than the corresponding mean transition density predicted by the WDR model. The orientational order at the transition
is somewhat higher than that for the purely steric model. We
might be able to understand this behavior by considering more
closely the mechanism responsible for ordering in this system.
In particular, as the system is compressed from the isotropic
state, the hard core repulsions begin to force the molecules into
alignment. The presence of an incipient alignment will, given the
nature of the anisotropic interactions, enhance the tendency towards alignment arising from the attractive forces. The overall
effect, relative to a purely steric model, would be to increase
the extent of alignment in the system as it moves through the
transition into the anisotropic phase. This effect, moreover,
would be most noticeable for attractive forces with considerable
anisotropy.

Also, concerning the anisotropy of the attractive interactions, there is some evidence in this study to suggest that
anisotropy may be a less important feature of the attractive
interactions with regard to determining the properties of the
first order transition than had been previously suspected; and,
for a fluid of rigid, rodlike molecules, treating the attractive
interactions as a uniform isotropic background ($\Phi = -\lambda\rho^n$) may, in
fact, not be too bad an approximation. We note in passing that
density discontinuities are somewhat smaller for isotropic than

TABLE II

RELATIVE DENSITY DISCONTINUITY, $\Delta\rho/\bar{\rho}$ ($\ell = 10$)

(Λ, τ)	n = 1	n = 2	n = 3	n = 4
$(1, -\frac{1}{2})$	0.39	0.39	0.31	0.27
$(1, 0)$	0.36	0.36	0.31	0.27
$(1, \frac{1}{2})$	0.35	0.34	0.31	0.27
$(1, 1)$	0.33	0.33	0.30	0.27
$(2, -\frac{1}{2})$				0.32
$(2, 0)$			0.46	0.32
$(2, \frac{1}{2})$			0.46	0.31
$(2, 1)$			0.47	0.31
$(3, 0)$				0.43
$(3, \frac{1}{2})$				0.44

for anisotropic attractions, although the order and transition are not especially different.

On the whole, it seems that the anisotropic attractive inter-actions play some role in determining the extent of orientational alignment in a system of rigid rodlike molecules. The predicted isotropic to anisotropic transition (in particular, the critical density) appears to be determined primarily by the hard core steric forces; the properties of the concomitant first order phase transition, on the other hand, appear to be shaped largely by the attractive interactions.

Quantitatively, the predictions of the present model do not, unfortunately, agree well with the properties of real nematogenic systems. Some improvement would certainly be achieved by lower-ing, for instance, the effective axial ratio of the constituent molecules or by making some provision for molecular flexibility.[13]

Referring to Table VI of reference (13), realistic mesophase-like behavior is insured over a relatively small range of flexibility (y ≈ 4-5) and molecular length, ℓ = 10-20. In this range, moreover, the transition density lies in the range $0.60 \leq \rho \leq 0.80$ --liquid-like densities--and the density discontinuity (DD) is of the order of $\leq 1\%$, both in excellent agreement with experience.

It is remarked, based on the effect of attractive interactions in the present model, that the inclusion of attractions would for semiflexible molecules of fixed length, shift the transition DD to higher values--in the direction of rigid rod-like behavior. This tendency could be counteracted, for instance, by increasing molecular flexibility.

Also, for a fixed flexibility, y, attractions would be expected to shift the transition parameters to values more characteristic of longer semiflexible rods; a corresponding decrease of length, ℓ, would maintain the transition in the density regime and with the corresponding DD characteristic of real mesophase systems.

Attractive interactions, it is apparent, constrain the molecular architecture needed to insure mesophase-like behavior, and in the direction which insures a more realistic assignment of shape to nematogens. In short, rigid or semi-flexible rods, with local anisotropic interactions, must have their shape (flexibility, axial ratio) altered towards smaller axial ratios or, equivalently, greater flexibility in order to accommodate the presence of attractive forces which promote nematic alignment.

REFERENCES

1. W. Maier and A. Saupe, Z. Naturforchg. 12a, 668 (1957); 13a, 564 (1958); 14a, 882 (1959); 15a, 287 (1960).

2. L. Onsager, Ann. N. Y. Acad. Sci. 51, 627 (1949).

3. J. Kaplan, E. Drauglis, Chem. Phys. Lett. 9, 645 (1971); A. Wulf, J. Chem. Phys. 64, 104 (1976); R. G. Priest, Phys. Rev. Lett. 26, 423 (1971); T. D. Schultz, Mol. Cryst. Liq. Cryst. 14, 147 (1971).

4. For example, M. Cotter, Phys. Rev. A10, 625 (1974); P. Sheng, J. Chem. Phys. 59, 1942 (1974). See also, A. Wulf, Ref. (3).

5. J. R. McColl, C. S. Shih, Phys. Rev. Lett. 29, 85 (1972); also, Phys. Rev. Lett. A38, 55 (1972).

6. M. A. Cotter and D. E. Martire, Mol. Cryst. Liq. Cryst. 7, 295 (1969).

7. E. A. DiMarzio, J. Chem. Phys. $\underline{35}$, 658 (1961).

8. S. L. Brenner and D. A. McQuarrie, J. Chem. Phys. $\underline{61}$, 3090 (1974).

9. R. Zwanzig, J. Chem. Phys. $\underline{39}$, 1714 (1963).

10. M. Cotter, J. Chem. Phys. $\underline{66}$, 1098 (1977).

11. W. M. Gelbart and B. A. Baron, J. Chem. Phys. $\underline{66}$, 207 (1977).

12. These ideas were first described in a paper (no. 38) presented at the Second International Liquid Crystal Conference (1968) --"A Van der Waals Model for the Nematic Isotropic Transition," referred to in Ref. (13) and again made explicit at the Van der Waals Centennial Conference, Amsterdam, 1973.

13. A. Wulf and A. G. DeRocco, J. Chem. Phys. $\underline{55}$, 12, (1971).

14. M. S. Rapport, Ph. D. thesis, University of Maryland (1976).

15. M. A. Cotter, Mol. Cryst. Liq. Cryst. $\underline{39}$, 173 (1977).

16. R. L. Humphries, P. G. James and G. R. Luckhurst, J. Chem. Soc., Faraday Transactions II $\underline{68}$, 1031 (1972).

AN EPR INVESTIGATION OF A NEMATOGEN-LIKE NITROXIDE SPIN

PROBE IN THE SOLID MODIFICATIONS OF A LIQUID CRYSTAL

Arthur E. Stillman, Loretta L. Jones, and
Robert N. Schwartz

Department of Chemistry
University of Illinois at Chicago Circle
Chicago, Illinois   60680

Barney L. Bales

Centro de Fisica, I.V.I.C.
Apartado 1827, Caracas 101, Venezuela

ABSTRACT

    The nematogen-like spin probe N-(p-methoxybenzyli-
dene)-4-amino-2,2,6,6-tetramethylpiperidino-1-oxyl
(MBATPO) in the various solid modifications of the
liquid crystal MBBA was investigated by EPR spectroscopy. In particular, by cooling the solution in a
magnetic field of 6000 G, a nematic glass was prepared
and the orientation dependent EPR spectra were studied
as a function of the temperature. As the temperature
is raised the nematic glass converts irreversibly to
the polycrystalline state. The EPR spectra in both
solid modifications reveal motional effects which are
analyzed utilizing the methods of Freed and co-workers.

INTRODUCTION

It is a well known fact that the glassy state can be established by quenching many supercooled liquids below their glass transition temperatures without crystallization. Recently, Lydon and Kessler observed that when the nematic phase of the liquid crystal N-(p-methoxybenzylidene)-p-butylaniline (MBBA) is rapidly cooled by exposure to liquid nitrogen, a glassy solid is formed which has the nematic ordering frozen in.[1] As the temperature is increased, a metastable crystalline phase occurs at 259 K. This phase then converts into the nematic phase via a smectic mesophase or stable crystalline phase. These authors monitored the phase changes using X-ray diffraction and differential thermal analysis (DTA) techniques.

Sorai and Seki have also carried out a DTA investigation on MBBA and reported that rapid cooling (>10 K/ min) of the nematic mesophase leads to the formation of a nematic glassy state at 201 K.[2] Upon heating a glass-type transformation occurs at the same temperature. This transformation was irreversible and leads to a crystalline modification of MBBA.

A calorimetric, infra-red, and far infra-red study of the various solid modifications of MBBA has also recently been reported by Janik, et al.[3] Upon cooling the sample down from the nematic phase they observed the formation of a metastable solid which was crystalline (not glassy). This metastable modification spontaneously transforms into a stable crystalline form when kept for several hours at temperatures not far below the melting point.

The studies briefly discussed above have been concerned with the identification and thermodynamics of the solid modifications of MBBA. Information regarding the structural and dynamical characteristics of the liquid crystal molecules in the various solid forms can certainly provide molecular insight into the nature of these phases. Electron paramagnetic resonance (EPR) studies of spin probes are ideally suited to investigate these structural and dynamical aspects. In addition, nuclear magnetic resonance (NMR) studies of the pure liquid crystals can also provide structural as well as dynamical information at the molecular level.

Information regarding the anisotropic pseudo-potential for nematic liquid crystals has been obtained from analysis of the EPR spectrum of paramagnetic probes aligned in the nematic glass phase. For example, James and Luckhurst found that the pseudo-potential for the spin probe vanadyl acetylacetonate dissolved in a binary mixture of nematogens and then cooled to the nematic glass state is mainly a result of the repulsive forces, whereas, for a single component of the mixture, the pseudo-potential for the spin probe in the nematic mesophase is determined by dispersion forces.[4] These authors attributed the difference to the close proximity of the molecules at low temperatures in the glassy nematic state which increases the repulsive contribution to the pseudo-potential.

Another recent EPR investigation of spin probes partially oriented in nematic glasses was reported by Ovchinnikov, et al.[5] They observed that in addition to the absorption lines typical of isotropic glassy systems, the EPR spectra of partially aligned spin probes in nematic glasses exhibit additional lines with an angular dependence typical of single crystals.

A nitroxide spin label study of 4,4'-diocta-decyloxyazoxybenzene (ODOAB) which exhibits smectic B and C phases was recently reported by Poldy, et al.[6] The spin labels used in this study were specially synthesized to resemble the host molecule as closely as possible. By attaching the nitroxide moiety at different positions along the alkyl end chains, these workers were able to investigate the state of the chains in the different phases of ODOAB. One notable feature of the EPR data was that in the solid phase the chain motions are of small amplitude and inter-mediate rate, whereas in the smectic mesophases these motions are of large amplitude and rapid. Therefore, the transition SOLID → SMECTIC B is an example of a chain melting transition. Poldy and co-workers also suggested the possibility of molecular rotation of the spin probe (and probably the liquid crystal molecules) about the long axes in the solid phase below the SOLID-SMECTIC B transition.

Freed and co-workers have carried out an important EPR study of ordering and spin relaxation as a function of pressure for the perdeuterated Tempone nitroxide radical in Phase V liquid crystal solvent.[7] They observed that the rotational correlation time τ_R of

the spin probe decreases as the pressure is increased
in the solid phase which is indicative that the probe
is located in a cavity (or clathrate) whose structure
is similar to that in the nematic mesophase except
that the pressure freezes out the motion of the solvent
molecules and hence the motion of the spin probe is
less hindered. These authors also suggested that at
still higher pressures the free volume of the cavity
should be reduced resulting in an increase in τ_R.

A fairly detailed proton spin relaxation study of
the alkoxybenzoic acid series liquid crystals in the
solid phase has been reported by Thompson and Pintar.[8]
At low temperatures (77-200 K) the proton spin relaxation is dominated by methyl group reorientation. At
higher temperatures reorientation of the molecules
around their long molecular axes as well as self-
diffusion are a source of spin relaxation in the solid
phase.

Kornberg and Gilson have very recently carried
out a wide line NMR study of the stable and metastable
solid phases of MBBA.[9] These samples were quenched
in liquid nitrogen before being placed in the spectrometer probe. From the observed temperature dependence
of the second moments of the proton NMR absorption
lines these authors conclude that: (1) methyl group
rotation plus rotation of the methoxy group are
responsible for the low temperature transition (77-140 K)
in the second moment versus temperature plot, (2) motion
within the n-butyl chain is responsible for the high
temperature transition in both solid modifications, and
(3) the difference in the measured second moments for
the metastable and stable phases is due to reorientation
of the benzylidene rings about the para axes.

In this paper we report the results of a preliminary study of the solid modifications of MBBA using
the nematogen-like nitroxide radical N-(p-methoxybenzylidene)-4-amino-2,2,6,6-tetramethylpiperidino-1-oxyl
(MBATPO) as a monitor of the microenvironment (cf.
Figure 1). The use of nitroxide free radicals to probe
the structural and dynamical properties of anisotropic
systems is a well established technique. Information
regarding the microscopic and macroscopic molecular
organization of the anisotropic system is reflected in
the ordering tensor of the nitroxide spin probe. Dynamical
properties of the molecular environment surrounding the
spin probe may be indireetly obtained from a careful

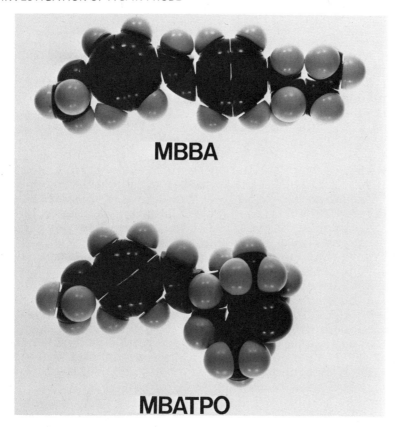

Figure 1.   Pauling-Corey-Koltun models of MBBA
             and MBATPO.

analysis of the EPR relaxation data which is a manifestation, for example, of the rotational and translation
diffusion of the paramagnetic probe.[10]   It should be
clear that even though we are studying the dynamics
of the spin probe, there is ample evidence that it
is sensitive to the dynamics of surrounding environment.[10] This is because the reorientational and
translational correlation times and the ordering
potential (anisotropic pseudo-potential) depend on the
intermolecular interactions provided by the local
molecular environment. If one is judicious in his
choice of the size and shape of the spin probe one
can expect dynamical behavior which more accurately
reflects that of the molecular surroundings.

EXPERIMENTAL METHODS

The nitroxide free radical MBATPO was prepared
according to the method of Pudzianowski, et al.
The liquid crystal MBBA was obtained from Eastman
Kodak and used without further purification. Samples
were sealed under approximately 1 atm of nitrogen
in 2.0 mm I.D. quartz sample tubes after removal of
oxygen by the freeze-pump-thaw technique.

EPR spectra at X-band were recorded with a homebuilt reflection-type spectrometer incorporating 6 KHz
magnetic field modulation. A Varian Mark II Fieldial
and 13 KW regulated magnet power supply was used to
deliver the stable magnetic field. Variable temperature experiments were carried out with a homebuilt
gas-flow cryostat capable of regulating the temperature over the active region of the sample to within
± 1°C. Experiments at liquid nitrogen temperatures
utilized a quartz insertion dewar in which helium gas
was passed through the liquid nitrogen to minimize the
nitrogen bubbling. The frequency of the spectrometer
was accurately determined using a Micro-Now Instrument
Co. Model 101C frequency multiplier chain and a Hewlett-
Packard 5248M frequency counter. Magnetic field
strengths were measured with a Systron-Donner NMR
Gaussmeter.

RESULTS

Figure 2 shows EPR spectra of MBATPO in MBBA at
77 K. The sample denoted nematic glass was prepared
by quenching to 77 K in a magnetic field of 6000 G.
The polycrystalline sample was prepared in the absence
of a magnetic field by first quenching the sample to
77 K, then warming to the polycrystalline state and
finally, cooling to 77 K. It is clear that the EPR
spectra of MBATPO in the nematic glass exhibit an
angular dependence. Another notable feature of the
spectra in the nematic glass state is that the high
field extremum absorption line is almost absent in the
spectrum with the magnetic field parallel to the
direction of alignment (optic axis), $\hat{n}$.

Figures 3 and 4 show spectra in the nematic glass
with the magnetic field parallel and perpendicular,
respectively, as a function of the temperature. The

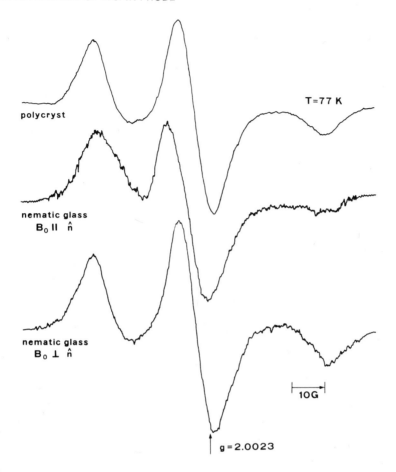

T=77 K

polycryst

nematic glass
$B_0 \parallel \hat{n}$

nematic glass
$B_0 \perp \hat{n}$

10G

g=2.0023

Figure 2.   First derivative EPR spectra of MBATPO
in the polycrystalline and nematic glass
phases of MBBA at 77 K.

nematic glassy state was prepared by rapidly cooling
(~40 K/min) the samples in a magnetic field of 6000 G
from the nematic mesophase. These spectra reveal that
as the temperature is increased from 108 to 203 K the
lines narrow resulting in additional structure.
Furthermore, the separation of the low and high field
hyperfine extrema of the spectra with B_0 perpendicular
to $\hat{n}$ (cf. Figure 4) decreases as the temperature is
increased. These splittings are listed in Table 1.

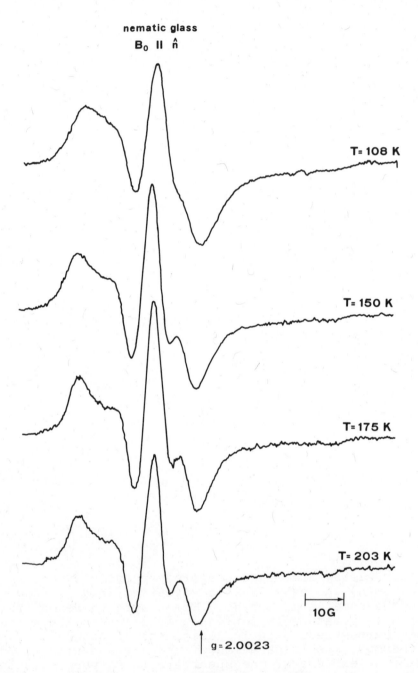

Figure 3.   EPR spectra of MBATPO in the nematic glass
            phase of MBBA with the magnetic field
            parallel to the optic axis.

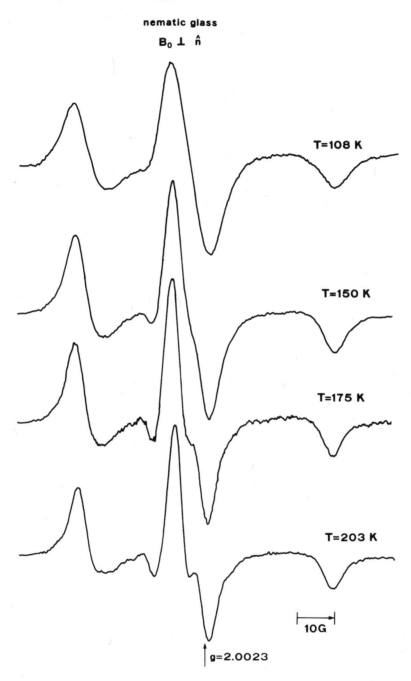

Figure 4.   EPR spectra of MBATPO in the nematic glass
             phase of MBBA with the magnetic field
             perpendicular to the optic axis.

Table 1. A_{zz} values as a function of temperature
           for MBATPO in the nematic glass.

Temperature	$A_{zz}^{\dagger}$
77 K	34.86 ± 0.05 G
108	34.70
150	34.60
175	34.29
203	33.92

$^{\dagger}A_{zz}$ =   one-half the separation of the outer ^{14}N
        hyperfine extrema (cf. Figure 4).

If the nematic glassy sample is cooled from 203 K
back down to 108 K one obtains the original spectra.
However, if the glassy sample is warmed above approximately 208 K and then cooled back to 108 K the original
spectra are not obtained. The spectral differences
observed for the irreversibly transformed samples are
moderate (parallel and perpendicular spectra are still
observed) as long as the sample is not warmed above
250 K. It should be pointed out that above approximately
208 K the sample is no longer semi-transparent.

Figure 5 show spectra of a sample which has been
warmed from the glassy state to 249.5 K. A striking
feature of these traces is that there is no difference
in the spectra taken with B_0 parallel and perpendicular
to n. Furthermore, the spectra at 135 K are similar to
those at 249.5 K except that the lines are broadened at
the lower temperatures.

At still higher temperatures the spectrum of MBATPO
in the solid phase is like that in viscous liquids,
i.e., the spectra are typical of probes in the slow-
tumbling region.[7,10,11,12] Some experimental results
are shown in Figure 6. The spectrum at 281.5 K is
especially indicative of a radical undergoing slow-
motional anisotropic reorientation.[7,11,13]

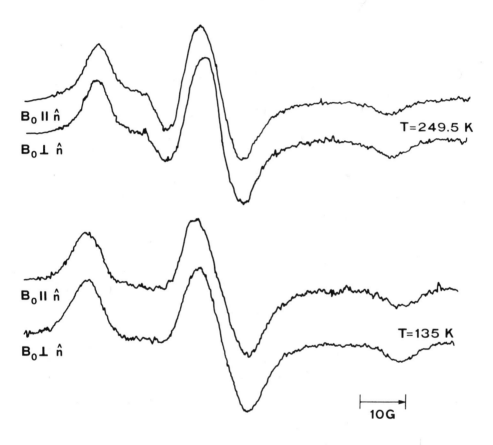

Figure 5.   EPR spectra of MBATPO in the polycrystalline
           phase of MBBA.

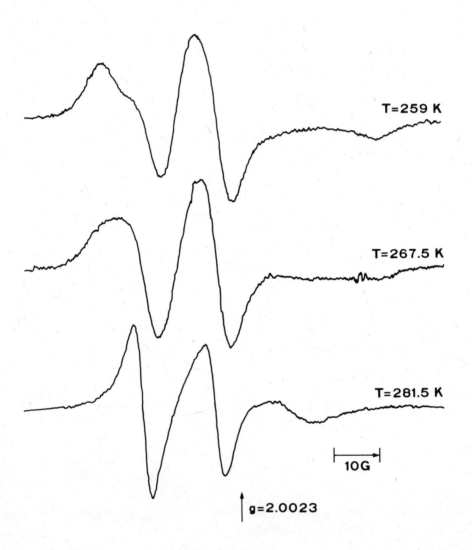

Figure 6.   EPR spectra of MBATPO in the polycrystalline
             phase of MBBA.

DISCUSSION

In the nematic mesophase the long axes of the molecules remain correlated with one another over a large number of molecules.[14] This is a manifestation of the anisotropic pseudo-potential $\mathcal{U}(\Omega)$ which in its most general form is given as[12,13]

$$\mathcal{U}(\Omega) = \sum_{L,k,m} \mathcal{E}_{km}^{(L)} \mathcal{D}_{k,m}^{(L)}(\Omega), \tag{1}$$

where $\mathcal{D}_{k,m}^{(L)}(\Omega)$ are the matrix elements of the Wigner rotation matrix and $\mathcal{E}_{k,m}^{(L)}$ are the general expansion coefficients. Knowledge of the pseudo-potential is essential in order to calculate the various static and dynamic properties of the ordered fluid. For example, the static property $<A>$ can be calculated by taking the Boltzmann average

$$<A> = \int d\Omega P_O(\Omega) A, \tag{2}$$

where $P_O(\Omega)$ is the equilibrium distribution of Euler angles Ω between the molecule-fixed coordinate system and the director frame and is given by

$$P_O(\Omega) = \exp(-\mathcal{U}/kT) / \int d\Omega \, \exp(-\mathcal{U}/kT). \tag{3}$$

In the above equation k is the Boltzmann constant and T is the absolute temperature.

The EPR spectra of MBATPO dissolved in a sample quenched by rapid cooling to a solid in the presence of a magnetic field exhibits an angular dependence, whereas those samples similarly prepared but in the absence of an external field do not. These observations are indicative of long-range macroscopic ordering for those samples prepared in the magnetic field. One may expect, however, that the microscopic ordering is the same for both types of samples.

An important feature of the orientational dependence of the EPR spectra in the ordered glass is that it can provide information regarding the form of $\mathcal{U}$.[4] As a first step we have computer simulated EPR spectra of the nitroxide probe using the rigid limit methods described by Lefebvre and Maruani.[15] This computer program was modified by Dr. C. F. Polnaszek in order to take into account the anisotropic properties of the nematic glass.[13] The calculated EPR intensities have been weighted by

an angular dependent probability distribution function
given by

$$P_o(\beta) = C \exp(\gamma_2 \cos^2 \beta), \qquad\qquad (4)$$

where C and γ_2 are constants and β is the angle between
the N-O bond axis (approximately along the long molecular axis) and the optic axis, n. The above distribution function is related to that found in the mean-
field theory of Maier and Saupe.[16]

Figure 7 shows computer simulated nitroxide spectra
for the case where B_o is parallel to $\hat{n}$ as a function
of the parameter γ_2 or equivalently, of the ordering
of the spin probe, $S_{zz}^{(P)}$. It is evident that as the
ordering increases the computed spectra resemble more
closely spectra corresponding to radicals whose N-O
bond is parallel to B_o.[17] Figure 8 shows a comparison
of a simulated spectrum with the corresponding experimental spectrum for the parallel orientation at 205 K. The
fit is reasonably good. The differences between the
theoretical and experimental spectra may be the result
of the presence of residual motion, unresolved proton
hyperfine interactions[18] and possibly the inadequacy of
Equation (4) to accurately describe the intermolecular
interactions.[12,13] Another important source of error
may be ascribed to the fact that the theoretical spectra
are computed assuming that there is a distribution of
molecular orientations only about a single optic axis
which is parallel to the external magnetic field, B_o.
The simulations may be improved by incorporating in
the calculation a distribution function for the directors
with respect to B_o.

It is interesting to note that the EPR spectrum
of an ordered nitroxide probe in the glassy nematic
phase may be a convenient method to obtain the elements
of the magnetic interaction tensors (cf. Fig. 9). For
example, from the perpendicular spectrum of MBATPO,
one-half the separation of the low and high field
hyperfine extrema is a measure of the principal z-
component of the ^{14}N hyperfine tensor, A_{zz}; the
principal z-component of the g-tensor is approximately
given by the mid-point between the low and high field
extrema. Assuming an axially symmetric hyperfine
tensor, the perpendicular components may be obtained
from the isotropic hyperfine coupling and the trace
of the hyperfine tensor. With the sample in
the parallel orientation the principal x-component

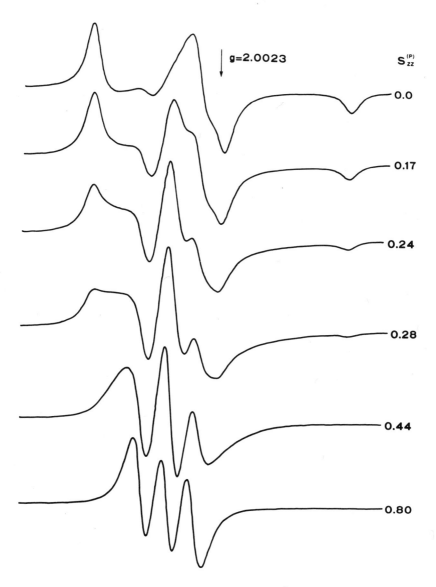

Figure 7. Computer simulated $B_0 \parallel \hat{n}$ EPR spectra for a nitroxide with $g_{xx} = 2.0102$, $g_{yy} = 2.0060$, $g_{zz} = 2.0021$, $A_{xx} = A_{yy} = 6.85$ G and $A_{zz} = 34.3$ G as a function of $S_{zz}^{(P)}$. An angular dependent line width of $(3.0 + 1.0 \cos^2 \beta)$ G was used in the simulations. These order parameters correspond to γ_2 values of 0.00, 1.20, 1.60, 1,87, 3.00, and 8.24, respectively.

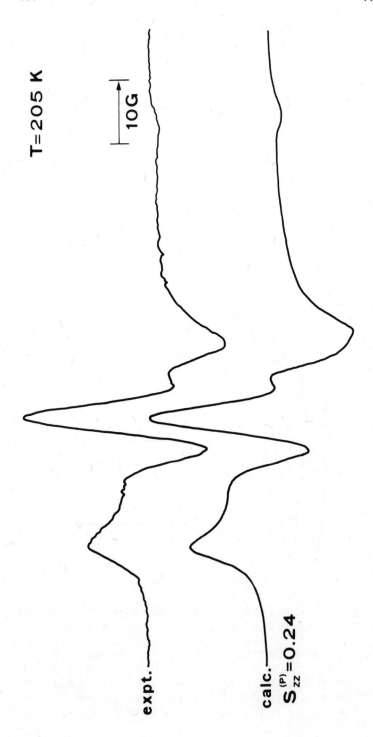

Figure 8.   A comparison of the experimental and
computed EPR spectra for MBATPO in the
nematic glass phase of MBBA at 285 K.
The parameters used in the simulations
are the same used in Figure 7.

of the g-tensor corresponds to the position of the
maximum of the central portion of the first derivative
representation of the EPR spectrum. From the relationship between the isotropic g-value and the trace of
the g-tensor the principal y-component is obtained.

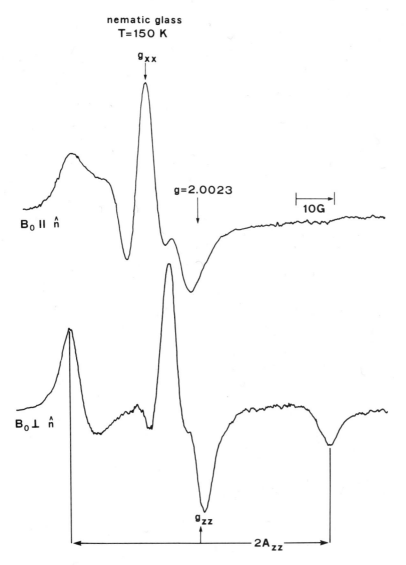

Figure 9.   Experimental EPR spectra of MBATPO in
            the nematic glass phase of MBBA. Also
            indicated are the assignments of some
            elements of the magnetic hyperfine and
            g-tensors.

As is evident in Figures 3-6 the line widths
narrow as the temperature is increased. Also apparent
in Figure 4 and Table 1 is that the separation of the
low and high field hyperfine extrema decreases with
increasing temperature. These observations are
indicative of motional effects associated with the
spin probe and possibly the solvent molecules.

Freed and co-workers have developed methods for
estimating reorientational correlation times τ_R for
nitroxide in the slow tumbling[19] and near-rigid limit
regimes[20] using stochastic Liouville methods (SLM).[21]
In the rigid limit the outer hyperfine extrema for an
isotropic distribution of nitroxide radicals arise
from those radicals whose $2p\pi$ orbital on the nitrogen
atom is parallel to B_o. Mason and Freed, using
SLM computed incipient motional effects on the line
widths of the outer hyperfine extrema near the rigid
limit.[20] These authors regard the motional broadening
in the near rigid limit as arising from lifetime
broadening due to motional effects which carry the
spin probe between different orientations corresponding
to different EPR resonance frequencies. This is
analogous to the well known phenomena of chemical or
Heisenberg exchange in which the magnetic resonance
absorption lines broaden before they shift as the exchange
rate increases.[22] However, in the rigid limit where
the spectrum is not influenced by motional effects,
unresolved inter- and intramolecular electron-nuclear
dipolar interactions dominate the line widths.[20] These
inhomogeneous contributions to the line widths are
quickly averaged out with the onset of molecular motion.
This is analogous to the case of exchange narrowing in
inhomogeneously broadened nitroxides[23] and in agreement
with our observations that the line widths sharpen at
higher temperatures.

We have estimated the slow-motional rotational
correlation times τ_R for the spin probe MBATPO in
the nematic glass using the parameter $S = A'_{zz}/A_{zz}$;
where A'_{zz} is one-half the separation of the outer ^{14}N
hyperfine extrema and A_{zz} is the rigid limit value
for the same quantity. As was shown by Goldman,
Bruno, and Freed, S depends upon the intrinsic line
width, hyperfine parameters, and models for rotational
diffusion.[19] Table 2 lists the estimated correlation
times as a function of temperature for the different
models for rotational diffusion. The important feature
of the data in Table 2 is that the correlational times

Table 2.   Rotational correlation times
estimated for MBATPO in a nematic
glass using the S parameter formalism.

Temperature	τ_R (Brownian Diffusion)	τ_R (Moderate Jump Diffusion)	τ_R (Strong Jump Diffusion)
108 K	7×10^{-7} sec	2×10^{-7} sec	7×10^{-8} sec
150	5×10^{-7}	1×10^{-7}	5×10^{-8}
175	2×10^{-7}	7×10^{-8}	3×10^{-8}
203	7×10^{-8}	4×10^{-8}	2×10^{-8}

fall in the range $7 \times 10^{-7} \lesssim \tau_R \lesssim 2 \times 10^{-8}$ sec over the
temperature range 108-203 K and that an Arrhenius-type
analysis yields an activation energy of approximately
1 kcal/mole. These values for the correlation times
are only approximate because the analysis based on
Freed and co-workers' S parameter formalism assumes
isotropic rotational reorientation. However, we do
feel that these results are significant because they
suggest that the probe is undergoing small amplitude
motions at an intermediate frequency.

NMR activation data for the alkoxybenzoic acid
series liquid crystals in the solid phase is on the
order of 3 kcal/mole for methyl group and ring reorientation.[8] One would expect similar NMR results for MBBA
in the solid phase. The discrepancy in the activation
energies obtained from the NMR and EPR data is most
likely due to enlarged cavities from perturbations
produced by the nitroxide portion of the spin probe.
The main point is that both types of experiments
indicate that there is some type of rotational or
librational motion about the long axes of the rod shaped
molecules in these systems. The EPR data, of course,
only yield direct information about the motion of the
spin probe and the motion of the solvent must be
inferred. One might expect that the reorientational
motion of the spin probe in the solid phase is concertive, that is, one may envision the probe's motion

correlated with the random reorientational motion of the liquid crystal molecules in the immediate vicinity. This type of reorientational behavior can be roughly analyzed in terms of Freed's fluctuating torque theory.[24] The concertive motion is manifested in a frequency dependent rotational diffusion tensor and can in principle be studied by a careful line shape analysis. Such a study is currently in progress using deuterated MBATPO to test this possibility as well as to quantify some of the previously discussed results.

ACKNOWLEDGEMENTS

Acknowledgement is made to the N.S.F. Grant No. DMR-76-00368 and the University of Illinois at Chicago Circle Research Board for partial support of this work. The U.I.C.C. Computer Center is acknowledged for its donation of computer time. The authors wish to thank Dr. C. F. Polnaszek for helpful discussions and for copies of his computer programs.

REFERENCES

1.   J. E. Lydon and J. O. Kessler, J. Physique (Paris) 36, C1-153 (1975).

2.   M. Sorai and S. Seki, Bull. Chem. Soc. Japan 44, 2887 (1971).

3.   J. A. Janik, J. M. Janik, J. Mayer, E. Ściesińska, J. Ściesiński, J. Twardowski, T. Waluga, and W. Witko, J. Physique (Paris) 36, C1-159 (1975).

4.   P. G. James and G. R. Luckhurst, Mol. Phys. 19, 489 (1970).

5.   I. V. Ovchinnikov, I. B. Bikchantaev, and N. E. Domracheva, Sov. Phys. Solid State 18, 2081 (1976).

6.   F. Poldy, M. Dvolaitzky and C. Taupin, J. Physique (Paris), 36, C1-27 (1975).

7.   J. S. Hwang, K.V.S. Rao, and J. H. Freed, J. Phys. Chem. 80, 1490 (1976).

8.   R. T. Thompson and M. M. Pintar, J. Chem. Phys. 65, 1787 (1976).

9.   B. Kornberg and D.F.R. Gilson, Chem. Phys. Lett. 47, 503 (1977).

10. J. H. Freed, in Spin Labeling Theory and Applications (L. J. Berliner, ed.), pp. 53-132, Academic Press, New York, 1976.

11. A. T. Pudzianowski, A. E. Stillman, R. N. Schwartz, B. L. Bales, and E. S. Lesin, Mol. Cryst. Liq. Cryst. 34 (Letters), 33 (1976).

12. C. F. Polnaszek and J. H. Freed, J. Phys. Chem. 79, 2283 (1975).

13. C. F. Polnaszek, Ph.D. Thesis, Cornell University, 1976.

14. P. G. de Gennes, The Physics of Liquid Crystals, Oxford University Press, New York, N. Y., 1974.

15. R. Lefebvre and J. Maruani, J. Chem. Phys. 42, 1480 (1965).

16. W. Maier and A. Saupe, Z. Naturforsch. A 14, 882 (1959).

17. L. J. Libertini and O. H. Griffith, J. Chem. Phys. 53, 3185 (1973).

18. J. S. Hwang, R. P. Mason, L. P. Hwang, and J. H. Freed, J. Phys. Chem. 79, 489 (1975).

19. S. A. Goldman, G. V. Bruno, and J. H. Freed, J. Phys. Chem. 76, 1858 (1972).

20. R. P. Mason and J. H. Freed, J. Phys. Chem. 78, 1324 (1974).

21. J. H. Freed, G. V. Bruno, and C. F. Polnaszek, J. Phys. Chem. 75, 3385 (1971).

22. P. W. Anderson, J. Phys. Soc. Jap. 9, 316 (1954).

23. A. E. Stillman and R. N. Schwartz, J. Mag. Resonance 22, 269 (1976).

24. L. P. Hwang and J. H. Freed, J. Chem. Phys. 63, 118 (1975).

CONDUCTION CURRENTS ASSOCIATED WITH THE VARIABLE GRATING MODES IN NEMATIC LIQUID CRYSTALS

P.K. Watson, J.M. Pollack and J.B. Flannery

Webster Research Center
Xerox Corporation
Rochester, New York   14644

INTRODUCTION

Electrohydrodynamic (EHD) instabilities in nematic liquid crystals (NLC) induced by dc or low-frequency excitation are of several kinds. The earliest discovered[1], and perhaps most widely known, is the turbulent mode, known in recent years as Dynamic Scattering (DSM)[2].

There have been additional discoveries and studies of non-turbulent EHD modes, which have come to be known as Variable Grating Modes (VGM)[3]. Several experimental studies since about 1970 have delineated the characteristics of at least three such modes[4,5,6,7]. While it appears in hindsight, that at least two of the confirmed three modes might have been observed by earlier workers[4,5] the some-- what subtle lines of observable distinction have only emerged of late[7,8,9,10]. Furthermore, a boundary value problem describing the generalized VGM effect, based on a conduction induced alignment mechanism[11], was solved earlier[3], but does not in its present form admit to immediate predictive extension to the three modes presently known[12].

While it is the intent of this paper to focus on the correlated optical and electrical transients precursor to two of these modes, it is appropriate in this introduction to establish the broader context within which the present results should be viewed. The VGM phenomena, are dc (<10 Hz) EHD effects in NLCs with negative dielectric anisotropy, and high electrical resistivity ($\rho > 10^{9\text{--}10}\Omega$cm). These general characteristics are common to all. With incoherent normal illumination, the NLC adopts a texture appearing as parallel striations resembling the well known Williams domains[13]. Photomicrographs of these textures are shown in several of the previously cited references. These parallel or "striped" domains consist of ordered regions of vortical flow, and are generally most apparent in polarized light because of the optical anisotropy of the NLC, and the spatial regularity of optical path length variations within the regions of flow. Under coherent illumination, these domains act in aggregate as a sinusoidal phase grating, as described earlier[6].

The onset of domain structure occurs at a characteristic voltage threshold in the range of 5-10V, independent of cell thickness, ℓ. The domain patterns are generally stationary at a fixed voltage, but the domain width decreases with increasing voltage.

While all three varieties of VGM have the above features in common, there are critical points of distinction which are correlated principally to the experimental geometry (*viz*, cell thickness, L), or quiescent state alignment. Table 1 summarizes these key points with respect to a three-dimensional cell coordinate system (x and y define the plane of the liquid crystal layer, and z the direction of applied voltage and optical observation). The first two modes, which we call VGM$\perp$ and VGM$||$, exhibit a mutually orthogonal domain orientation with respect to the quiescent state director orientation. These differences are illustrated schematically in Figure 1. The stationary state domain width is never greater than the cell thickness for either mode, and, for each, the domain width decreases linearly with increasing applied voltage. Furthermore, the only observable parameter determining which of these two modes prevails for a given NLC is the cell thickness. That critical thickness, L^*, for the material used in this study is about 5μm. For another similar material, L^* was about 10μm[7]. While the dominance of VGM$||$ below, and VGM$\perp$ above the critical thickness, are always observed, there is a region of measured thicknesses within ~0.5 - 1μm of L^*

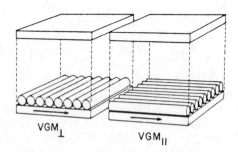

Figure 1. Schematic illustration of the domain axial orientations for VGM$\perp$ and VGM$||$. The arrows on the faces of the substrates indicate the director orientation in the homogeneous quiescent state.

Table 1. CHARACTERISTICS OF VARIABLE GRATING MODES

EHD MODE	Experimental Conditions		Characteristics of Steady-State EHD Mode		References
	Quiescent State Director(n) Orientation	Cell Thickness(L) Regime	Domain Axial Orientation	Maximum Domain Width at Threshold	
$VGM_\perp$	$n\|\|x$	>5-10μm†	$\perp x$	~ℓ	4,5,6,7
$VGM_\|\|$	$n\|\|x$	<5-10μm†	$\|\|x$	~ℓ	7,8,10
VGM_θ	$n\perp x$ $\theta_{xz}=30°$	10-100μm	$\|\|x$	>ℓ	9

† depends on material. Range represents extremes presently known.

for which either of the two stationary-state modes can be observed, even within a single cell.

In contrast to the above, the third mode, which we call VGM$_\theta$ (extraordinary domains[9]) (to reflect the non-zero tilt angle for the quiescent state director) appears not directly related to, and has significantly different characteristics from VGM$_\perp$ and VGM$_{||}$. It occurs in moderately thick cells, and although the steady-state domain orientation is similar to that of VGM$_{||}$, the domain width is greater than the cell thickness, and the dependence of domain width on applied voltage is substantially sublinear. We have not examined VGM$_\theta$ in more than cursory fashion, but cite its several distinct features here to mitigate the obvious confusion which might arise on comparison with VGM$_{||}$ on the single common point of domain axial orientation.

Our principal interest, reported here, was in gaining a mechanistic understanding of the physical processes which must be precursor to the establishment of either steady-state VGM$_\perp$ or VGM$_{||}$ in a given NLC. To this end we examined experimentally the correlated optical and conduction transients which follow application of a voltage pulse to VGM cells of systematically varied thickness. We also developed an isotropic one-dimensional conduction model which allows a simulation of the various features observed in the conduction transients. One essential characteristic of this model, which describes the growth of a space charge layer in the NLC, is a direct relationship between the time required for unipolar carrier sweepout and the thickness of the NLC cell.

While we cannot, in the context of the present conduction model, determine the causal relationships between the development of this depletion layer and the natural selection of a preferred mode of molecular orientation in VGM, this new understanding of electrical carrier dynamics is seen as essential to further elucidation of the more complex molecular dynamics of VGM, perhaps through a three-dimensional hydrodynamic analysis.

EXPERIMENTAL

The material used in these studies was a room temperature NLC. It consisted of a mixture of terminally-substituted aromatic azoxy compounds, obtained from E. Merck, and is referred to as N5. This material is a mixture of two eutectics: the first, designated N4, is the eutectic of

$$CH_3O\!-\!\bigcirc\!-\!N(O) \;=\; N\!-\!\bigcirc\!-\!C_4H_9$$

and

$$CH_3O\langle O \rangle N = N(O)\langle O \rangle C_4H_9$$

The second component in the mixture is the eutectic of

$$CH_3O\langle O \rangle N(O) = N\langle O \rangle C_2H_5$$

and

$$CH_3O\langle O \rangle N = N(O)\langle O \rangle C_2H_5$$

This material, used as received, is characterized electrically by a resistivity in the range 10^9 to $10^{10}\Omega$cm, a negative dielectric anisotropy ($\varepsilon|| - \varepsilon\perp = -0.2$) and a dielectric constant of 5.5.

Parallel-plate test cells were used for the measurements. The cells were constructed from 0.25 inch plate glass coated with a transparent layer of indium oxide. Both electrodes were selectively etched to produce a 1 cm^2 circular active area.

For all transient measurements, homogeneous alignment was induced by orientation rubbing. (Obliquely coated silicon monoxide was used for alignment in steady-state experiments.) The alignment and tilt angle of the NLC layers in the zero-field states were checked by conoscopic techniques. Rubbed samples exhibited uniform alignment with a small tilt angle. Samples oriented using silicon monoxide provided superior homogeneous alignment with zero tilt angle. Both techniques allowed the formation of VGM domains regardless of the zero-field tilt angle.

Cell thicknesses were established through the use of evaporated spacers placed outside the active area. Coplanarity of the confining electrodes was determined by observation of optical fringes using a helium lamp illuminator. The observed variation in cell thickness was less than 0.15μm. Measurement of cell thickness was made by capacitive techniques using a General Radio Impedance Bridge Model 1615A, both before and after filling with NLC. The sealant for the cell was an epoxy resin ("Epoxypatch," Dexter Corporation). All measurements were made at ambient temperature without additional temperature control of the NLC cell.

The diffraction efficiency of voltage-induced phase gratings was determined through measurement of the relative decrease of the intensity of the non-polarized zero-order beam of a Spectra Physics 133M 5mw He-Ne laser used for normal illumination of the NLC cell. The optical transients were measured by observing the change in intensity of the zero-order beam; this was detected by an Edmunds Scientific solar cell and displayed on an oscilloscope. Triggering the NLC cell was accomplished with a high slewing speed switch using

TTL logic with a switching time of less than 120μsec between 0-60V dc.

The current transients were obtained from the voltage drop across a 10 KΩ resistor in series with the test cell. This voltage was displayed on an oscilloscope; a photograph of the trace was digitized and used for subsequent analysis. In all cases the current transients exhibited a capacitive displacement peak, but of time duration very short relative to that of the phenomena analyzed here.

RESULTS

Current transients and the accompanying optical transients were measured for three cell thicknesses of the NLC. These were 3.7, 4.7 and 7.4μm; the applied voltages ranged from 10 to 60 volts.

Three main features emerged from these measurements:

i) The current transient for a VGM⊥ cell was characterized by a broad, primary current peak, which in turn was followed by a smaller, secondary peak or shoulder. Current transients for a VGM∥ cell were similar, but the secondary peak was much more pronounced. Figure 2 shows typical transient response characteristics for VGM⊥ and VGM∥ cells, respectively.

ii) In all cases the optical transient was characterized by a delay interval, followed by a rise to a maximum at or slightly above the diffraction efficiency of the steady state grating. The onset of the optical transients correlated closely with the evolution of the secondary current transients. A typical example is shown in Figure 3. The position of the secondary current peak was measured as a function of the applied voltage in each of the VGM∥ cells. It was found that the maximum of this current peak occurred when the optical transient reached 85% of its maximum, as shown in Figure 4.

iii) The initial value of current, J_o, defined as in Figure 3, scaled linearly with applied voltage for each sample, though there was considerable variation from sample to sample, as may be seen from the results plotted in Figure 5. However, from the linearity of current with voltage it was inferred that the conductivity was largely ionic in character. The wide variation of conductivity between samples, suggests that the ions were present in the original material, caused by the dissociation of impurities, or were produced by the electrode surfaces.

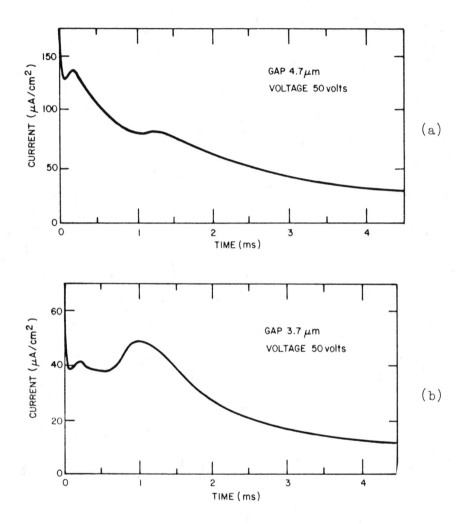

Figure 2.   Current transient in a VGM⊥ cell(a) and a VGM|| cell (b)

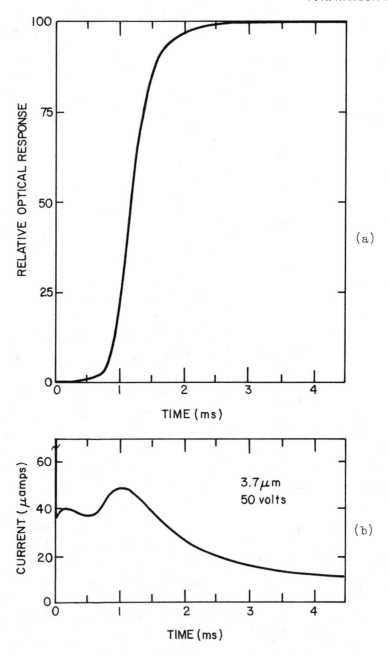

Figure 3.   Plots of (a) diffraction efficiency (%) versus time, and
   (b) current versus time for a typical VGM|| cell. The capaci-
   tive displacement current is shown in (b) as a vertical marker
   on the time zero axis. The initial current, J_o, is defined as
   the current at t = o.

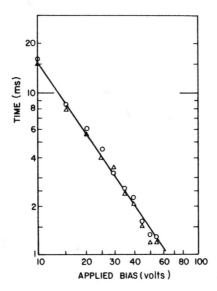

Figure 4. Plot showing correlation of optical transient response
with secondary current peak for a VGM∥ cell: O, time to secondary current maximum; Δ, time to 85% of maximum diffraction
efficiency.

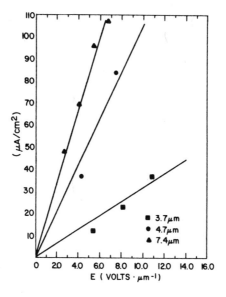

Figure 5. Plot of initial current, J_o, versus applied electric
field for VGM∥ and VGM⊥ cells.

DISCUSSION

The essentially featureless optical transients found for both VGM|| and VGM| are formally the same as reported earlier[6]. Taken alone, they allow little insight at the mechanistic level as to the root physical phenomena driving the VGM modes. The new correlation, however, with the secondary maximum in the feature-rich current transients, is amenable to analysis, as are the current transients themselves. We, therefore, approached the problem from the latter viewpoint, attempting to construct a plausible conduction model which would generate a suitable simulation of the experimental data. Several possible origins of charge in the NLC were considered, as well as conduction mechanisms themselves.

In the following discussion, allusion is made to alternate mechanisms which led to inadequate quantitative models, but the focus is on development of the one which ultimately led to the most reasonable description of all but one of the experimental observations, including the characteristics of the current transients, the correlation of the optical maximum with the secondary current maximum, and the existence of a critical cell thickness determining the transition from VGM|| to VGM|. The sole exception is a satisfactory analysis of the molecular alignment mechanism.

The linear relationship observed between initial current, J_o, and the applied voltage suggests that the conduction process in NLCs is ionic, in contrast to a non-linear process such as carrier injection. Therefore, the transient current is probably caused by the drift of ions in the applied field, modified by an ionic space charge. A kinetic analysis of the transient space charge problem for two sets of mobile carriers is very difficult; however, there is a simplifying assumption which can be made in the present case: the form of the measured current transients at low fields is strongly reminiscent of the $sech^2 (t/\tau)$ current transient which accompanies the formation of a depletion layer. Such transients are observed where one species of charge carrier is far more mobile than the countercharge. Moreover, in the limit of zero mobility for the slow carrier, the analysis reduces to a particularly simple form[14].

In most nonpolar dielectric fluids the positive and negative ions have similar mobilities[15], but it is reasonable to hypothesize that there are ion-dipole interactions in NLCs which give rise to widely divergent positive and negative ion mobilities. One possibility is that the localized negative charge on the oxygen-atom of the N-O dipole creates a potential well for a positive ion in the NLC, whereas the corresponding positive charge of the dipole is delocalized, leaving a negative ion relatively free. Therefore it is suggested that the current transients in these NLCs are a consequence of this energetic and kinetic inequality of positive and negative ions. For the purpose of developing a model it is assumed that the negative ion is the mobile species, while the positive ion is relatively immobilized.

The Formation of a Depletion Layer

After the application of voltage to this model dielectric the mobile negative ions will be swept out of part of the cell by the field. Thus, the interior of the dielectric is divided into two regions as shown in Figure 6. Region A is depleted of negative ions leaving behind the stationary, positive countercharge. If Q is the charge in region A, then the current in that region, which is purely capacitive, will be given by J_A = dQ/dt = N dλ/dt, where the co-ordinate λ defines the position of the interface between regions A and B (*i.e.*, λ defines the position of the mobile carrier front) and N is the initial concentration of charge carriers in the NLC.

In region B the current is entirely due to the drift of the mobile carriers in the field E (λ). Current in region B is therefore given by J_B = N eμE(λ), where μ is the mobility of the mobile carrier.

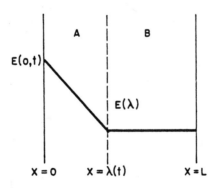

CARRIER SWEEPOUT

Figure 6. Schematic diagram showing spatial variation of electric field in NLC cell of thickness L as mobile ionic carriers are removed. The boundary between depletion region A and region B is at a position x = λ which moves in time as the space-charge region develops.

Applying Poisson's equation, $dE/dx = Ne/\varepsilon$, to the space charge in region A, the relationship between applied voltage V, the position coordinate λ, and initial ion concentration N, is given by

$$V = E(\lambda) L + Ne\lambda^2/2\varepsilon . \tag{1}$$

It follows that

$$E(\lambda) = [V - Ne\lambda^2/2\varepsilon]/L. \tag{2}$$

Equating J_A and J_B, and rearranging, yields

$$\frac{d\lambda}{dt} = \frac{Ne\mu}{2\varepsilon L} \left[\frac{2\varepsilon V}{Ne} - \lambda^2 \right] . \tag{3}$$

Integrating, with the boundary condition $\lambda = 0$ at $t = 0$, gives

$$\lambda(t) = \sqrt{\frac{2\varepsilon V}{Ne}} \tanh\left[\sqrt{\frac{NeV}{2\varepsilon}} \cdot \frac{\mu t}{L} \right] . \tag{4}$$

On further manipulation, the expression for ionic current is

$$J_1(t) = Ne\frac{d\lambda}{dt} = Ne\mu\frac{V}{L} \ \text{sech}^2\left(\frac{t}{\tau}\right) , \tag{5}$$

$$= J_0 \ \text{sech}^2\left(\frac{t}{\tau}\right) \tag{6}$$

where $J_0 = Ne\mu E_0$ is the current at zero time, and τ is a time constant given by

$$\tau = \sqrt{\frac{2\varepsilon L^2}{Ne\mu^2 V}} = \sqrt{\frac{2\varepsilon L}{\mu J_0}} . \tag{7}$$

There is a clear similarity between the curve of $\text{sech}^2 (t/\tau)$, shown in Figure 7, and the initial portion of the current transient in Figure 2a (*i.e.*, for VGM⊥; the similarity is less evident for VGM||). A direct comparison is shown in Figure 8.

The foregoing model requires that J_0 increase linearly with field, and the time constant τ be proportional to $\sqrt{L/J_0}$. The experimental results for larger cell thicknesses and lower applied fields were in reasonable agreement with these requirements, and will be considered further in a later section.

Further analysis, however, shows that the model is incomplete.

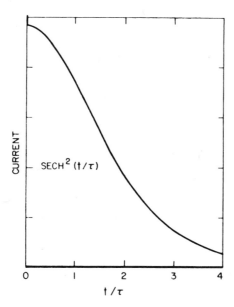

Figure 7. Functional dependence of current, $J(t)$, on (t/τ) according to Equation 6.

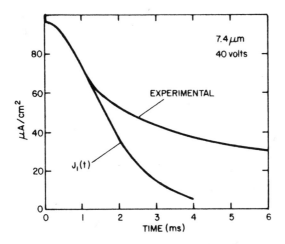

Figure 8. Plot of current versus time for a VGM cell, comparing experimental data with the computed result of Equation 6.

In particular, it does not lead to a shoulder or secondary peak in the current transients. It is the second current peak which correlates well with the onset of the optical effect.

Clearly, a second source of current must be identified, which becomes operative as the depletion layer is established. There are at least two processes which could provide this additional current. One is that of field emission from the cathode as a result of the high field created by the space charge in the depletion region. This possibility cannot be ruled out entirely, but it would require a highly non-linear relationship between applied voltage and the magnitude of the second current peak. The latter relationship is very nearly linear.

This approximate linearity prompts consideration of ionic processes as the source of the secondary current. The mechanism which was examined in detail and appears most suitable is that of ionic dissociation in the depletion layer.

The Effect of Ionic Dissociation in the Depletion Region

In formulating the depletion layer model it was assumed that the number of ions in the system is constant and equal to the initial value N. However, ion generation is continuous, and the concentration of ions is determined by both the rate of generation, α, and the rate of recombination. Ion recombination in the bulk is bimolecular. This rate is given by $\beta N_+ N_-$, where N_+ and N_- are the concentration of positive and negative ions, respectively, and β is the Langevin recombination coefficient $e\mu/\epsilon$. The equilibrium carrier concentration is therefore given by $N_+ = N_- = N = \sqrt{\alpha/\beta}$.

Returning again to Figure 6, it is seen that in region B both sets of carriers are present at their initial concentrations and equilibrium is maintained. In region A, however, fields are generated which are so high that mobile carriers are swept out more rapidly than recombination can occur. In the depletion region, therefore, in addition to the capacitive current, $Ne(d\lambda/dt)$, there is a current attributable to the generation of new charge carriers. These new carriers contribute to the total current, and their countercharge contributes to the space charge in the depletion region.

This space charge problem is not simply tractable. However, an upper bound to the contribution of these excess carriers to the total current can be set by neglecting both their countercharge and their recombination. Thus, if all the new carriers are swept out of the depletion region before recombination can occur, and their countercharge is negligible in comparison to eN, then the extra con-

tribution to the current is given by $J_2(t) \simeq \alpha e \lambda(t)$. The total current is then given by

$$J(t) = J_o \operatorname{sech}^2 \left(\frac{t}{\tau}\right) + \alpha e \lambda(t) \tag{8}$$

$$= J_o \operatorname{sech}^2 \left(\frac{t}{\tau}\right) + J_o \tanh\left(\frac{t}{\tau}\right), \tag{9}$$

where τ is the time constant defined in equation 7. (The extra
space charge has the further effect of reducing the time constant,
τ. This will be discussed below.) The current components according to equation 9 and their additive functional behavior are illustrated in Figure 9.

Equation 9 has the correct form, but gives too high a value for
the current component J_2. However, if the effect of ion recombination in the depletion layer is included, the resulting value of J_2
is reduced to about $0.6 \, J_o$ for $t > \tau$. This modified equation fits
the experimental results well for large and intermediate cell thicknesses. A typical example of an experimental curve for $J(t)$, and
a computed curve of this form is shown in Figure 10.

Experimental data for the 3.7μm cell, are not well replicated
by this model. The secondary current peak rises above the calculated value. However, the computed fit to the data for 4.7 and 7.4μm
cells is very satisfactory.

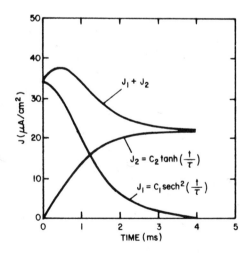

Figure 9. Graphical display of the temporal evolution of current
components J_1 and J_2, and their composite, $J = J_1 + J_2$, according to equation 9.

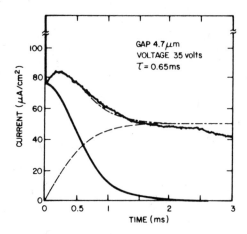

Figure 10. Comparison of computed and experimental current transients: $-J_1$; $---J_2$; $-\cdot-J_1 + J_2$;. Superimposed on $J_1 + J_2$ is an
experimental trace for $J(t)$.

Estimate of Charge Carrier Mobility.

According to equation 7 a plot of time constant, τ, versus
$\sqrt{L/J_0}$ should be linear, with slope proportional to $\sqrt{2\varepsilon/\mu}$. Figure
11 shows such a plot of the time constants, τ, generated from com-
puted fits of experimental data according to equation 9. Although
there is considerable scatter among the data the results are reasonably consistent with the model. The carrier mobility calculated
on the basis of equation 9, however, would be too high. This is
because of neglect of the contribution of extra space charge in the
depletion layer. This extra space charge decreases the time constant τ. To calculate exactly the charge carrier mobility it is
necessary to use a modified time constant $\tau^* = \sqrt{2\varepsilon L^2/N^* e\mu^2 V}$, where
N^* is the augmented space charge density.

The fact that J_2 is comparable to J_0 indicates that the extra
space charge must be of the same order as N. The magnitude of this
extra space charge can be estimated directly from J_2, thus automatically taking into account the effect of carrier recombination in
region A. As given above, $J_2 \simeq 0.6\,J_0\,\tanh(t/\tau)$. This current,
which is caused by the mobile species, leaves behind an equal and
opposite countercharge. The total space charge density in region A
after time τ must, therefore, be given approximately by

$$N^* \simeq N + \frac{0.6\,J_0\,\tanh(t/\tau^*)}{e\,\lambda^*(t)} \tag{10}$$

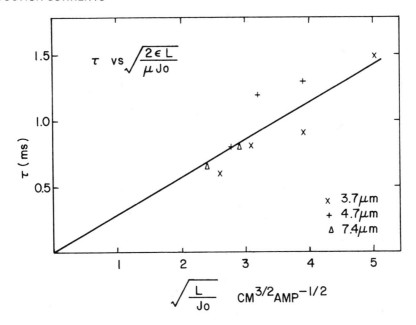

Figure 11. Plot of time constant, τ, versus $(L/J_o)^{1/2}$ for three VGM cells.

where

$$\lambda^*(t) = \sqrt{\frac{2\epsilon V}{N^* e}} \ \tanh\left(\frac{t}{\tau^*}\right). \tag{11}$$

In equations 10 and 11 the starred variables indicate that the initial space charge density N has been replaced by the augmented space charge N*.

Substituting for λ* and J_o in equation 10 it is found that the space charge density N* at time t = τ* is given by N* ≃ N + 0.6N. Thus, the modified time constant is related to the initial space charge density by the equation

$$\tau^* = \sqrt{\frac{2\epsilon L^2}{1.6 N e \mu^2 V}} = \sqrt{\frac{1.25 \epsilon L}{\mu J_o}}. \tag{12}$$

Using equation 12 and the data in Figure 11, the mobility of the mobile charge carrier in this NLC is found to be μ=7.1 ± 0.9 x 10^{-6} cm^2/volt sec.

It is interesting to compare this carrier mobility with ion mobilities in homogeneous liquids. The viscosity of Merck N5 is 2.9 poise at room temperature. Adamczewski[15] shows negative ion mobilities in hydrocarbon liquids of 3 poise viscosity ranging from 1 to 3×10^{-6} cm^2/volt sec. Positive ion mobilities are about thirty percent lower. The mobility estimated above for a negative ion in N5 is therefore quite reasonable for an ordered fluid of this viscosity.

Length of Fully Developed Depletion Layer

Having an estimate of the carrier mobility in the NLC, it is possible to address the question of how far the depletion layer moves in the cell, and ultimately to ask whether this has any significance in distinguishing between the VGM$\perp$ and VGM$\parallel$ modes.

The length of the depletion layer is given by equation 4, modified to take account of the augmented space charge N*.

$$\lambda(t) = \sqrt{\frac{2\varepsilon V}{N^* e}} \ \tanh\left(\frac{t}{\tau^*}\right) . \tag{13}$$

For $t \gg \tau^*$ this function saturates. Thus, the length of the fully developed space charge is given by

$$\lambda_{max} = \sqrt{\frac{2\varepsilon V}{N^* e}} . \tag{14}$$

Since $N^* \simeq 1.6N$, it follows that

$$\lambda_{max} = \sqrt{\frac{1.25\varepsilon V}{Ne}} = \sqrt{\frac{1.25\mu\varepsilon V^2}{J_o L}} . \tag{15}$$

Substituting values of the experimental variables into equation 15, leads to the following:

i)   For the 7.4μm cell, λ_{max} is less than half the cell thickness for all applied voltages. This cell operates only in the VGM$\perp$ mode for all values of applied voltage;

ii) For the 4.7μm cell λ_{max} is somewhat less than the cell thickness for all voltages except the maximum, at which it exceeds the cell thickness. At this voltage a significant increase in J_2 also occurs. This cell is one which may operate spontaneously in either VGM$\perp$ or VGM$\parallel$ mode;

iii) For the 3.7μm cell λ_{max} exceeds the cell thickness for all values of applied voltage. The transient currents always

exhibit a high contribution from J_2. This cell operates exclusively
in the VGM$||$ mode for all values of applied voltage.

From the foregoing it appears that there is a clear distinction
between VGM modes on the basis of the length of a fully developed
charge depletion layer. It is probable that this change in dynamic
electrical boundary conditions leads to the different hydrodynamic
effects which are observed optically.

Condition of Hydrodynamic Stability

So far in this analysis it has been assumed that the ions of
the positive space charge are stationary, but it is unlikely that
these ions can be totally immobile given the observable fact of
hydrodynamic flow.

The motion of the positive ion space charge has interesting consequences for the hydrodynamic stability of the system. It is known
from the work of Schneider and Watson[16], and of Atten and Moreau[17],
that a unipolar space charge current is hydrodynamically unstable
when the voltage across the space charge exceeds a critical voltage.
At the space charge limit (*i.e.*, with a perfect injecting contact)
this critical voltage is given by

$$V_c = \frac{99\mu\eta}{\varepsilon} .$$
(16)

The application of these electrohydrodynamic concepts to the stability of nematic liquid crystals has been discussed by Koelmans
and Van Boxtel[18], and by Lacroix and Tobazéon[19] both of whom used
injecting contacts to study these phenomena.

In the present instance, the operative phenomena are not injection processes, but rather a sweep-out process. However, from
the space charge point of view the process of sweeping out the mobile negative species has a similar effect to injecting the relatively immobile positive species. Referring again to Figure 6 the
interface λ between regions A and B can be regarded as equivalent to
an ohmic contact which is injecting positive ions into region A.
Then, the condition for EHD stability can be applied to region A.

Experimentally, it is found that the threshold voltage at which
EHD effects are observed is about 10 volts, regardless of cell thickness. Moreover, at threshold, there probably exists a fully developed depletion layer, with the entire applied voltage dropped across
the space charge. Knowing the voltage across the space charge
allows use of equation 16 to estimate the mobility of the positive
ions in this system.

Substituting for the threshold voltage (10 volts), the viscosity (3 poise) and the dielectric constant $(5.5\varepsilon_o)$, an approximate value for the positive carrier mobility, is obtained:

$$\mu_p \sim 2 \times 10^{-7} \text{ cm}^2/\text{volt sec}. \tag{17}$$

This is an order of magnitude less than the mobility which we deduced for the negative ion in N5. As pointed out earlier, there is reason to suppose that the positive ion will be trapped in this material. Unfortunately, in the absence of data over a range of temperature it is not possible to assess this trap depth.

CONCLUSIONS

The currents in NLCs which lead to the VGM can be explained in terms of an ionic space charge transient with the assumption that the carrier of one sign is much more mobile than the countercharge.

The current has two components. The first results from the sweepout of a mobile ionic species, and the consequent establishment of a depletion layer. The second component is due to the generation of excess ions within the depletion layer. These two components give rise to an initial peak and a shoulder or secondary peak in the current trace.

The depletion region, which is in effect, a high field region adjacent to one of the electrodes, is established before the hydrodynamic effects set in to generate the VGM modes. The latter occur only when a critical voltage is exceeded.

The fast carrier may be a negative ion. The mobility of the fast carrier is about 7×10^{-6} cm^2/volt sec.

In thick cells at low fields (VGM$\perp$), the fully developed depletion layer is less than the cell thickness; whereas for thin cells and high fields (VGM$\parallel$), the depletion layer crosses the cell. This changes the dynamic electrical boundary conditions. It is, however, not clear how these changes precipitate the molecular orientational dynamics which are observed optically.

REFERENCES

1. V. Frederiks and I.V. Tsvetkov, Dokl. Akad. Nauk SSSR, 4, 123 (1935).

2. G.H. Heilmeier, L.A. Zanoni, and L.A. Barton, Proc. IEEE, 56, 1162 (1968).

3. P.A. Penz and G.W. Ford, Phys. Rev. A, 6, 414 (1972).

4. L.K. Vistin, Kristallografiya, 15, 594 (1970) [Sov. Phys. Crystallogr., 15, 514 (1970)].

5. W. Greubel and U. Wolff, Appl. Phys. Lett., 19, 213 (1971).

6. J.M. Pollack and J.B. Flannery, Liquid Crystals and Ordered Fluids, eds. J.F. Johnson and R.S. Porter, Plenum Press, New York, 1974, pp. 557-71.

7. I.G. Chistyakov and L.K. Vistin, Kristallografiya, 19, 195 (1974) [Sov. Phys. Crystallogr., 19, 119 (1974)].

8. J.M. Pollack, Domain Formation in Nematic Liquid Crystals, Thesis submitted in partial fulfillment of the requirements for the M.S. degree, 1974, The University of Rochester, Rochester, New York.

9. S. Pikin, G. Ryschenkow and W. Urbach, J. Phys., 37, 241 (1976).

10. J.M. Pollack and J.B. Flannery, 1976 SID International Symposium, Digest of Technical Papers, Lewis Winner, New York, 1976, pp. 142-3; J.M. Pollack, J.B. Flannery and P.K. Watson, Sixth International Conference on Liquid Crystals, Kent, Ohio, 1976; P.K. Watson, J.M. Pollack and J.B. Flannery, Conference on Anisotropic Dielectrics, Cambridge, England, 1977; J.M. Pollack, P.K. Watson and J.B. Flannery, Symposium on Ordered Fluids and Liquid Crystals, 174th American Chemical Society National Meeting, Chicago, Ill., 1977, Coll 114.

11. W. Helfrich, J. Chem. Phys., 51, 4092 (1969).

12. P.A. Penz, private communication.

13. R. Williams, J. Chem. Phys., 39, 384 (1963).

14. A. von Hippel, E.P. Gross, J.G. Jelatis, and M. Geller, Phys. Rev., 91, 568 (1953).

15. I. Adamczewski, Ionization, Conductivity and Breakdown in Dielectric Liquids, Taylor and Francis Ltd., London, 1969, pp. 185-232.

16. J.M. Schneider and P.K. Watson, Physics of Fluids, 13, 1948 (1970).

17. P. Atten and R. Moreau, Compt. Rend., A270, 415 (1970).

18. H. Koelmans and A.M. Van Boxtel, Molec. Cryst. Liquid Cryst., 12, 185 (1971).

19. J.C. Lacroix and R. Tobazéon, Appl. Phys. Lett., 20, 251 (1972).

MICROPOLAR THEORY OF LIQUID CRYSTALS[1]

A. Cemal Eringen

Princeton University

Princeton, New Jersey 08540

ABSTRACT

A unified, nonlinear, hydrodynamic theory of liquid crystals is presented. The theory makes use of the balance laws of micropolar continuum mechanics. New strain and rate measures appropriate to liquid crystals and suspensions are introduced. The nonlinear constitutive equations are obtained for the nematic and cholesteric liquid crystals employing these measures and the micro-inertia tensor. The thermodynamic restrictions are studied in detail. The relation of the theory to the director theory is established. The solution is given for the shear flow of nematic liquids in a channel.

1. INTRODUCTION

There exists now a fairly extensive literature on the theory of liquid crystals. After the initial approaches of Oseen [1929, 1933], Anzelius [1931] and Frank [1958] in a series of papers Ericksen [1961-1969] and Leslie [1968a, 1968b] proposed a continuum theory of liquid crystals. A number of other approaches to the subject were made by other authors, e.g., Davison [1967, 1969], Martin et al. [1970], Helfrich [1969-1972], Aero and Bulygin [1970,1971], Lee and Eringen [1971a,b,c], Eringen and Lee [1973]. An assessment of the subject matter was recently made by Stephen and Straley [1974] where extensive references to the literature on the subject are to be found. Together with de Gennes' [1974]

---

[1]The present work was supported by the National Science Foundation.

book on the subject it hardly seems necessary to comment on the
nature and the differences of various theories available today.

Any theory of liquid crystals must necessarily rely on an extra
basic concept, namely the orientability, which is foreign to the
concept of fluidity in the realm of the classical continuum
mechanics. This is also borne out by the physics of liquid crystals.
Thus many of the continuum theories proposed, in one way or another,
make use of the concept of the director.

If one prefers a statistical, mechanical approach to the
subject, provision will have to be made in the nature of material
points of Newtonian mechanics through the introduction of polar
molecules or attached particles. This implies the introduction of
the concept of subcontinua (e.g., rigid or flexible rods) as the
material points of the bodies. In this case it is clear that the
fundamental concept that arises is the inertia per unit mass
(microinertia) rather than the director. Indeed this is in perfect
harmony with classical continuum mechanics, where the mass density
takes the place of Newtonian discrete mass. Although a microinertia tensor can always be associated with a director, the
converse is not necessarily true since it requires a very special
kind of mass distribution over small regions of the space.

In the molecular approach to the subject matter a fundamental
concept that naturally arises is the order parameter which is a
measure of the degree of alignment of the molecules in a small
volume element (cf. de Gennes [1974, p. 24]. This expression is
reminiscent of the deviatoric micro-inertia concept. When the
order parameter vanishes the liquid crystal is considered to be
isotropic. From the continuum viewpoint this may be achieved by
equating the inertia tensor to its trace, just as in the order
parameter. This fact too directs our attention to the fundamental
nature of the concept of microinertia. In fact, since the local
continuum theory can never hope to explain the behavior of
materials in the molecular and atomic scales, the use of the
director to replace an aggregate of molecules, however simplifying
it may be, is, at best, an approximation philosophically not
satisfying as a model. Moreover, it lacks a generalization in the
way to deal with liquid crystals and suspensions whose molecules
(or their aggregates) are not slender, rod-like, and/or they have
perfect geometrical distributions as idealized by the usual
schematic sketches drawn to visualize the structures of nematic,
smectic and cholesteric liquids.[1]

---

[1]In this regard, de Gennes' [1974, p.59] statement for the raison
d'être of a director-type continuum theory, based on the
hypothesis of perfect alignment and straight thread-like elements,
is illuminating.

Clearly a rational continuum theory employing the inertia tensor should reduce, in the limit when the microelements of the body are thin rigid rods, to the director theory. This, in fact, turns out to be the case under various assumptions (see Section 9 here).

With the introduction of the director Ericksen [1961] had to borrow an equation of motion suggested by Oseen [1933] for the director which we have shown to be a special case of the law of balance of moment of momentum of the micropolar continuum mechanics (cf. Eringen [1964a], Lee and Eringen [1973]. Moreover, the law of conservation of inertia has been lacking (or trivially satisfied) in the Ericksen-Leslie theory, because of the very nature of the director.

The raison d'être of the present article stems from these observations and the desire to obtain a unified approach to the subject of liquid crystals. We believe that the theory developed here can also be used profitably in the discussion of mechanics of suspensions. It represents a generalization of our previous work, valid for nonlinear deformations and motions. In the process, not only the basic concepts underlying all liquid crystals come to the surface, but also the fact that the liquid crystal theory is indeed a branch of micropolar continuum mechanics. As a consequence it is clear that the well-established principles and tools of micropolar continuum theories are available for the discussion of many physical phenomena hitherto not cultivated satisfactorily (e.g., electromagnetic effects [2]).

In contrast to our previous work, the approach here is novel in that the theory is hydrodynamic in character at the start, and no need arises for the use of the concept of the natural state hypothesis which seems to have caused some confusion in its interpretations (cf. Shahinpoor [1975], Lee and Eringen [1975]).

In Section 2 new strain and rate measures are introduced which appear to be most appropriate for the mechanics of orientable fluent media. Section 3 contains a summary of the basic laws of micropolar continuum mechanics. In Sections 4 to 8 we obtain the constitutive equations of nematic and cholesteric liquid crystals and study the restrictions arising from the second law of thermodynamics. In Section 9 we obtain the special case when the inertia tensor acquires the special form appropriate to a slender rod. This provides a link between the constitutive equations based on the director and the inertia concepts. Section 10 deals with a shear flow problem based on the present theory. The predictions are gratifyingly in accord with the observed phenomena.

[2]Separate publications are planned for E-M effects.

2. KINEMATICS

A micropolar continuum is a collection of orientable material points. Referred to a rectangular frame of reference X_K, $K=1,2,3$, in the reference state, a material point X of a micropolar continuum is characterized by its position vector $\underset{\sim}{X}_K$ and a director $\underset{\sim}{\Xi}$ attached to the point. The motion, at time t carries $\underset{\sim}{X}$ to a spatial point x with rectangular coordinate x_k and rotates the director Ξ to a new direction $\underset{\sim}{\xi}$. Thus

$$x_k = x_k(\underset{\sim}{X},t), \qquad \xi_k = \chi_{kK}(\underset{\sim}{X},t)\, \Xi_K \qquad (2.1)$$

express the motion and the rotation of a micropolar point. It is assumed that the motion x_k and the microrotation χ_{kK} possess continuous partial derivatives with respect to X_K and t, and they possess unique inverses, i.e.,

$$X_K = X_K(\underset{\sim}{x},t), \qquad \chi_{Kk}^{-1} = \chi_{kK} \qquad (2.2)$$

Consequently,

$$x_{k,K}X_{K,\ell} = \delta_{k\ell} \;,\quad X_{K,k}x_{k,L} = \delta_{KL}$$
$$\chi_{kK}\chi_{\ell K} = \delta_{k\ell} \quad,\quad \chi_{kL}\chi_{kK} = \delta_{KL} \qquad (2.3)$$

where an index following a comma represents partial differentiation, e.g.,

$$x_{k,K} \equiv \partial x_k/\partial X_K \;,\quad X_{K,k} \equiv \partial X_K/\partial x_k$$

and the repeated indices are summed over the range (1,2,3).

By use of $(2.3)_4$ we can solve for Ξ_K:

$$\Xi_K = \chi_{kK}\xi_k \qquad (2.4)$$

By taking the material derivative of $(2.1)_2$ and using (2.4) we obtain

$$\dot{\xi}_k = \nu_{k\ell}\xi_\ell, \qquad \nu_{k\ell}(\underset{\sim}{x},t) \equiv \dot{\chi}_{kK}\chi_{\ell K} = -\nu_{\ell k} \qquad (2.5)$$

where a superposed dot or D/Dt indicates the material time derivative. Here $\nu_{k\ell}$ is the gyration tensor which is equivalent to a vector ν_k called the angular velocity vector

$$\nu_k = -\tfrac{1}{2}\, \varepsilon_{k\ell m}\nu_{\ell m}, \qquad \nu_{k\ell} = -\varepsilon_{k\ell m}\nu_m \qquad (2.6)$$

where $\varepsilon_{k\ell m}$ is the permutation symbol.

For a fluent body it is necessary to consider the relative motion and relative rotations from the present configuration. The relative motion $\underset{\sim}{x}(t)$, the relative director $\underset{\sim}{\xi}(t)$ at time τ with respect to those at time t ($\underset{\sim}{x}$ and $\underset{\sim}{\tau}$) are defined by

$$\underset{\sim}{x}(t) = \underset{\sim}{x}(t) \; (\underset{\sim}{x},\tau) = \underset{\sim}{x}[\underset{\sim}{X}(\underset{\sim}{x},t),\tau]$$

$$\underset{(t)k}{\xi} = \chi_{kK}(\underset{\sim}{X},\tau)\underset{K}{\Xi} = \chi_{kK}(\underset{\sim}{X},\tau)\chi_{\ell K}(\underset{\sim}{X},t)\underset{\ell}{\xi} = \chi_{(t)k\ell}(\underset{\sim}{x},\tau)\underset{\ell}{\xi} \tag{2.7}$$

where we have introduced the relative microrotation

$$\chi_{(t)k\ell}(\underset{\sim}{x},\tau) \equiv \chi_{kK}[\underset{\sim}{X}(\underset{\sim}{x},t),\tau]\chi_{\ell K}[\underset{\sim}{X}(\underset{\sim}{x},t),t] \tag{2.8}$$

Note that $\chi_{(t)k\ell}$ is a <u>two-point</u> tensor, the index k refers to the configuration at time τ and ℓ to that at time t. Thus, this function depends on both t and τ.

Two basic deformation measures of a micropolar continuum are those introduced by Cosserat

$$\mathcal{E}_{KL} \equiv x_{k,K}\chi_{kL} \; , \qquad \Gamma_{KL} \equiv \tfrac{1}{2}\varepsilon_{KMN}\chi_{kM,L}\chi_{kN} \tag{2.9}$$

Of these the first is the generalization of the deformation tensor accustomed in classical elasticity, and the second represents the strain measure arising from the local twist of the directors. Conducive to the study of isotropy are the spatial deformation measures which we now introduce:

$$\mathcal{C}_{k\ell} \equiv X_{K,k}\chi_{\ell K} \; , \qquad \gamma_{k\ell} \equiv \tfrac{1}{2}\varepsilon_{kmn}\chi_{mK}\chi_{nK,\ell} = \Gamma_{KL}\chi_{kK}\chi_{L,\ell} \tag{2.10}$$

For fluent bodies one needs also the deformation rate measures. These are derivable from the relative strain measures defined by (cf., Eringen and Kafadar [1976, p. 10]).

$$\mathcal{C}_{(t)k\ell}(\underset{\sim}{x},\tau) \equiv x_{(t)r,k}\chi_{(t)r\ell}$$

$$\Gamma_{(t)k\ell}(\underset{\sim}{x}\;\tau) \equiv \tfrac{1}{2}\varepsilon_{kmn}\chi_{(t)rm,\ell}\chi_{(t)rn} \tag{2.11}$$

In fact the material time rates of these tensors at $\tau=t$ give all the deformation rate measures, e.g.,

$$\underset{\sim n}{a}(t) \equiv \frac{D^n}{D\tau^n}\mathcal{C}_{(t)}(\tau)\Big|_{\tau=t}, \underset{\sim n}{b} \equiv \frac{D^n}{D\tau^n}\Gamma_{(t)}(\tau)\Big|_{\tau=t} \tag{2.12}$$

For the present investigation we only need the first order rates (n=1), i.e.

$$a_{1k\ell} \equiv a_{k\ell} = v_{\ell,k} + v_{k\ell} \; , \qquad b_{1k\ell} \equiv b_{k\ell} = v_{k,\ell} \tag{2.13}$$

It can be shown that (cf. Eringen [1970, p. 17]) a proper orthogonal tensor $\underset{\sim}{\chi}$ may be expressed in terms of an axial vector $\underset{\sim}{\phi}$ by

$$\chi_{kK} = [\cos \phi \delta_{k\ell} + (1-\cos \phi)n_k n_\ell - \sin \phi \varepsilon_{k\ell m} n_m]\delta_{\ell K} \tag{2.14}$$

where $\delta_{\ell K}$ are the direction cosines (shifters) between x_ℓ and X_K and

$$n_k \equiv \phi_k/\phi, \qquad \phi \equiv (\phi_k \phi_k)^{\frac{1}{2}} \tag{2.15}$$

are respectively the axis of rotation and the angle of rotation of $\underset{\sim}{\chi}$. By means of (2.14) we can express $\gamma_{k\ell}$ in terms of ϕ:

$$\gamma_{k\ell} = n_k \phi_{,\ell} + \sin \phi\, n_{k,\ell} - (1-\cos \phi)\varepsilon_{kmn} n_m n_{n,\ell} \tag{2.16}$$

For computation purposes, also useful is the expression of $\underset{\sim}{\nu}$ given in terms of ϕ (Kafadar and Eringen [1971, p. 274]).

$$\nu_k = \Lambda_{k\ell}\, \dot{\phi}_\ell \tag{2.17}$$

where

$$\Lambda_{k\ell} \equiv \frac{\sin\phi}{\phi}\, \delta_{k\ell} + (1 - \frac{\sin\phi}{\phi})n_k n_\ell - (1-\cos \phi)\varepsilon_{k\ell m} n_m \tag{2.18}$$

We note that when $\phi \ll 1$, then

$$\gamma_{k\ell} \simeq \phi_k, \quad , \quad \nu_k \simeq \dot{\phi}_k \tag{2.19}$$

a result well-known from the linear micropolar theory of Eringen [1966].

3. Balance Laws

The balance laws of liquid crystals are the same as those of the micropolar continua. They are (cf. Eringen [1966] and Eringen and Kafadar [1976])

Mass:

$$\frac{\partial \rho}{\partial t} + (\rho v_k)_{,k} = 0 \text{ in } V - \sigma \tag{3.1}$$

$$[\rho(v_k - u_k)]n_k = 0 \text{ on } \sigma$$

Microinertia:

$$\frac{Dj_{k\ell}}{Dt} - \nu_{km} j_{\ell m} - \nu_{\ell m} j_{km} = 0 \quad \text{in } V - \sigma$$

$$[\rho j_{k\ell}(v_r - u_r)]n_r = 0 \qquad \text{on } \sigma \tag{3.2}$$

Momentum:

$$t_{k\ell,k} + \rho(f_\ell - \dot{v}_\ell) = 0 \quad \text{in } V - \sigma$$

$$[\underset{\sim}{t}_{k\ell} - \rho v_\ell(v_k - u_k)]\, n_k = 0 \qquad \text{on } \sigma$$

(3.3)

Moment of momentum:

$$m_{k\ell,k} + \varepsilon_{\ell mn} t_{mn} + \rho(\ell_\ell - \dot{\sigma}_\ell) = 0 \quad \text{in } V - \sigma$$

$$[\underset{\sim}{m}_{k\ell} - \rho\sigma_\ell(v_k - u_k)]n_k = 0 \qquad \text{on } \sigma$$

(3.4)

Energy:

$$\rho\dot{\varepsilon} - t_{k\ell}a_{k\ell} - m_{k\ell}b_{\ell k} - q_{k,k} - \rho h = 0 \quad \text{in } V - \sigma$$

$$[(\rho\varepsilon + \tfrac{1}{2}\rho\underset{\sim}{v}\cdot\underset{\sim}{v} + \tfrac{1}{2}\rho\underset{\sim}{\sigma}\cdot\underset{\sim}{v})(v_k - u_k) - t_{k\ell}v_\ell - m_{k\ell}\nu_\ell - q_k]n_k = 0$$

on σ

(3.5)

Entropy inequality (the second law of thermodynamics):

$$\rho\dot{\eta} - (q_k/\theta)_{,k} - (\rho h/\theta) \geq 0 \qquad \text{in } V - \sigma$$

$$[\rho\eta(v_k - u_k) - \theta^{-1}q_k]n_k \geq 0 \qquad \text{on } \sigma$$

(3.6)

For our purpose, a useful form of $(3.6)_1$ is the generalized Clausius-Duhem inequality obtained by eliminating h between $(3.5)_1$ and $(3.6)_1$:

$$-\frac{\rho}{\theta}(\dot{\psi} + \eta\dot{\theta}) + \frac{1}{\theta}t_{k\ell}a_{k\ell} + \frac{1}{\theta}m_{k\ell}b_{\ell k} + \frac{q_k}{\theta^2}\theta_{,k} \geq 0$$

(3.7)

where $\psi \equiv \varepsilon - \theta\eta$ is the Helmholtz free energy and

ρ	$\equiv$ mass density,	v_k	$\equiv$ velocity vector
$j_{k\ell}$	$\equiv$ microinertia tensor	$\nu_{k\ell}$	$\equiv$ gyration tensor
$t_{k\ell}$	$\equiv$ stress tensor,	f_ℓ	$\equiv$ body force density
$m_{k\ell}$	$\equiv$ couple stress tensor	ℓ_ℓ	$\equiv$ body couple density
ε	$\equiv$ internal energy density	q_k	$\equiv$ heat vector
η	$\equiv$ entropy density,	θ	$\equiv$ absolute temperature
h	$\equiv$ the heat source,		

Equations $(3.1)_1$ to $(3.5)_1$ are respectively, the local balance laws of mass, microinertia, momentum, moment of momentum and energy. The inequality $(3.6)_1$ is the expression of the second law of thermodynamics for the micropolar continua. The spin inertia is defined by

$$\dot{\sigma}_k = \frac{D}{Dt}(j_{k\ell}\nu_\ell) = j_{k\ell}\dot{\nu}_\ell - \varepsilon_{kmr}j_{\ell m}\nu_r\nu_\ell$$

(3.8)

Accompanying each one of the balance laws are the jump conditions

$(3.1)_2$ to $(3.6)_2$ at a discontinuity surface σ which may be sweeping the body with a velocity u_k in the direction of the positive unit normal n_k (not to be confused with n_k introduced by (2.15)) of σ.

For future use we note the integral of $(3.2)_1$ which was first derived by Eringen [1964b]

$$j_{k\ell} = J_{KL} \chi_{kK} \chi_{\ell L} \tag{3.9}$$

where J_{KL} is the microinertia tensor at the natural state of the body.

4. Constitutive Equations

The state of a fluent micropolar continuum is determined by the characterization of the dependent constitutive variables (response functions)

$$t_{k\ell} \; , \; m_{k\ell} \; , \; q_k \; , \; \Psi \; , \; \eta \tag{4.1}$$

as functions of certain independent variables that characterize the constitution of the body in motion. For the first order rate dependent, orientable, fluent materials, independent variables may be established according to axioms of causality and objectivity as (see Eringen [1966], Eringen and Kafadar [1976])

$$\rho^{-1} \; , \; j_{k\ell} \; , \; \gamma_{k\ell} \; , \; a_{k\ell} \; , \; b_{k\ell} \; , \; \theta, \; \theta_{,k} \tag{4.2}$$

Note the absence from this list of the strain measure $\varsigma_{k\ell}$ which would be included in the case of viscoelastic micropolar bodies (Eringen [1967]).

For the constitutive equations of liquid crystals, we write equations of the form

$$\psi = \psi(\rho^{-1}, \theta, \underset{\sim}{j}, \underset{\sim}{\gamma}, \underset{\sim}{a}, \underset{\sim}{b}, \nabla\theta) \tag{4.3}$$

Similar equations containing the same list of variables are valid for other members of the response functions (4.1), according to the principle of equipresence. The constitutive equations are restricted by:

(i) The axiom of objectivity (material frame-indifference)
(ii) The second law of thermodynamics.

The axiom of objectivity requires that the response functions remain form-invariant under time-dependent rigid motions of the spatial frame of reference (with ρ^{-1} fixed for fluids) as described by

$$\bar{\underset{\sim}{x}}(\underset{\sim}{X}, t) = \underset{\sim}{Q}(t)\underset{\sim}{x}(\underset{\sim}{X}, t) + \underset{\sim}{c}_o(t)$$

$$\bar{\chi}_K(\underset{\sim}{X}, t) = \underset{\sim}{Q}(t) \, \chi_K(\underset{\sim}{X}, t) \tag{4.4}$$

where $c_0(t)$ is an arbitrary time dependent translation and $\{Q(t)\}$ represent the proper group of orthogonal transformations, i.e.,

$$QQ^T = Q^TQ = 1, \text{ det } Q = 1 \tag{4.5}$$

For nematic and smectic liquid crystals $\{Q\}$ is the full group of orthogonal transformations so that the center of symmetry is also included,[1] i.e.,

$$\text{det } Q = \pm 1 \tag{4.6}$$

The axiom of objectivity requires that scalar, vector and tensor response functions transform as

$$\Psi(\rho^{-1},\theta,\bar{j},\bar{\gamma},\bar{a},\bar{b},\bar{\nabla}\theta) = \Psi(\rho^{-1},\theta,j,\gamma,a,b,\nabla\theta)$$

$$q(\rho^{-1},\theta,\bar{j},\bar{\gamma},\bar{a},\bar{b},\bar{\nabla}\theta) = \bar{q}(\rho^{-1},\theta,j,\gamma,a,b,\nabla\theta) \tag{4.7}$$

$$t(\rho^{-1},\theta,\bar{j},\bar{\gamma},\bar{a},\bar{b},\bar{\nabla}\theta) = \bar{t}(\rho^{-1},\theta,j,\gamma,a,b,\nabla\theta)$$

where

$$\{\bar{j},\bar{a},\bar{t}\} = Q(t) \{j,a,t\} Q^T(t)$$

$$\{\bar{\gamma},\bar{b},\bar{m}\} = Q(t)\{\gamma,b,m\} Q^T(t) \text{ det } Q \tag{4.8}$$

$$\bar{q} = Q(t)q$$

$$\bar{\nabla}\theta = Q(t)\nabla\theta$$

The constitutive independent variables obeying (4.8) are called objective and the dependent variables obeying (4.7) are called hemitropic if $\{Q\}$ is the proper group and isotropic if it is the full group. It is clear that objective vectors and tensors are the proper candidates for the characterization of the material properties.

We first investigate the consequence of the entropy inequality (3.7). To this end, we note

---

[1]This dual use of the axiom of objectivity, one with center of symmetry one without, raises a question that remains unsettled to date. While it is not supposed to have any connection with material symmetry regulation but only connected with the form-invariance of the response functions under the change of the spatial frame of reference it is a statement of the fact that the configurations obtained by 180° rotations of the elements of nematic and smectic liquid crystals are identical to their original states. This, however, is not the case for cholesteric liquid crystals.

$$\frac{D\rho^{-1}}{Dt} = -\rho^{-2}\dot{\rho} = \rho^{-1}v_{k,k}$$

$$\dot{\gamma}_{k\ell} = b_{k\ell} + v_{kr}\gamma_{r\ell} + v_{\ell r}\gamma_{kr} - \gamma_{kr}a_{\ell r}$$

(4.9)

In $(4.9)_1$ we employed $(3.1)_1$. In deriving $(4.9)_2$, we employed $X_{kK} = v_{k\ell}X_{\ell K}$ obtained from $(2.5)_2$, $DX_{K,k}/Dt = -X_{K,r}v_{r,k}$ and (2.13).

Upon substituting (4.3) into (3.7) and using (4.9) we obtain

$$-\frac{\rho}{\theta}\left(\frac{\partial\psi}{\partial\theta} + \eta\right)\dot{\theta} + \frac{1}{\theta}\left(t_{k\ell} - \frac{\partial\psi}{\partial\rho^{-1}}\delta_{k\ell} + \rho\frac{\partial\psi}{\partial\gamma_{rk}}\gamma_{r\ell}\right)a_{k\ell}$$

$$+\frac{1}{\theta}\left(m_{k\ell} - \frac{\rho\partial\psi}{\partial\gamma_{\ell k}}\right)b_{\ell k} - \frac{\rho}{\theta}\left(\frac{\partial\psi}{\partial j_{kr}}j_{\ell r} + \frac{\partial\psi}{\partial j_{rk}}j_{r\ell} - \frac{\partial\psi}{\partial\gamma_{kr}}\gamma_{\ell r} - \frac{\partial\psi}{\partial\gamma_{rk}}\gamma_{r\ell}\right)v_{k\ell}$$

$$-\frac{\rho}{\theta}\frac{\partial\psi}{\partial a_{k\ell}}\dot{a}_{k\ell} - \frac{\rho}{\theta}\frac{\partial\psi}{\partial b_{k\ell}}\dot{b}_{k\ell} - \frac{\rho}{\theta}\frac{\partial\psi}{\partial\theta,k}\dot{\theta},_k + \frac{1}{\theta^2}q_k\theta,_k \geq 0$$

(4.10)

This inequality is posited not to be violated for all independent variations of $\theta, a, b, v, \dot{a}, \dot{b}, \overline{\theta,_k}$, and $\theta,_k$. It is linear in $\dot{\theta}, \dot{a}, \dot{b}$, and $\overline{\theta,_k}$. For arbitrary and independent variations of these quantities this inequality cannot be maintained unless

$$\eta = -\frac{\partial\psi}{\partial\theta}, \quad \frac{\partial\psi}{\partial a_{k\ell}} = \frac{\partial\psi}{\partial b_{k\ell}} = 0, \qquad \frac{\partial\psi}{\partial\theta,_k} = 0$$

(4.11)

so that ψ is independent of a, b, and $\nabla\theta$.

Since $\psi(\rho^{-1}, \theta, \underset{\sim}{j}, \underset{\sim}{\gamma})$ must satisfy $(4.7)_1$ for arbitrary rigid motions, it is simple to show that

$$\frac{\partial\psi}{\partial j_{kr}}j_{\ell r} + \frac{\partial\psi}{\partial j_{rk}}j_{r\ell} - \frac{\partial\psi}{\partial\gamma_{kr}}\gamma_{\ell r} - \frac{\partial\psi}{\partial\gamma_{rk}}\gamma_{r\ell}$$

$$= \frac{\partial\psi}{\partial j_{\ell r}}j_{kr} + \frac{\partial\psi}{\partial j_{r\ell}}j_{rk} - \frac{\partial\psi}{\partial\gamma_{\ell r}}\gamma_{kr} - \frac{\partial\psi}{\partial\gamma_{r\ell}}\gamma_{rk}$$

(4.12)

This is the skew-symmetric part of the coefficient of $v_{k\ell}$ in (4.10). Conversely it can be shown that the solution of (4.12) obeys $(4.7)_1$. With this, (4.10) reduces to

$$\frac{1}{\theta}D^t_{k\ell}a_{\ell k} + \frac{1}{\theta}D^m_{k\ell}b_{\ell k} + \frac{1}{\theta^2}q_k\theta,_k \geq 0$$

(4.13)

where

$$D^t_{k\ell} = t_{\ell k} - E^t_{\ell k}, \qquad D^m_{k\ell} = m_{k\ell} - E^m_{k\ell}$$

(4.14)

and

$$E^t_{k\ell} = -\pi\delta_{k\ell} - E^m_{kr}\,\gamma_{r\ell}\,, \qquad E^m_{k\ell} = \rho\,\frac{\partial\psi}{\partial\gamma_{\ell k}}$$

$$\pi = -\frac{\partial\psi}{\partial\rho^{-1}}\,, \qquad \psi = \psi(\rho^{-1},\theta,j,\gamma) \tag{4.15}$$

Here π is the thermodynamic pressure, $\underset{\sim}{E^t}$ and $\underset{\sim}{E^m}$ are the equilibrium parts of the stress and couple stress, respectively.

If we assume that $\underset{\sim}{D^t},\underset{\sim}{D^m}$ and q are continuous functions of a, b and $\nabla\theta$, from (4.13) it follows that

$$\underset{\sim}{D^t} = \underset{\sim}{D^m} = 0,\ q=0 \quad\text{when } a = b = 0,\ \nabla\theta = 0 \tag{4.16}$$

We have therefore proved:

Theorem: Constitutive equations of liquid crystals are thermodynamically admissible if and only if they are of the forms $(4.11)_1$, (4.14), (4.15) subject to (4.13) (and consequently (4.16)).

5. Reduction of Constitutive Equations

The axion of objectivity (4.7) places restrictions on the constitutive functions $\psi,\underset{\sim}{D^t},\underset{\sim}{D^m}$ and q. For the scalar function ψ of two second order tensors j and $\gamma(4.7)_1$ implies that ψ will be a function of certain minimal number invariants of j and γ. The complete irreducible sets of invariants (the function basis) of two symmetric tensors $\underset{\sim}{j}$, $\underset{\sim}{\gamma}_S$ and one antisymmetric tensor γ_A, defined by

$$\gamma_S \equiv \frac{1}{2}(\underset{\sim}{\gamma} + \underset{\sim}{\gamma}^T)\,, \quad \gamma_A \equiv \frac{1}{2}(\underset{\sim}{\gamma} - \underset{\sim}{\gamma}^T)$$

are given by (cf., Wang [1970]).

$$I_1 \equiv \mathrm{tr}\gamma_S\,, \qquad I_2 \equiv \mathrm{tr}\,\gamma_S^2, \qquad I_3 \equiv \mathrm{tr}\,\gamma_S^3$$

$$I_4 \equiv \mathrm{tr}\,\underset{\sim}{j} \qquad I_5 \equiv \mathrm{tr}\,\underset{\sim}{j}^2, \qquad I_6 \equiv \mathrm{tr}\,\underset{\sim}{j}^3,$$

$$I_7 \equiv \mathrm{tr}\,\gamma_A^2, \qquad I_8 \equiv \mathrm{tr}\,(\gamma_S\underset{\sim}{j}) \qquad I_9 \equiv \mathrm{tr}(\gamma_S^2\underset{\sim}{j}),$$

$$I_{10} \equiv \mathrm{tr}(\gamma_S\underset{\sim}{j}^2), \qquad I_{11} \equiv \mathrm{tr}\,(\gamma_S^2\underset{\sim}{j}^2), \qquad I_{12} \equiv \mathrm{tr}\,(\gamma_S\gamma_A^2).$$

$$I_{13} \equiv \mathrm{tr}(\gamma_S^2\gamma_A^2), \qquad I_{14} \equiv \mathrm{tr}(\gamma_S^2\gamma_A\gamma_S\gamma_A),\ I_{15} \equiv \mathrm{tr}\,(\underset{\sim}{j}\gamma_A^2),$$

$$I_{16} \equiv tr(\underset{\sim}{j}^2 \underset{\sim}{\gamma}_A^2), \qquad I_{17} \equiv tr(\underset{\sim}{j}^2 \underset{\sim}{\gamma}_A^2 \underset{\sim}{j} \underset{\sim}{\gamma}_A),$$

$$I_{18} \equiv tr(\underset{\sim}{\gamma}_S \underset{\sim}{j} \underset{\sim}{\gamma}_A), \qquad I_{19} \equiv tr(\underset{\sim}{\gamma}_S^2 \underset{\sim}{j} \underset{\sim}{\gamma}_A),$$

$$I_{20} \equiv tr(\underset{\sim}{\gamma}_S \underset{\sim}{j}^2 \underset{\sim}{\gamma}_A), \qquad I_{21} \equiv tr(\underset{\sim}{\gamma}_S \underset{\sim}{\gamma}_A^2 \underset{\sim}{j} \underset{\sim}{\gamma}_A), \qquad (5.1)$$

$$I_{22} \equiv tr(\underset{\sim}{j} \underset{\sim}{\gamma}_S \underset{\sim}{\gamma}_A^2)$$

Thus ψ is a function of ρ^{-1}, θ and 22 invariants I_α. By differentiation, we obtain

$$2\underset{\sim}{E}^m/\rho = 2\frac{\partial\psi}{\partial\underset{\sim}{\gamma}^T} = \frac{\partial\psi}{\partial I_1}\underset{\sim}{I} + 2\frac{\partial\psi}{\partial I_2}\underset{\sim}{\gamma}_S + 3\frac{\partial\psi}{\partial I_3}\underset{\sim}{\gamma}_S^2 + 2\frac{\partial\psi}{\partial I_7}\underset{\sim}{\gamma}_A + \frac{\partial\psi}{\partial I_8}\underset{\sim}{j}$$

$$+ \frac{\partial\psi}{\partial I_9}(\underset{\sim}{\gamma}_S\underset{\sim}{j} + \underset{\sim}{j}\underset{\sim}{\gamma}_S) + \frac{\partial\psi}{\partial I_{10}}\underset{\sim}{j}^2 + \frac{\partial\psi}{\partial I_{11}}(\underset{\sim}{\gamma}_S\underset{\sim}{j}^2 + \underset{\sim}{j}^2\underset{\sim}{\gamma}_S)$$

$$+ \frac{\partial\psi}{\partial I_{12}}(\underset{\sim}{\gamma}_A^2 + \underset{\sim}{\gamma}_A\underset{\sim}{\gamma}_S + \underset{\sim}{\gamma}_S\underset{\sim}{\gamma}_A) + \frac{\partial\psi}{\partial I_{13}}(\underset{\sim}{\gamma}_S\underset{\sim}{\gamma}_A^2 + \underset{\sim}{\gamma}_A^2\underset{\sim}{\gamma}_S + \underset{\sim}{\gamma}_A\underset{\sim}{\gamma}_S^2 + \underset{\sim}{\gamma}_S^2\underset{\sim}{\gamma}_A)$$

$$+ \frac{\partial\psi}{\partial I_{14}}(\underset{\sim}{\gamma}_S\underset{\sim}{\gamma}_A^2\underset{\sim}{\gamma}_S\underset{\sim}{\gamma}_A + \underset{\sim}{\gamma}_A^2\underset{\sim}{\gamma}_S\underset{\sim}{\gamma}_A\underset{\sim}{\gamma}_S + \underset{\sim}{\gamma}_A\underset{\sim}{\gamma}_S\underset{\sim}{\gamma}_A^2 + \underset{\sim}{\gamma}_A\underset{\sim}{\gamma}_S\underset{\sim}{\gamma}_A\underset{\sim}{\gamma}_S + \underset{\sim}{\gamma}_S\underset{\sim}{\gamma}_A\underset{\sim}{\gamma}_S\underset{\sim}{\gamma}_A^2 + \underset{\sim}{\gamma}_S^2\underset{\sim}{\gamma}_A\underset{\sim}{\gamma}_S)$$

$$+ \frac{\partial\psi}{\partial I_{15}}(\underset{\sim}{\gamma}_A\underset{\sim}{j} + \underset{\sim}{j}\underset{\sim}{\gamma}_A) + \frac{\partial\psi}{\partial I_{16}}(\underset{\sim}{\gamma}_A\underset{\sim}{j}^2 + \underset{\sim}{j}^2\underset{\sim}{\gamma}_A) \qquad (5.2)$$

$$+ \frac{\partial\psi}{\partial I_{17}}(\underset{\sim}{\gamma}_A\underset{\sim}{j}\underset{\sim}{\gamma}_A\underset{\sim}{j}^2 + \underset{\sim}{j}\underset{\sim}{\gamma}_A\underset{\sim}{j}^2\underset{\sim}{\gamma}_A + \underset{\sim}{j}^2\underset{\sim}{\gamma}_A^2\underset{\sim}{j}) + \frac{\partial\psi}{\partial I_{18}}(\underset{\sim}{j}\underset{\sim}{\gamma}_A + \underset{\sim}{\gamma}_S\underset{\sim}{j})$$

$$+ \frac{\partial\psi}{\partial I_{19}}(\underset{\sim}{\gamma}_S\underset{\sim}{j}\underset{\sim}{\gamma}_A + \underset{\sim}{j}\underset{\sim}{\gamma}_A\underset{\sim}{\gamma}_S + \underset{\sim}{\gamma}_S^2\underset{\sim}{j}) + \frac{\partial\psi}{\partial I_{20}}(\underset{\sim}{j}^2\underset{\sim}{\gamma}_A + \underset{\sim}{\gamma}_S\underset{\sim}{j}^2)$$

$$+ \frac{\partial\psi}{\partial I_{21}}(\underset{\sim}{\gamma}_A^2\underset{\sim}{j}\underset{\sim}{\gamma}_A + \underset{\sim}{\gamma}_A\underset{\sim}{j}\underset{\sim}{\gamma}_A\underset{\sim}{\gamma}_S + \underset{\sim}{j}\underset{\sim}{\gamma}_A\underset{\sim}{\gamma}_S\underset{\sim}{\gamma}_A + \underset{\sim}{\gamma}_S\underset{\sim}{\gamma}_A\underset{\sim}{j})$$

$$+ \frac{\partial\psi}{\partial I_{22}}(\underset{\sim}{\gamma}_A^2\underset{\sim}{j} + \underset{\sim}{\gamma}_A\underset{\sim}{j}\underset{\sim}{\gamma}_S + \underset{\sim}{j}\underset{\sim}{\gamma}_S\underset{\sim}{\gamma}_A)$$

For the equilibrium part of the stress tensor we have:

$$\underset{\sim}{E}^t = -\pi\underset{\sim}{I} - \underset{\sim}{E}^m\underset{\sim}{\gamma} \qquad (5.3)$$

The complete irreducible invariants of $\underset{\sim}{j}, \gamma, a, b$ and $\nabla\theta$, that appear in the arguments of $\underset{\sim}{D}^t, \underset{\sim}{D}^m$ and $\underset{\sim}{q}$ and the final forms of these dissipative parts are much too lengthy to produce here. In the following section, we shall study some special cases, e.g., equations linear in $\underset{\sim}{a}, \underset{\sim}{b}$, and $\nabla\theta$. Two special cases of interest are:

(i) <u>Non-heat conducting materials.</u> In this case $\nabla\theta$ is absent in the constitutive equations and the Clausius-Duhem inequality (4.13) gives

$$\underset{\sim}{q} = \underset{\sim}{0} \tag{5.4}$$

(ii) <u>Dissipation due to intrinsic rotations is absent.</u> In this case $\underset{\sim}{b}$ is not a constitutive variable and (4.13) gives

$$\underset{\sim}{D}^m = \underset{\sim}{0} \tag{5.5}$$

Of great interest is the combination of cases (i) and (ii). In this case we have

$$\underset{\sim}{t} = -\pi \underset{\sim}{I} - \underset{E\underset{\sim}{\sim}}{m\gamma} + \underset{D}{t}^T(\rho^{-1},\theta,\underset{\sim}{j},\underset{\sim}{\gamma},\underset{\sim}{a}) \tag{5.6}$$

$$\underset{\sim}{m} = \underset{E}{m}, \qquad \underset{\sim}{q} = \underset{\sim}{0}$$

where $\underset{E}{m}$ is given by (5.2).

6. Special Constitutive Equations (Nematic Liquid Crystals)

From the exact constitutive equations one can obtain various approximate theories in an obvious way by expanding the response functions into power series in some of the argument vectors and tensors and retaining various terms up to a fixed degree. Important among these is the case in which $\underset{D}{t},\underset{D}{m}$ and $\underset{\sim}{q}$ depend on $\underset{\sim}{a},\underset{\sim}{b}$, and $\nabla\theta$, linearly and independent of $\underset{\sim}{\gamma}$. This means that the dissipative power is quadratic in $\underset{\sim}{a},\underset{\sim}{b}$ and $\nabla\theta$ with directional dependence arising from the microinertia tensor $\underset{\sim}{j}$. Considering the fact that $\underset{D}{t}$ is an absolute tensor, $\underset{D}{m}$, is a relative tensor and $\underset{\sim}{q}$ is an absolute vector, we can write down the forms of these tensors by employing the tables provided by Wang [1970]. Thus, for the liquid crystals with center of symmetry (nematic and smectic), after some lengthy manipulations we arrive at

$$\underset{D}{t} = \alpha_1 \underset{\sim}{I} + \alpha_2 \underset{\sim}{j} + \alpha_3 \underset{\sim}{j}^2 + \alpha_4 \underset{\sim}{a} + \alpha_5 \underset{\sim}{a}^T + \alpha_6 \underset{\sim}{ja} + \alpha_7 \underset{\sim}{aj}$$

$$+ \alpha_8 \underset{\sim}{ja}^T + \alpha_9 \underset{\sim}{a}^T j + \alpha_{10} \underset{\sim}{j}^2 a + \alpha_{11} \underset{\sim}{aj}^2 + \alpha_{12} \underset{\sim}{j}^2 a^T \tag{6.1}$$

$$+ (\alpha_{10} + \alpha_{11} - \alpha_{12})\underset{\sim}{a}^T j^2 + \alpha_{13}(\underset{\sim}{jaj}^2 - \underset{\sim}{j}^2 aj + \underset{\sim}{ja}^T j^2 - \underset{\sim}{j}^2 a^T j),$$

$$\underset{\sim}{q} = (\kappa_1 \underset{\sim}{I} + \kappa_2 \underset{\sim}{j} + \kappa_3 \underset{\sim}{j}^2)\nabla\theta \tag{6.2}$$

The constitutive equation for $_D\tilde{m}$ is identical to (6.1) with α_i
and $\underset{\sim}{a}$ replaced by β_i and b, respectively. The material moduli α_r
and β_r, $r=1,2,3$ have the forms

$$\alpha_r = \tilde{\alpha}_{r1} \text{ tr } \underset{\sim}{a} + \tilde{\alpha}_{r2} \text{ tr } \underset{\sim\sim}{ja} + \tilde{\alpha}_{r3} \text{ tr } \underset{\sim}{j}^2\underset{\sim}{a}, \quad r = 1,2,3 \qquad (6.3)$$

Here $\tilde{\alpha}_{rs}$ and $\tilde{\beta}_{rs}$, $(r,s = 1,2,3)$ and α_2 to α_{13}, β_2 to β_{13} and κ_1
to κ_3 are functions of ρ^{-1}, $\text{tr } \underset{\sim}{j}$, $\text{tr } \underset{\sim}{j}^2$, and $\text{tr } \underset{\sim}{j}^3$. In arriving
at these results we also used (4.16).

We reiterate that the second law of thermodynamics (inequality
(4.13)) places restrictions on these moduli. We study these
restrictions for the special case in which <u>constitutive equations</u>
<u>are linear in j</u> also. In this case we have the simpler forms

$$D_{\sim}^t = (\alpha_1 \text{ tr } \underset{\sim}{a} + \alpha_2 \text{tr } \underset{\sim}{j} \text{ tr } \underset{\sim}{a} + \alpha_3 \text{ tr } \underset{\sim}{j} \underset{\sim}{a}) I + \alpha_4 \underset{\sim}{j} \text{ tr } \underset{\sim}{a}$$

$$+ (\alpha_5 + \alpha_6 \text{ tr } \underset{\sim}{j}) \underset{\sim}{a} + (\alpha_7 + \alpha_8 \text{ tr } \underset{\sim}{j}) \underset{\sim}{a}^T + \alpha_9 \underset{\sim\sim}{ja}$$

$$+ \alpha_{10} \underset{\sim\sim}{aj} + \alpha_{11} \underset{\sim\sim}{ja}^T + \alpha_{12} \underset{\sim}{a}^T\underset{\sim}{j}, \qquad (6.4)$$

$$\underset{\sim}{q} = [(\kappa_1 + \kappa_2 \text{ tr } \underset{\sim}{j}) I + \kappa_3 \underset{\sim}{j}] \nabla\theta \qquad (6.5)$$

where α_i (different from α_i used in (6.1)), β_i (material moduli
for $_D\tilde{m}$, and κ_i are now functions of ρ^{-1} and θ only.

In the spirit of this special case we also obtain special
constitutive equations for the equilibrium parts of the stress
and couple stress tensors. To this end we only need to retain
the invariants $I_1, I_2, I_4, I_7, I_8, I_9, I_{15}$ and I_{18} from the list
(5.1). With a new nomenclature

$$J_1 \equiv \text{tr } \underset{\sim}{\gamma}, J_2 \equiv \text{tr } \underset{\sim}{\gamma}^2, J_3 \equiv \text{tr } (\underset{\sim\sim}{\gamma\gamma}^T),$$

$$J_4 \equiv \text{tr } \underset{\sim}{j}, J_5 \equiv \text{tr } (\underset{\sim\sim}{\gamma j}), J_6 \equiv \text{tr } (\underset{\sim}{\gamma}^2\underset{\sim}{j}), \qquad (6.6)$$

$$J_7 \equiv \text{tr } (\underset{\sim}{\gamma}^T\underset{\sim\sim}{\gamma j}), J_8 \equiv \text{tr } (\underset{\sim\sim}{\gamma\gamma}^T\underset{\sim}{j})$$

We write

$$\psi = \frac{1}{2} A_1 J_1^2 + \frac{1}{2} A_2 J_1^2 J_4 + A_3 J_1 J_5 + \frac{1}{2} A_4 J_2 + \frac{1}{2} A_5 J_2 J_4$$

$$+ \frac{1}{2} A_6 J_3 + \frac{1}{2} A_7 J_3 J_4 + A_8 J_6 + \frac{1}{2} A_9 J_7 + \frac{1}{2} A_{10} J_8 \qquad (6.7)$$

where A_1 to A_{10} are functions of ρ^{-1} and θ. Hence

$$E_{\sim}^{m\,T}/\rho = (A_1 \operatorname{tr} \gamma + A_2 \operatorname{tr} j \operatorname{tr} \gamma + A_3 \operatorname{tr} \gamma j) \underset{\sim}{I} + A_3 j \operatorname{tr} \gamma$$

$$+ (A_4 + A_5 \operatorname{tr} j) \gamma^T + (A_6 + A_7 \operatorname{tr} j) \gamma \qquad (6.8)$$

$$+ A_8 (j\gamma^T + \gamma^T j) + A_9 \gamma j + A_{10} j \gamma$$

With this the stress constitutive equations are complete since the expression of $E_{\sim}^{t}$ follows from (5.3).

7.   Thermodynamic Restrictions

Since $\underset{\sim}{a}, \underset{\sim}{b}$ and $\nabla\theta$ are not coupled in the expressions of $_D t$, $_D m$ and $\underset{\sim}{q}$, it follows that the Clausius-Duhem inequality (4.13) is not violated for arbitrary and independent variations of $\underset{\sim}{a}, \underset{\sim}{b}$ and $\nabla\theta$ if and only if

$$\operatorname{tr} (_D \underset{\sim}{ta}) \geq 0 \quad , \quad \operatorname{tr} (_D \underset{\sim}{mb}) \geq 0 \quad , \quad \underset{\sim}{q} \cdot \nabla\theta \geq 0 \qquad (7.1)$$

Substituting $\underset{\sim}{q}$ from (6.5) into $(7.1)_3$ we find that

$$\underset{\sim}{q} \cdot \nabla\theta = \bar{\kappa}_{ij} \theta_{,i} \theta_{,j} \geq 0 \qquad (7.2)$$

where

$$\bar{\kappa}_{ij} = (\kappa_1 + \kappa_2 \operatorname{tr} j) \delta_{ij} + \kappa_3 j_{ij} \qquad (7.3)$$

The necessary and sufficient conditions for (7.2) not to be violated are

$$\bar{\kappa}_{11} \geq 0 \quad , \quad \begin{vmatrix} \bar{\kappa}_{11} & \bar{\kappa}_{12} \\ \bar{\kappa}_{21} & \bar{\kappa}_{22} \end{vmatrix} \geq 0 \quad , \quad \begin{vmatrix} \bar{\kappa}_{11} & \bar{\kappa}_{12} & \bar{\kappa}_{13} \\ \bar{\kappa}_{21} & \bar{\kappa}_{22} & \bar{\kappa}_{23} \\ \bar{\kappa}_{31} & \bar{\kappa}_{32} & \bar{\kappa}_{33} \end{vmatrix} \geq 0 \qquad (7.4)$$

If these inequalities shall not be violated for all non-negative matrices j including $(j_{ij}=0, \infty)$ then it is necessary and sufficient that

$$\kappa_1, \kappa_2, \kappa_3 \geq 0 \qquad (7.5)$$

To determine the implications $(7.1)_1$ we express it as

$$\operatorname{tr}(_D \underset{\sim}{ta}) = _D t_{ji} a_{ij} = \alpha_{ijk\ell} a_{ij} a_{k\ell} \qquad (7.6)$$

where

$$\alpha_{ijk\ell} = [(\alpha_1 + \alpha_2 \mathrm{tr}\underset{\sim}{j})\delta_{k\ell} + \alpha_3 j_{k\ell}]\delta_{ij} + \alpha_4 j_{ij}\delta_{k\ell}$$

$$+ (\alpha_5 + \alpha_6 \mathrm{tr}\underset{\sim}{j})\delta_{jk}\delta_{i\ell} + (\alpha_7 + \alpha_8 \mathrm{tr}\underset{\sim}{j})\delta_{ik}\delta_{j\ell} \qquad (7.7)$$

$$+ \alpha_9 j_{jk}\delta_{i\ell} + \alpha_{10} j_{i\ell}\delta_{jk} + \alpha_{11} j_{j\ell}\delta_{ik} + \alpha_{12} j_{ik}\delta_{j\ell}$$

By writing $a_{11}=a_1$, $a_{22}=a_2$, $a_{33}=a_3$, $a_{12}=a_4$, $a_{23}=a_5$, $a_{31}=a_6$, $a_{21}=a_7$, $a_{32}=a_8$, $a_{13}=a_9$ we can express (7.6) as:

$$\mathrm{tr}(\underset{D}{\underset{\sim}{t}}a) = \bar{\alpha}_{ij} a_i a_j \quad , \quad (i,j=1,2,\ldots,9) \qquad (7.8)$$

If we accept Onsager relations $\bar{\alpha}_{ij}=\bar{\alpha}_{ji}$ then

$$\alpha_3 = \alpha_4 \quad , \quad \alpha_9 = \alpha_{10} \qquad (7.9)$$

Referred to the principal axes of $j_{k\ell}$ the mixed components of $j_{k\ell}$ vanish and we obtain for the 9x9 symmetric matrix $\bar{\alpha}_{ij}$

$$\|\bar{\alpha}_{ij}\| = \begin{vmatrix} \bar{\alpha}_{11} & \bar{\alpha}_{12} & \bar{\alpha}_{13} & 0 & 0 & 0 & 0 & 0 & 0 \\ & \bar{\alpha}_{22} & \bar{\alpha}_{23} & 0 & 0 & 0 & 0 & 0 & 0 \\ & & \bar{\alpha}_{33} & 0 & 0 & 0 & 0 & 0 & 0 \\ & & & \bar{\alpha}_{44} & 0 & 0 & \bar{\alpha}_{47} & 0 & 0 \\ & & & & \bar{\alpha}_{55} & 0 & 0 & \bar{\alpha}_{58} & 0 \\ & & & & & \bar{\alpha}_{66} & 0 & 0 & \bar{\alpha}_{69} \\ & & & & & & \bar{\alpha}_{77} & 0 & 0 \\ & & & & & & & \bar{\alpha}_{88} & 0 \\ & & & & & & & & \bar{\alpha}_{99} \end{vmatrix} \qquad (7.10)$$

where

$$\bar{\alpha}_{ii}=\alpha_{iiii}=\alpha_1+\alpha_5+\alpha_7+(\alpha_2+\alpha_6+\alpha_8)\mathrm{tr}\underset{\sim}{j}+(2\alpha_3+2\alpha_9+\alpha_{11}+\alpha_{12})j_{ii} \quad ,$$

$$\bar{\alpha}_{ij}=\alpha_{iijj}=\alpha_1+\alpha_2\mathrm{tr}\underset{\sim}{j}+\alpha_3(j_{ii}+j_{jj}) \quad , \quad i,j=1,2,3$$

$$\bar{\alpha}_{44}=\alpha_{1212}=\alpha_7+\alpha_8 \mathrm{tr}\underset{\sim}{j}+\alpha_{11}j_{22}+\alpha_{12}j_{11} \quad ,$$

$$\bar{\alpha}_{55}=\alpha_{2323}=\alpha_7+\alpha_8 \mathrm{tr}\underset{\sim}{j}+\alpha_{11}j_{33}+\alpha_{12}j_{22} \quad ,$$

$$\bar{\alpha}_{66}=\alpha_{3131}=\alpha_7+\alpha_8 \mathrm{tr}\underset{\sim}{j}+\alpha_{11}j_{11}+\alpha_{12}j_{33} \quad , \qquad (7.11)$$

$$\bar{\alpha}_{77}=\alpha_{2121}=\alpha_7+\alpha_8 \mathrm{tr}\underset{\sim}{j}+\alpha_{11}j_{11}+\alpha_{12}j_{22} \quad ,$$

$$\bar{\alpha}_{88}=\alpha_{3232}=\alpha_7+\alpha_8 \mathrm{tr}\underset{\sim}{j}+\alpha_{11}j_{22}+\alpha_{12}j_{33} \quad ,$$

$$\bar{\alpha}_{99}=\alpha_{1313}=\alpha_7+\alpha_8 \mathrm{tr}\underset{\sim}{j}+\alpha_{11}j_{33}+\alpha_{12}j_{11} \quad ,$$

$$\bar{\alpha}_{47}=\alpha_{1221}=\alpha_5+\alpha_6 \mathrm{tr}\underset{\sim}{j}+\alpha_9(j_{11}+j_{22}) \quad ,$$

$$\bar{\alpha}_{58}=\alpha_{2332}=\alpha_5+\alpha_6 \mathrm{tr}\underset{\sim}{j}+\alpha_9(j_{22}+j_{33}) \quad ,$$

$$\bar{\alpha}_{69}=\alpha_{3113}=\alpha_5+\alpha_6 \mathrm{tr}\underset{\sim}{j}+\alpha_9(j_{33}+j_{11}) \quad ,$$

The necessary and sufficient conditions for (7.8) to be a non-negative definite form are that all subdeterminants along the main diagonal of (7.10) must be non-negative. Hence

$$\bar{\alpha}_{11}\geq 0 \quad , \quad \begin{vmatrix} \bar{\alpha}_{11} & \bar{\alpha}_{12} \\ \bar{\alpha}_{21} & \bar{\alpha}_{22} \end{vmatrix}\geq 0 \quad , \quad \begin{vmatrix} \bar{\alpha}_{11} & \bar{\alpha}_{12} & \bar{\alpha}_{13} \\ \bar{\alpha}_{21} & \bar{\alpha}_{22} & \bar{\alpha}_{23} \\ \bar{\alpha}_{31} & \bar{\alpha}_{32} & \bar{\alpha}_{33} \end{vmatrix}\geq 0$$

$$\bar{\alpha}_{44}\geq 0 \quad , \quad \bar{\alpha}_{55}\geq 0 \quad , \quad \bar{\alpha}_{66}\geq 0 \quad , \qquad (7.12)$$

$$\begin{vmatrix} \bar{\alpha}_{44} & \bar{\alpha}_{47} \\ \bar{\alpha}_{74} & \bar{\alpha}_{77} \end{vmatrix}\geq 0 \quad , \quad \begin{vmatrix} \bar{\alpha}_{55} & \bar{\alpha}_{58} \\ \bar{\alpha}_{85} & \bar{\alpha}_{88} \end{vmatrix}\geq 0 \quad , \quad \begin{vmatrix} \bar{\alpha}_{66} & \bar{\alpha}_{69} \\ \bar{\alpha}_{96} & \bar{\alpha}_{99} \end{vmatrix}\geq 0$$

Recalling that $j_{ii}\geq 0$, if we wish that the inequalities (7.12) should not be violated for all j_{ii} ($j_{ii}=0$ and $j_{ii}=\infty$ included) then we must have

$$\alpha_5+\alpha_7\geq 0, \quad \alpha_7-\alpha_5\geq 0, \quad 3\alpha_1+\alpha_5+\alpha_7\geq 0 \quad ,$$

$$\alpha_6+\alpha_8\geq 0, \quad \alpha_8-\alpha_6\geq 0, \quad 3\alpha_2+\alpha_6+\alpha_8\geq 0 \qquad (7.13)$$

$$\alpha_8+\alpha_{11}\geq 0, \quad \alpha_8+\alpha_{12}\geq 0, \quad -2\alpha_6+2\alpha_8-2\alpha_9+\alpha_{11}+\alpha_{12}\geq 0,$$

$$\alpha_2+2\alpha_3+\alpha_6+\alpha_8+2\alpha_9+\alpha_{11}+\alpha_{12}\geq 0 \quad ,$$

$$(\alpha_8 + \alpha_{11})(\alpha_8 + \alpha_{12}) - (\alpha_6 + \alpha_9)^2 \geq 0 \quad ,$$

$$(\alpha_6 + \alpha_8 + 2\alpha_9 + \alpha_{11} + \alpha_{12})(2\alpha_2 + \alpha_6 + \alpha_8) - (\alpha_2 + 2\alpha_3)(\alpha_6 + \alpha_8) - 2\alpha_3^2 \geq 0$$

Thus we have proved:

Theorem: The necessary and sufficient conditions for the non-
equilibrium parts of the stress and heat vector not to violate
the second law of thermodynamics for all independent variations
of a, $\nabla\theta$ and for all positive definite matrices j are (7.4) and
(7.12). The conditions on β_i (the rotational viscosities) are
identical to (7.12).

If the above conditions are not to be violated for all non-
negative j (including $j_{ii} = 0, \infty$) then (7.5) and (7.13) must not
be violated. Whether (7.5) and (7.13) are both necessary and
sufficient needs further investigation.

Physically it is expected that as the twisting increases the
free energy will also increase, i.e. $2\psi = E^m_{ji} \gamma_{ij}$ must be a non-
negative function of γ. Classically this is equivalent to the
material stability condition for the elastic solids. To obtain
these conditions we write

$$E^m_{ji} = \rho A_{ijk\ell} \gamma_{k\ell} \tag{7.14}$$

where

$$\begin{aligned}
A_{ijk\ell} = A_{k\ell ij} &= (A_1 + A_2 \operatorname{tr} j)\delta_{ij}\delta_{k\ell} + A_3(j_{ij}\delta_{k\ell} + j_{k\ell}\delta_{ij}) \\
&+ (A_4 + A_5 \operatorname{tr} j)\delta_{jk}\delta_{i\ell} + (A_6 + A_7 \operatorname{tr} j)\delta_{ik}\delta_{j\ell} \\
&+ A_8(j_{jk}\delta_{i\ell} + j_{i\ell}\delta_{jk}) + A_9 j_{j\ell}\delta_{ik} + A_{10} j_{ik}\delta_{j\ell}
\end{aligned} \tag{7.15}$$

By comparing this with (7.7) we make identifications:

$$\begin{aligned}
&\rho A_1 \sim \alpha_1 \quad , \quad \rho A_2 \sim \alpha_2 \quad , \quad \rho A_3 \sim \alpha_3 \quad , \quad \rho A_4 \sim \alpha_5 \quad , \quad \rho A_5 \sim \alpha_6 \quad , \\
&\rho A_6 \sim \alpha_7 \quad , \quad \rho A_7 \sim \alpha_8 \quad , \quad \rho A_8 \sim \alpha_9 \quad , \quad \rho A_9 \sim \alpha_{11} \quad , \quad \rho A_{10} \sim \alpha_{12}
\end{aligned} \tag{7.16}$$

The inequalities on ρA_i can now be written down immediately by
substituting from (7.16) ρA_i for α_i into (7.12) and (7.13). We
reproduce only the results corresponding to (7.13) here.

$$\rho\,(A_4 + A_6) \geq 0 \quad,$$

$$\rho\,(A_6 - A_4) \geq 0 \quad,$$

$$\rho\,(3A_1 + A_4 + A_6) \geq 0 \quad,$$

$$\rho\,(A_5 + A_7) \geq 0 \quad,$$

$$\rho\,(A_7 - A_5) \geq 0 \quad,$$

$$\rho\,(3A_2 + A_5 + A_7) \geq 0 \quad, \tag{7.17}$$

$$\rho\,(A_7 + A_9) \geq 0 \quad,$$

$$\rho\,(A_7 + A_{10}) \geq 0 \quad,$$

$$\rho\,(-2A_5 + 2A_7 - 2A_8 + A_9 + A_{10}) \geq 0 \quad,$$

$$\rho\,(A_2 + 2A_3 + A_5 + A_7 + 2A_8 + A_9 + A_{10}) \geq 0 \quad,$$

$$\rho^2(A_7 + A_9)(A_7 + A_{10}) - \rho^2(A_5 + A_8)^2 \geq 0 \quad,$$

$$\rho^2(A_5 + A_7 + 2A_8 + A_9 + A_{10})(2A_2 + A_5 + A_7) - \rho^2(A_2 + 2A_3)$$
$$\cdot (A_5 + A_7) - 2\rho^2 A_3^2 \geq 0$$

8. Cholesteric Liquid Crsytals

For the cholesteric liquid crystals the material frame indifference requires the use of the proper group of orthogonal transformations. In this case then axial and polar vectors and tensors are not distinguished. Therefore, the response functions $D_{\sim}^t, D_{\sim}^m$ and q can now depend on a, b and $\nabla\theta$ simultaneously. The constitutive equations (6.4) will now contain the following additional terms:

$$D_{\sim}^{t\,c} = (\alpha_1' \ \mathrm{tr}\ b + \alpha_2' \mathrm{tr}\ j\ \mathrm{tr}\ b + \alpha_3' \mathrm{tr}\ jb)I + \alpha_4' \ j\ \mathrm{tr}\ b$$
$$+ (\alpha_5' + \alpha_6'\ \mathrm{tr}\ j)b + (\alpha_7' + \alpha_8'\ \mathrm{tr}\ j)b^T + \alpha_9'\ jb$$
$$+ \alpha_{10}'bj + \alpha_{11}'jb^T + \alpha_{12}'b^Tj + \alpha_{13}'j\theta^d + \alpha_{14}'\theta^d j \tag{8.1}$$

$$q_i^c = \kappa_4'\theta\varepsilon_{ijk}a_{j\ell}j_{\ell k} + \kappa_5'\theta\varepsilon_{ijk}j_{j\ell}a_{\ell k}$$

where θ^d is the dual of $\nabla\theta$, i.e.,

$$\theta_{ij}^d \equiv \varepsilon_{ijk}\theta_{,k} \tag{8.2}$$

The expression $_D\underset{\sim}{m}$ will also contain terms identical to $(8.1)_1$ with α'_i and $\underset{\sim}{b}$ replaced by β_i and $\underset{\sim}{a}$, respectively. These additional material moduli α'_i, β'_i and κ'_i are, in general, functions of ρ^{-1} and θ.

In the literature the viscosity moduli β_i and β'_i are not accounted for. Such effects may be small; however, consistent with the physics of the matter they represent the internal friction against the intrinsic rotations of the molecular elements. Experimentally they should be detectable by means of experiments on the relaxation of twisted molecules.

The restrictions arising from the second law of thermodynamics can be studied in a similar fashion to Section 7. In order to simplify the matter here we ignore the dissipation due to intrinsic rotation, which according to (5.6) gives $_D\underset{\sim}{m} = 0$ and $_D\underset{\sim}{t}^c$ reduces to

$$D^t{}_{ij} = \alpha'_{13} j_{ik} \varepsilon_{kj\ell} \theta_{,\ell} + \alpha'_{14} \varepsilon_{ik\ell} \theta_{,\ell} j_{kj} \tag{8.3}$$

The contribution of these terms and those of q_i^c in $(8.1)_2$ to the entropy inequality (4.13) is

$$[(\alpha'_{13} - \kappa'_4)\varepsilon_{kj\ell} j_{ik} + (\alpha'_{14} - \kappa'_5)\varepsilon_{ik\ell} j_{kj}] a_{ji} \theta_{,\ell}$$

which must be non-negative independently. Hence it is sufficient that

$$\alpha'_{13} - \kappa'_4 = 0 \;, \; \alpha'_{14} - \kappa'_5 = 0 \tag{8.4}$$

Thus the constitutive equations of cholesteric liquid crystals are of the form

$$D^t{}_{ij} = (\alpha_1 \operatorname{tr} \underset{\sim}{a} + \alpha_2 \operatorname{tr} \underset{\sim}{j} \operatorname{tr} \underset{\sim}{a} + \alpha_3 \operatorname{tr} \underset{\sim}{ja})\delta_{ij} + \alpha_4 j_{ij} \operatorname{tr} \underset{\sim}{a}$$

$$+ (\alpha_5 + \alpha_6 \operatorname{tr} \underset{\sim}{j})a_{ij} + (\alpha_7 + \alpha_8 \operatorname{tr} \underset{\sim}{j})a_{ji} + \alpha_9 j_{ik} a_{kj}$$

$$+ \alpha_{10} a_{ik} j_{kj} + \alpha_{11} j_{ik} a_{jk} + \alpha_{12} a_{ki} j_{kj} + \tag{8.5}$$

$$+ \kappa'_4 j_{ik} \varepsilon_{kj\ell} \theta_{,\ell} + \kappa'_5 \varepsilon_{ik\ell} \theta_{,\ell} j_{kj} \tag{8.6}$$

$$q_i = (\kappa_1 + \kappa_2 \operatorname{tr} \underset{\sim}{j})\theta_{,i} + \kappa_3 j_{ik} \theta_{,k} + \kappa'_4 \theta \varepsilon_{ijk} a_{j\ell} j_{\ell k} + \kappa'_5 \theta \varepsilon_{ijk} j_{j\ell} a_{\ell k}$$

in the case when the rotational viscosities are negligible, hence $_D\underset{\sim}{m} = 0$. The material moduli α_i and κ_i are subject to (7.4), (7.5), (7.12) and (7.13).

9. Relation to the Director Theory

If it is assumed that the thin bar-like elements of the liquid crystals have vanishingly small thickness and they are perfectly aligned then a great deal of simplification can be

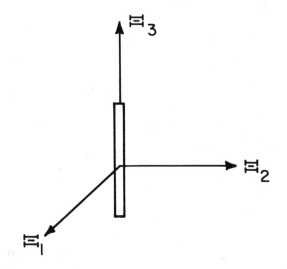

FIGURE 1:   THREAD LIKE ELEMENTS (NEMATIC LIQUIDS)

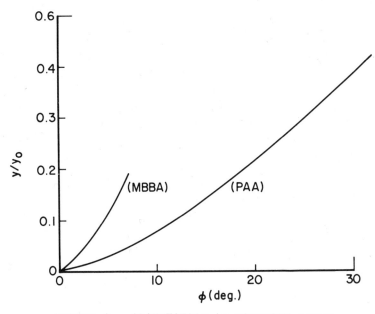

FIGURE 2:   ORIENTATION IN BOUNDARY LAYER

achieved because of the special approximate expression that the micro-inertia tensor takes. Selecting the Ξ_3-axis of the material frame of reference along a thread-like element of the liquid crystal and Ξ_1 and Ξ_2 through its center of mass, Fig. 1, we find that the inertia tensor J_{KL} at the natural state is given by

$$J_{KL} = I_o \left(\delta_{KL} - \delta_{K3}\delta_{L3} \right) \tag{9.1}$$

where I_o is the common constant value of the micro-inertia about Ξ_1 and Ξ_2 axes. Through (3.9) this gives for the spatial inertia tensor

$$j_{k\ell} = I_o \left(\delta_{k\ell} - d_k d_\ell \right) \tag{9.2}$$

where we set

$$\chi_{k3} = d_k \tag{9.3}$$

so that

$$d_k d_k = 1 \tag{9.4}$$

We employ (9.2) and (9.4) to simplify the constitutive equations. Thus for example (6.1) reduces to the form

$$
\begin{aligned}
{}_D t_{ij} = & (\bar{\alpha}_{11} \text{tr } \underset{\sim}{a} + \bar{\alpha}_{12} a_{k\ell} d_k d_\ell) \delta_{ij} + (\bar{\alpha}_{21} \text{tr } \underset{\sim}{a} + \bar{\alpha}_{22} a_{k\ell} d_k d_\ell) d_i d_j \\
& + \bar{\alpha}_3 a_{ij} + \bar{\alpha}_4 a_{ji} + \bar{\alpha}_5 a_{kj} d_k d_i + \bar{\alpha}_6 a_{ik} d_k d_j \\
& + \bar{\alpha}_7 a_{jk} d_k d_i + \bar{\alpha}_8 a_{ki} d_k d_j
\end{aligned} \tag{9.5}
$$

where

$$
\begin{aligned}
\bar{\alpha}_{11} &= \tilde{\alpha}_{11} + (\tilde{\alpha}_{12} + \tilde{\alpha}_{21}) I_o + (\tilde{\alpha}_{13} + \tilde{\alpha}_{22} + \tilde{\alpha}_{31}) I_o^2 + (\tilde{\alpha}_{23} + \tilde{\alpha}_{32}) I_o^3 + \tilde{\alpha}_{33} I_o^4 \\
\bar{\alpha}_{12} &= -[\tilde{\alpha}_{12} I_o + (\tilde{\alpha}_{13} + \tilde{\alpha}_{22}) I_o^2 + (\tilde{\alpha}_{23} + \tilde{\alpha}_{32}) I_o^3 + \tilde{\alpha}_{33} I_o^4] \\
\bar{\alpha}_{21} &= -[\tilde{\alpha}_{21} I_o + (\tilde{\alpha}_{22} + \tilde{\alpha}_{31}) I_o^2 + (\tilde{\alpha}_{23} + \tilde{\alpha}_{32}) I_o^3 + \tilde{\alpha}_{33} I_o^4] \\
\bar{\alpha}_{22} &= -\tilde{\alpha}_{22} I_o^2 + (\tilde{\alpha}_{32} - \tilde{\alpha}_{23}) I_o^3 + \tilde{\alpha}_{33} I_o^4 \\
\bar{\alpha}_3 &= \alpha_4 + (\alpha_6 + \alpha_7) I_o + (\alpha_{10} + \alpha_{11}) I_o^2 \\
\bar{\alpha}_4 &= \alpha_5 + (\alpha_8 + \alpha_9) I_o + (\alpha_{10} + \alpha_{11} - \alpha_{12}) I_o^2 \\
\bar{\alpha}_5 &= -\alpha_6 - \alpha_{10} I_o^2 \ , \quad \bar{\alpha}_6 = \alpha_7 I_o - \alpha_{11} I_o^2 \\
\bar{\alpha}_7 &= -\alpha_8 I_o - \alpha_{12} I_o^2 \\
\bar{\alpha}_8 &= -\alpha_9 I_o - (\alpha_{10} + \alpha_{11} - \alpha_{12}) I_o^2
\end{aligned} \tag{9.6}
$$

Equations (9.5) can be reduced to the form given by Ericksen and Leslie by noticing the following identities among the variables of the present theory and those of Ericksen [1969] and Leslie [1968 a]:

$$A_{ij} = a_{(ij)}, \quad N_i = a_{[ij]}d_j$$
$$a_{[ij]} = a_o(d_i N_j - d_j N_i), \quad N_i d_i = 0 \tag{9.7}$$

where A_{ij} and N_i are the independent variables in the Ericksen-Leslie theory, i.e.

$$A_{ij} = v_{(i,j)} \equiv \frac{1}{2}(v_{i,j} + v_{j,i})$$
$$N_i = \dot{d}_i + w_{ki}d_k, \quad w_{ki} = v_{[k,i]} \equiv \frac{1}{2}(v_{k,i} - v_{i,k}) \tag{9.8}$$

The expression of $a_{[ij]}$ follows from the fact that the axial vector $\underset{\sim}{a}$ coresponding to $a_{[ij]}$ is normal to $\underset{\sim}{N}$ and $\underset{\sim}{d}$ hence $\underset{\sim}{a} = a_o \underset{\sim}{d} x \underset{\sim}{N}$ where a_o is a constant. Employing (9.7) and incorporating the coefficient of δ_{ij} in (9.5) to the unknown pressure p in the stress tensor, for the incompressible case (for which tr $\underset{\sim}{a} = 0$), we obtain

$$D^t_{ij} = \mu_1 d_k d_p A_{kp} d_i d_j + \mu_2 d_j N_i + \mu_3 d_i N_j + \mu_4 A_{ij} +$$
$$\mu_5 d_j d_k A_{ki} + \mu_6 d_i d_k A_{kj} \tag{9.9}$$

where

$$\mu_1 = \bar{\alpha}_{22}, \mu_2 = a_o(\bar{\alpha}_8 - \bar{\alpha}_6), \qquad \mu_3 = a_o(\bar{\alpha}_5 - \bar{\alpha}_7),$$
$$\mu_4 = \bar{\alpha}_3 + a_o \bar{\alpha}_4, \qquad \mu_5 = \bar{\alpha}_6 + \bar{\alpha}_8, \quad \mu_6 = \bar{\alpha}_5 + \bar{\alpha}_6 \tag{9.10}$$

Similarly, the "director-form" of (6.2) is obtained to be

$$q_i = (\bar{\kappa}_1 - \bar{\kappa}_2 d_i d_j)\theta_{,j} \tag{9.11}$$

where

$$\bar{\kappa}_1 = \kappa_1 + \kappa_2 I_o + \kappa_3 I_o^2, \qquad \bar{\kappa}_2 = \kappa_2 I_o + \kappa_3 I_o^2 \tag{9.12}$$

There is no particular difficulty in transforming the additional terms in the constitutive equations (8.1) of the cholesteric liquid crystals to those of Leslie [1968b].

This establishes the relations of the dynamical parts of the constitutive equations of the Ericksen-Leslie and the present theory. Note that the first term in (9.9) is of fourth degree in $\underset{\sim}{d}$ and would be left out had we used our equation (6.4) which is linear in $\underset{\sim}{j}$.

Transformations of the constitutive equations for the equilibrium part of the couple stresses $\underset{\sim}{E}m$ is somewhat more involved since it requires the development of the relationship between our strain measure γ_{ij} and $(d_{i,j}, d_i)$ used in the Ericksen-Leslie theory. The use of $d_{i,j}$ alone does not seem to correspond directly to γ_{ij} for we notice that $d_{i,jk} = d_{i,kj}$ but $\gamma_{ij,k} \neq \gamma_{ik,j}$. However, since χ_{kK} admits a representation in terms of a single vector (see 2.14) it is clear that γ_{ij} must be expressible in terms of such a vector and its gradient. From the theory of invariants it then follows that the free energy used here as a function of γ_{ij} and j_{ij} is reducible to that used in the Ericksen-Leslie theory, when j_{ij} is of the form (9.2). Together with our previous discussion on the balance laws (Eringen [1964a], Lee and Eringen [1973] it is clear that under the assumptions of straight thread-like elements, perfect alignment and incompressibility the Ericksen-Leslie theory results as a special case of the present theory.

10.   Channel Flow of Nematic Liquids

Here we consider the solution of the steady flows of a nematic liquid in a channel with parallel walls. We take the x-axis of the rectangular coordinates in the direction of the velocity field, which is parallel to the walls and the y-axis perpendicular to the walls. We leave the orientations of the elements (e.g. the inertia tensor at the natural state) to be prescribed later. The assumption of the channel flow is that

$$v_1 = u(y), \quad v_2 = 0, \quad v_3 = 0$$
$$\phi_1 = 0, \quad \phi_2 = 0, \quad \phi_3 = \phi(y) \tag{10.1}$$

Thus the flow is in the direction of the x-axis and the directors are constrained to rotate about the z-axis only. Whether such a steady flow of an incompressible nematic liquid is possible or not according to the present theory, and the results predicted would be crucial tests for the theory developed here.

From (2.15) we have $n_1 = n_2 = 0$, $n_3 = 1, \phi = \phi_3$. Hence, (2.16) to (2.18) give:

$$\gamma_{k\ell} = \phi' \delta_{\ell 2} \delta_{k3}, \qquad \nu_k = \dot{\phi}\,\delta_{k3} \tag{10.2}$$

where $\phi' \equiv \partial\phi/\partial y$. Under the assumption of steady flow $\dot{\phi} \equiv (\partial\phi/\partial y)\dot{y} = 0$ so that $\nu_k = 0$. From (2.13) then we have:

$$a_{k\ell} = u' \delta_{1\ell} \delta_{2k}, \quad b_{k\ell} = 0 \tag{10.3}$$

The microrotation tensor χ_{kK} follows from (2.14):

$$\chi_{kK} = [\cos\phi\,\delta_{k\ell}+(1-\cos\phi)\delta_{k3}\delta_{\ell3} - \sin\phi\,\varepsilon_{k\ell m}\delta_{m3}]\,\delta_{\ell K} \qquad (10.4)$$

so that the inertia tensor $j_{k\ell}$ given by (3.9) takes the form:

$$j_{k\ell} = J_{KL}\chi_{kK}\chi_{\ell L}=(J^{O}_{k\ell} -J^{O}_{3\ell}\delta_{k3}-J^{O}_{k3}\delta_{\ell3}+J^{O}_{33}\delta_{k3}\delta_{\ell3})\cos^2\phi$$

$$+ (J^{O}_{3\ell}\delta_{k3}+J^{O}_{k3}\delta_{\ell3}-2J^{O}_{33}\delta_{k3}\delta_{\ell3})\cos\phi+J^{O}_{33}\delta_{k3}\delta_{\ell3}$$

$$+ (-\varepsilon_{kr3}J^{O}_{r\ell}+\varepsilon_{kr3}J^{O}_{r3}\delta_{\ell3}-\varepsilon_{\ell r3}J^{O}_{kr}+\varepsilon_{\ell r3}J^{O}_{3r}\delta_{k3})\sin\phi\cos\phi \qquad (10.5)$$

$$+ (-\varepsilon_{kr3}\delta_{\ell3}J^{O}_{r3}-\varepsilon_{\ell r3}\delta_{k3}J^{O}_{3r})\sin\phi$$

$$+ \varepsilon_{kr3}\varepsilon_{\ell s3}J^{O}_{rs}\sin^2\phi$$

where

$$J^{O}_{k\ell} \equiv J_{KL}\delta_{kK}\delta_{\ell L} \qquad (10.6)$$

After some tedious calculations the constitutive equations (6.8), (5.3) and (6.4) are obtained to be:

$$m_{ji}/\rho = A_{ji}\phi' \quad, \quad {}_{E}t_{ij} = -p\,\delta_{ij}-\rho A_{i3}\delta_{j2}\phi'^2$$

$$D^{t}_{ij} = \alpha_{ij}u' \qquad (10.7)$$

where p is the unknown pressure and

$$A_{ji} \equiv A_3 j_{23}\delta_{ij}+(A_4+A_5 trj)\delta_{i2}\delta_{j3}+(A_6+A_7 trj)\delta_{i3}\delta_{j2}$$

$$+ A_8(j_{i2}\delta_{j3}+\delta_{i2}j_{3j})+A_9 j_{2j}\delta_{i3}+A_{10}j_{i3}\delta_{j2} \quad, \qquad (10.8)$$

$$\alpha_{ij} \equiv \alpha_3 j_{12}\delta_{ij}+(\alpha_5+\alpha_6 trj)\delta_{2i}\delta_{1j}+(\alpha_7+\alpha_8 trj)\delta_{1i}\delta_{2j}$$

$$+ \alpha_9(j_{i2}\delta_{1j}+j_{1j}\delta_{2i})+\alpha_{11}j_{i1}\delta_{2j}+\alpha_{12}\delta_{1i}j_{2j}$$

Substituting m_{ji} and $t_{ij}={}_{E}t_{ij}+{}_{D}t_{ji}$ into the equations of motion (3.3), and (3.4), we obtain six equations for the determination of three unknowns u,ϕ and p:

$$-p_{,i} + (-\rho A_{23}\delta_{i2}\phi'^2+\alpha_{i2}u')' = 0 \qquad (10.9)$$

$$(\rho A_{2i}\phi')'+\varepsilon_{ijk}(-\rho A_{j3}\delta_{k2}\phi'^2+\alpha_{kj}u') = 0 \qquad (10.10)$$

The fact that we are dealing with nematic liquids whose long axis is constrained to remain in the (x,y)-plane implies through (10.5) that $j_{13}=j_{23}=0$. With this we find that the first two equations of (10.10), (for i=1,2) are satisfied identically. Also the third equation of (10.9), (for i=3) gives $p_{,3}=0$ so that $p=p(x,y)$. Using this in the second equation of (10.9), (for i=2) we obtain:

$$p = -\rho [A_6 + A_7 \operatorname{tr} \underset{\sim}{j} + A_9 \, j_{22} + A_{10} \, j_{33}]\phi'^2$$
$$+ (\alpha_3 + \alpha_9 + \alpha_{11}) \, j_{12} u' + p_o(x) \tag{10.11}$$

where $p_o(x)$ is an arbitrary function of integration (an arbitrary pressure).   This leaves two additional equations one from (10.9), and one from (10.10). They are given below:

$$\alpha(\phi)u' - p_{o,x} y - C_1 = 0 \tag{10.12}$$

$$[f(\phi)\phi']' - g(\phi)(\phi')^2 + \beta(\phi)u' = 0 \tag{10.13}$$

where

$$\alpha(\phi) = \alpha_7 + \alpha_8 \operatorname{tr} \underset{\sim}{j} + \alpha_{11} \, j_{11} + \alpha_{12} \, j_{22}$$
$$= \frac{1}{2} (\lambda_o - \lambda_2 \cos 2\phi - \lambda_3 \sin 2\phi)$$

$$\beta(\phi) = \alpha_5 - \alpha_7 + (\alpha_6 - \alpha_8)\operatorname{tr} \underset{\sim}{j} + (\alpha_9 - \alpha_{11})j_{11} + (\alpha_9 - \alpha_{12})j_{22}$$
$$= \frac{1}{2} (\lambda_1 + \lambda_2 \cos 2\phi + \lambda_3 \sin 2\phi) \quad,$$

$$f(\phi) = \rho(A_6 + A_7 \operatorname{tr} \underset{\sim}{j} + A_9 j_{22} + A_{10} j_{33})$$
$$= \frac{1}{2} (f_o + f_2 \cos 2\phi + f_3 \sin 2\phi) \quad,$$

$$g(\phi) = \rho A_9 j_{21} = \frac{1}{2} \frac{df}{d\phi} \quad, \tag{10.14}$$

$$\lambda_o = 2\alpha_7 + (2\alpha_8 + \alpha_{11} + \alpha_{12})\operatorname{tr}\underset{\sim}{J}^o - (\alpha_{11} + \alpha_{12})J_{33}^o \quad,$$

$$\lambda_1 = 2(\alpha_5 - \alpha_7) + (2\alpha_6 - 2\alpha_8 + 2\alpha_9 - \alpha_{11} - \alpha_{12})\operatorname{tr}\underset{\sim}{J}^o - (2\alpha_9 - \alpha_{11} - \alpha_{12})J_{33}^o ,$$

$$\lambda_2 = (\alpha_{12} - \alpha_{11})(J_{11}^o - J_{22}^o) \quad,$$

$$\lambda_3 = 2(\alpha_{11} - \alpha_{12})J_{12}^o \quad,$$

$$f_o = \rho[2A_6 + (2A_7 + A_9)\operatorname{tr}\underset{\sim}{J}^o + (2A_{10} - A_9)J_{33}^o] \quad,$$

$$f_2 = \rho A_9 (J_{22}^o - J_{11}^o)$$

$$f_3 = 2\rho A_9 J_{12}^o$$

To obtain the right hand sides of the second equal signs in $\alpha(\phi)$ to $g(\phi)$ we employed (10.5). By use of (10.14), equations (10.12) and (10.13) may be written as:

$$u' = (C_1 + p_{o,x}y)\alpha^{-1}(\phi) \quad , \tag{10.15}$$

$$2f(\phi)\phi'' + \frac{df}{d\phi}(\phi')^2 + 2\beta(\phi)u' = 0 \tag{10.16}$$

In the case of thread-like elements that are oriented in the direction of x-axis, at the natural state, we have $J^o_{22} = J^o_{33} = J^o$ and all other $J^o_{KL} = 0$ so that

$$2\alpha(\phi) = \lambda_o - \lambda_2\cos2\phi \quad ,$$

$$2\beta(\phi) = \lambda_1 + \lambda_2\cos2\phi \quad ,$$

$$f(\phi) = k_{11}\cos^2\phi + k_{33}\sin^2\phi \quad ,$$

$$2g(\phi) = \frac{df}{d\phi} \tag{10.17}$$

$$\lambda_o = 2\alpha_7 + (4\alpha_8 + \alpha_{11} + \alpha_{12})J^o \quad ,$$

$$\lambda_1 = 2(\alpha_5 - \alpha_7) + (4\alpha_6 - 4\alpha_8 + 2\alpha_9 - \alpha_{11} - \alpha_{12})J^o \quad ,$$

$$\lambda_2 = (\alpha_{11} - \alpha_{12})J^o \quad , \quad \lambda_3 = 0$$

$$k_{11} = \rho[A_6 + (2A_7 + A_9 + A_{10})J^o],$$

$$k_{33} = \rho[A_6 + (2A_7 + A_{10})J^o]$$

In this special case (10.15) has the same form as eq. (6.11) of Leslie [1968a] and (10.16) coincides with his eq. (6.13).

Shear Flow. In the case of shear flow with no pressure gradient we give the solution of (10.12) and (10.13) which is valid for molecules having an arbitrary inertia tensor. A natural set of boundary conditions are

$$u = 0 \quad , \quad \phi = \phi_1 \quad \text{at} \quad y = -h$$

$$u = u_o \quad , \quad \phi = \phi_o \quad \text{at} \quad y = h \tag{10.18}$$

To integrate (10.13) we change the independent variable to ϕ by setting $(\phi')^2 = \psi(\phi)$ then the first integral of (10.13) is obtained to be

$$(\phi')^2 = f^{-1}(\phi)[C_2 - 2C_1\int^\phi \beta\alpha^{-1}d\phi] \tag{10.19}$$

Let ϕ_c be a critical angle at which $\phi'=0$, then (10.19) may be written as

$$(\phi')^2 = C_1F(\phi;\phi_c) \tag{10.20}$$

where

$$F(\phi;\phi_c) = 2f^{-1}(\phi) \int_\phi^{\phi_c} \beta\alpha^{-1}d\phi \qquad (10.21)$$

Carrying $\alpha(\phi)$ and $\beta(\phi)$ from (10.14), the integration in (10.21) may be performed leading to:

$$F(\phi;\phi_c) = f^{-1}(\phi)[F_1(\phi_c)-F_1(\phi)] \qquad (10.22)$$

where

$$F_1(\phi) = \phi + \frac{\lambda_o+\lambda_1}{\Delta} \tan^{-1}[\frac{(\lambda_o+\lambda_2)\tan\phi-\lambda_3}{\Delta}] \quad , $$
$$\Delta = (\lambda_o^2-\lambda_2^2-\lambda_3^2)^{1/2} \qquad (10.23)$$

provided of course $\Delta^2>0$. From the inequalities (7.13) and (7.17) it can be seen that (10.23) is valid at least for thread-like elements. If the sign of Δ^2 is reversed for some materials, the expression corresponding to (10.23) is not difficult to work out.

We now integrate (10.12) and use the two boundary conditions (10.18) on u to obtain

$$u/u_o = \pm\, U(\phi;\phi_1,\phi_c)/U(\phi_o;\phi_1,\phi_c) \qquad (10.24)$$

where we set

$$U(\phi;\phi_1,\phi_c) = \int_{\phi_1}^\phi \alpha^{-1}F^{-1/2}d\phi \;,\; C_1 = u_o^2/U^2(\phi_o;\phi_1,\phi_c) \qquad (10.25)$$

From $(10.25)_2$, it is clear that $C_1>0$. It is necessary to investigate the region in which $F(\phi;\phi_c)$ is non-negative. A close scrutiny of the matrix $||\bar{a}_{ij}||$, $i,j=1,2,\ldots,9$ in the general case ($\bar{a}_{ij}\neq0$ for $i\neq j$), with the identifications (7.16), reveals that $f(\phi)\geq0$, $\alpha(\phi)\geq0$, $\beta/\alpha=(\bar{a}_{47}/\bar{a}_{77})-1$. Thus $F(\phi;\phi_c)\geq0$ in the regions

a) $\phi\leq\phi_c$ when $\beta/\alpha\geq0$,

b) $\phi>\phi_c$ when $\beta/\alpha<0$

$$(10.26)$$

Equation (10.20) can now be integrated to give

$$\pm y = \frac{U(\phi_o;\phi_1,\phi_c)}{2u_o} (\int_{\phi_o}^\phi F^{-1/2}d\phi + \int_{\phi_1}^\phi F^{-1/2}d\phi) \qquad (10.27)$$

where ϕ_c is determined by the equation

$$\pm h = \frac{U(\phi_o;\phi_1,\phi_c)}{2u_o} \int_{\phi_1}^{\phi_o} F^{-1/2}d\phi \qquad (10.28)$$

Since $\alpha(\phi)\geq0$ it follows that

$$U(\phi;\phi_1,\phi_c) \geq 0 \quad \text{when } \phi > \phi_1$$

$$U(\phi;\phi_1,\phi_c) < 0 \quad \text{when } \phi < \phi_1 \tag{10.29}$$

The pressure p is determined from (10.11), i.e.,

$$p = p_0(x) - [f(\phi)F(\phi,\phi_c) - (\alpha_3 + \alpha_9 + \alpha_{11})j_{12}(\phi)\alpha^{-1}(\phi)]$$

$$\times [u_0^2/U^2(\phi_0;\phi_1,\phi_c)] \tag{10.30}$$

The formal solution is now complete.

In the case of _thread-like molecules_ that adhere to the walls with their long axes remaining parallel to the walls at the walls, we have $\phi_0 = \phi_1 = 0$. In this case it is reasonable to assume that $\phi(-y) = \phi(y)$ and therefore $\phi'(0) = 0$. This implies that $\phi_c = \phi(0)$. Hence the solution is given by

$$u/u_0 = U(\phi;0,\phi_c)/U(0;0,\phi_c) \tag{10.31}$$

$$y = \pm \frac{U(0;0,\phi_c)}{u_0} \int_\phi^{\phi_c} F^{-1/2}d\phi \quad , \quad y \geq 0$$

$$-y = \pm \frac{U(0;0,\phi_c)}{u_0} \int_\phi^{\phi_c} F^{-1/2}d\phi \quad , \quad y < 0 \tag{10.32}$$

The unknown constant ϕ_c is obtained by solving

$$\pm \frac{U(0;0,\phi_c)}{u_0} \int_0^{\phi_c} F^{-1/2}d\phi = h \tag{10.33}$$

In (10.32) and (10.33) (+) sign is valid when $\phi_c \geq \phi$ and (−) sign when $\phi_c < \phi$.

For thread-like molecules, $F_1(\phi)$ is simplified to:

$$F_1(\phi) = \phi + \frac{\lambda_0 + \lambda_1}{(\lambda_0^2 - \lambda_2^2)^{1/2}} \tan^{-1} [(\frac{\lambda_0 + \lambda_2}{\lambda_0 - \lambda_2})^{1/2} \tan\phi] \tag{10.34}$$

In this case from the thermodynamic inequalities it can be seen that $\lambda_0^2 - \lambda_2^2 > 0$ so that (10.34) is valid. It appears that the evaluation of the integrals in (10.27) to (10.32) to determine $\phi(y)$ and $u(y)$ will have to be made numerically.

11. Boundary Layer Near A Wall

For some nematic liquids $|k_{11}| + |k_{33}|$ may be sufficiently small so as to make the effect of couple stress on rotations of molecules negligible. In this case, the direction of the long axes of molecules near a wall changes from $\phi = 0$ to $\phi = \phi_\delta$ at some distance $y = \delta$. Afterwards the molecular axes remain at this

angle. The "boundary layer thickness" is determined from (10.13)
by dropping the first term. For the thread-like elements this
gives

$$\frac{y}{y_o} = \int_0^\phi \left[\pm\left(\frac{1}{\lambda+\mu\cos2\phi}-1\right)\sin2\phi\right]^{1/2}d\phi \qquad (11.1)$$

where (+) sign is for $C_1/A_9 \leq 0$ and (−) sign is for $C_1/A_9 \geq 0$, and

$$y_o = (-2C_1/\rho A_9 J^o)^{1/2} \quad ,$$

$$\lambda = \lambda_1/(\lambda_o+\lambda_1) \quad , \quad \mu = \lambda_2/(\lambda_o+\lambda_1) \; . \qquad (11.2)$$

From the thermodynamic inequalities (7.13) it is not difficult to
show that $\lambda_o \geq 0$ and $\lambda_1 \leq 0$.

Using Table 5.1 of de Gennes [1974] and Helfrich [1969] we
find that λ_o, λ_1 and λ_2 have the following values in centipoise.

MBBA at 25°C		PAA at 120°C	
$\lambda_o = 127 \pm 5.5$	,	$\lambda_o = 11.6 \pm 0.6$	,
$\lambda_1 = -76.3 \pm 1.6$	,	$\lambda_1 = -3.0 \pm 0.4$	,
$\lambda_2 = 78.7 \pm 1.6$	,	$\lambda_2 = 6.8 \pm 0.4$	,

(11.3)

Hence we have

$$\lambda \simeq -1.50 \quad , \quad \mu \simeq 1.55 \quad \text{(MBBA)}$$

$$\lambda \simeq 0.35 \quad , \quad \mu \simeq 0.80 \quad \text{(PAA)}$$

(11.4)

Computer calculations were carried out to determine y/y_o as a
function of ϕ. For the values of parameters given by (11.4)
only (+) sign in (11.1) will give real y/y_o which are plotted as
functions of ϕ in Fig. 2. These curves stop at $\phi_\delta = 7°3$ for (MBBA)
and $\phi_\delta = 32°0$ for PAA. Thus "the boundary layer thickness" is
about $\delta/y_o = 0.19$ for MBBA and $\delta/y_o = 0.42$ for PAA. Beyond these
points the long axes of nematic molecules remain at constant
angles 7°3 and 32° respectively. Because of the experimental
inaccuracies that exist in the values of λ_o, λ_1 and λ_2, of the
order of magnitude 13%(especially on λ_1 of PAA), clearly the
boundary layer thicknesses obtained above are approximate.

If $|k_{11}|$ and $|k_{33}|$ are not sufficiently small then the
molecular elements may rotate many times with y and the nematic
structure is highly distorted. This possibility will have to
be verified by carrying out computations for the full solution.
It must be noted, however, that the rotatory motions cannot

continue forever because of the damping against rotations. Such a damping mechanism, while not included in the present calculations, is in fact, accounted for with the expression of $_{D}m$ which does not vanish when the rate measure b is present. In this case, as we know $_{D}m$ has the form (6.4) with α_i and a replaced by β_i and b, respectively.

Acknowledgement

The author is indebted to Dr. T. Kelley for proofreading and checking the analysis, which led to the improvement of this work. Dr. Balta computed y/y_o.

REFERENCES

Oseen, C.W. [1929]: Die anisotropen Flüssigkeiten. Tatsachen und Theorien, Berlin.

Oseen, C.W. [1933] Trans. Faraday Soc. 29, 883.

Anzelius, A. [1931] Arssker, Mat. Ocb. Natl., 1.

Frank, F.C. [1958] Discussion Faraday Soc. 25, 1928.

Ericksen, J.L. [1962] Arch. Ration. Mech. Anal. 9, 371-378.

Ericksen, J.L. [1967] Arch. Ration. Mech. Anal. 23, 266.

Ericksen, J.L. [1969] Molecular Crystals and Liquid Crystals, 7, 153-164.

Leslie, F.M. [1968a] Arch. Ration. Mech. Anal. 28, 265-283.

Leslie, F.M. [1968b] Proc. Royal Soc. A 307, 359-372.

Davison, L. [1967] Phys. Fluid, 10, 2333-2338.

Davison, L. [1969] Phys. Review 180, 232-237.

Martin, P.C., D.S. Pershan and J. Swift [1970]. Phys. Review 25, 844-848.

Helfrich, W.J. [1969] J. Chem. Phys. 51, 4092.

Helfrich, W.J. [1970] J. Chem. Phys. 53, 2267.

Helfrich, W.J. [1972] J. Chem. Phys. 56, 3187.

Aero, E.L. and A.N. Bulygin [1970] Ikv Akad. Nauk SSR, MZLG, 3.

Aero, E.L. and A.N. Bulygin [1971] PMM, 35, 879-891.

Lee, J.D. and A.C. Eringen [1971a] J. Chem. Phys. 54, 5027-5034.

Lee, J.D. and A.C. Eringen [1971b] J. Chem. Phys. 55, 4504-4508.

Lee, J.D. and A.C. Eringen [1971c] J. Chem. Phys. 55, 4509-4512.

Eringen, A.C. and J.D. Lee [1973] Liquid Crystals and Ordered Fluids 2, 315-330, Plenum Publishing Co.

Stephen, M.J. and J.P. Straley [1974] Reviews of Modern Physics 46, 617-704.

P.G. de Gennes [1974] The Physics of Liquid Crystals, Clarendon Press, Oxford.

Eringen, A.C. [1964a] Proc. 11th Intern. Congress of Appl. Mech. Munich, Germany.

Lee, J.D. and A.C. Eringen [1973] Liquid Crystals and Ordered Fluids 2, 315-370, Plenum Publishing Co.

Shahinpoor, M. [1975] J. Chem. Phys. 63, 1319-1320.

Lee, J.D. and A.C. Eringen [1975] J. Chem. Phys. 63, 1321-1322.

Eringen, A.C. and C.B. Kafadar [1976] Continuum Physics Vol. 4 (edit. A.C. Eringen) Academic Press, New York.

Eringen, A.C. [1970] "Foundations of Micropolar Thermoelasticity" Springer-Verlag, Vienna.

Kafadar, C.B. and A.C. Eringen [1971] Int. J. Engng. Sci. 9, 271-305.

Eringen, A.C. [1966] J. Math. and Mechanics 16, 1-18.

Eringen, A.C. [1964b] Int. J. Engng. Sci. 2, 205-217.

Eringen, A.C. [1967] Int. J. Engng. Sci. 5, 191-204.

Wang, C.C. [1970] Arch. Ration. Mech. Analysis 36, 166-197, 198-223.

PHOTOCHEMICAL AND THERMAL STABILITY STUDIES ON A LIQUID CRYSTAL MIXTURE OF CYANOBIPHENYLS

Frederick G. Yamagishi, Deborah S. Smythe,
Leroy J. Miller, and J. David Margerum

Hughes Research Laboratories

3011 Malibu Canyon Road, Malibu, CA. 90265

INTRODUCTION

Liquid crystals (LCs) have unique properties that allow them to be used in information display devices.[1] In particular, LCs are useful for displays used in wrist watches, calculators, message boards, flat-panel television, and large-screen projection systems.[2] The choice of materials is often dictated by the desired application. The use of LCs in electro-optical devices exposed to high-intensity light must meet not only the usual criteria for these devices such as birefringence, dielectric and conductivity anisotropy, and alignment qualities, but must be photochemically stable to long exposures of visible light and thermally stable to heat generated by exposure to visible and near infrared light. Large-screen projection displays that use liquid crystal light valves[3] have more severe photochemical stability requirements to achieve long lifetime displays than many other devices. Even very small absorption "tails" of the LCs that extend into the visible region of the spectrum can cause lifetime problems (should photo-decomposition occur) because of the high intensity of light used. Figure 1 shows a light valve projection system.[3]

The LC used in these devices must show long-term photo and thermal stability in both composition, which determines the LCs physical properties (e.g., nematic range, $\Delta\varepsilon$, Δn, etc.), and in surface alignment stability. Alignment is usually the first thing to be affected by small amounts of degradation products, especially if those products are polar.

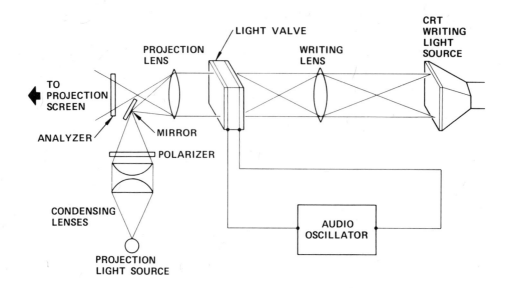

Figure 1.   Diagram of a light valve projection system.

We have chosen for study the commercial LC E7 which is a mixture of three cyanobiphenyl compounds and a cyanoterphenyl. LCs such as anils, azobenzenes, and azoxybenzenes are all expected to be more susceptible to photochemical degradation because of their longer wavelength chromophores (see Figure 2). Preliminary experiments indicated that cyanobiphenyls show higher stability than anils and esters when exposed to high intensity light, and, therefore, this class of materials was chosen for this study.

The purpose of these studies is to determine the effects of wavelength (UV cutoff), heating, high light intensity, oxygen, and impurities on the stability of E7 in hybrid field effect cells.[4] With a knowledge of these effects and an understanding of the failure mechanisms we expect to optimize conditions for long lifetimes with high-intensity lights.

For these studies, the technical approach was to monitor the changes in the electro-optical effect and to analyze and identify chemical products produced by light and heat.

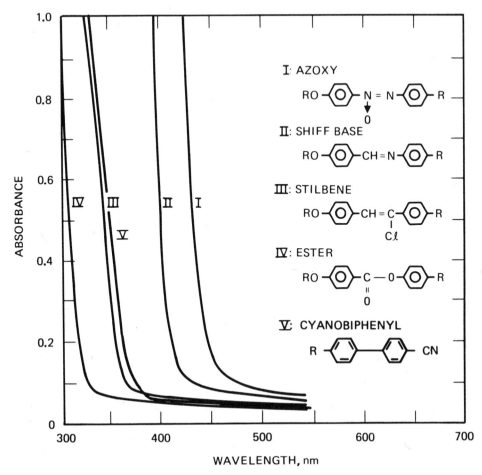

Figure 2.   Absorption and scattering of various classes of liquid
            crystals in 12.7 μm thick test cells.

EXPERIMENTAL SECTION

      The liquid crystal used in this study was E7 (BDH Chemicals,
Ltd.) which is a four-component mixture consisting of 4-cyano-4'-
pentylbiphenyl (K15), 4-cyano-4'-heptylbiphenyl (K21), 4-cyano-4'-
octyloxybiphenyl (M24), and 4-cyano-4"-pentyl-p-terphenyl. The LC
was used without further purification. The purity of each batch
was checked by high performance liquid chromatography (HPLC) after
opening. Figure 3 shows a typical chromatogram of the starting
material.

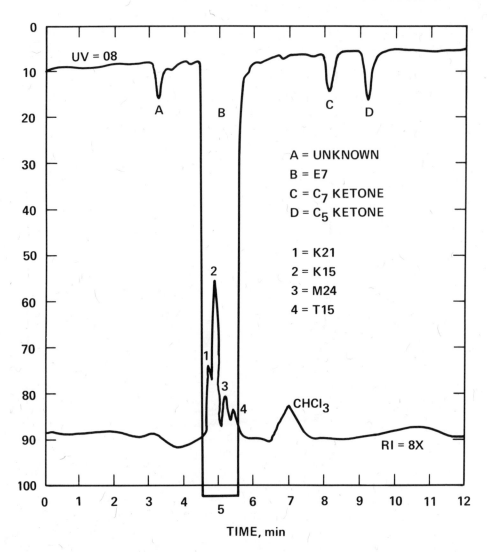

Figure 3. Liquid chromatogram of E7 as received. Upper trace is
           from a UV detector at high sensitivity; the lower trace
           is from a refractive index detector.

     Two xenon short arc lamps were used in this study: a Christie
Electric Corp. CSL-900-0 900 W lamp powered by an Electro Powerpacs
Corp. Model 354 power supply; and an Optical Radiation Corp. 1600 W
lamp operated at 1000 W for most runs and at 1200 W for high-
intensity tests. Figure 4 shows the arrangement of the filtering
system for each lamp. A Corning Laboratory Products' CS-3-73
sharp-cut filter was used for λ > 420 nm. No sharp-cut filter was

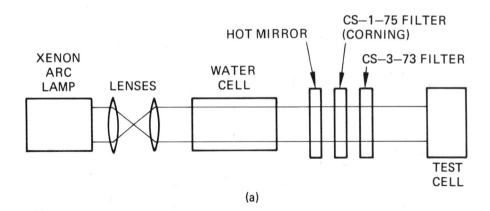

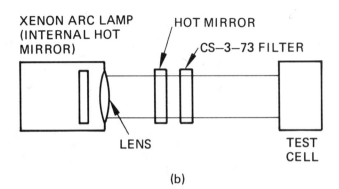

Figure 4. Arrangement of filters for photochemical stability
studies: (a) Christie 900 W Xenon lamp; (b) Optical
radiation 1600 W Xenon lamp.

used for λ > 360 nm (Christie) nor for λ > 385 nm (Optical Radiation) since that was the short wavelength cutoff of the hot mirror
being employed for that experiment. Figure 5 is a spectrum of the
lamp with a CS-3-73 filter in the system.

Light intensity measurements were made with a Hewlett-Packard
Model 8330A Radiant Flux Meter equipped with a Model 8334A Radiant
Flux Detector, which was factory-calibrated (traceable to NBS
standards). Measurements were made through a Balzers 1.1% transmission filter. Intensities are reported without regard to

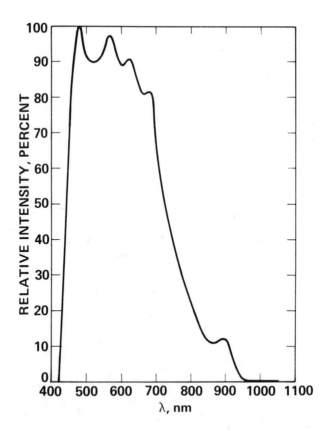

Figure 5.   Spectrum of a Xenon lamp with a CS-3-73, a CS-1-75, and
             a 6 in. Pyrex water cell filtering system.

spectral distribution of the lamp and assumes that the filter is
linear throughout the spectrum of interest.

    All experiments were carried out on neat LC. For thermal runs,
a weighed amount of LC was placed in several small tubes, all of
which were contained in a large tube protected by a calcium sulfate
drying tube. The system was immersed in a constant temperature
bath (±0.1°) and samples were removed periodically, diluted to a
known concentration and quantitatively analyzed by HPLC. All
photochemical cells had a 1 in.^{2} aperture which was completely filled
with the exposing light beam. An external, first-surface aluminum
mirror was used behind the test cell. Four of the cells (see
Table 1) were made from 1/8 in. thick Nesatron glass, which was
coated with indium-tin oxide (ITO), from PPG Industries. Mylar
spacers (6.4 μm) were used. The other two cells were made from
1/2 in. optical flats which were coated with an ITO layer. The

Table 1. Photochemical Lifetimes of E7 in 45° Twist Test Cells

CELL	THICKNESS, μm	LIQUID CRYSTAL	LAMP INTENSITY, mW/cm²	TEMP, °C	SHORT WAVELENGTH CUTOFF, nm	LIFETIME, hr	LIFETIME W.hr/cm²	CALCULATED LIFETIME IN SYSTEM[5], hr
1	6.4[1]	E7	418	30 – 33	420	1839	769	3076
2	6.4	E7	812	38 – 44	420	618	486	1944
3	2.2[2]	E7	427 1345[3]	32 – 39	420	774	486	1944
4	6.4	E7K[4]	409	30 – 32	420	876	358	1432
5	4	E7	408	35 – 37	385	145	59	236
6	6.4	E7K	508	30	360	43	22	88

[1] MYLAR SPACER USED

[2] SPACERS WERE PADS OF SiO_x

[3] THIS CELL WAS EXPOSED AT 1345 mW/cm² FOR 169 hr

[4] E7 DOPED WITH 0.37% OF 4-CYANO-4'-HEPTANOYLBIPHENYL

[5] CALCULATED LIFETIME FOR AN OPERATING SYSTEM WITH A LAMP INTENSITY OF 250 mW/cm². INCREASING SHORT WAVELENGTH CUTOFF WILL YIELD FURTHER SUBSTANTIAL INCREASE IN LIFE (SEE TEXT)

cells were assembled in air and were not hermetically sealed. In
all cases, the substrate electrodes were overcoated with SiO_2 and
shallow-angle ion-beam etched[5] to give surface parallel alignment.
They were positioned at 45° with respect to each other. Spacers
for the 1/2 in. glass were provided by pads of evaporated SiO_x.

Temperatures were measured by taping a thermocouple to the
backside of the mirror. With 1/8 in. glass, control experiments
showed that the measured temperature outside the cell was no more
than 1° lower than the internal temperature. No such control
experiments were made with 1/2 in. glass, but the external temperature was probably no more than 5° lower than the internal
temperature.

The cells were monitored by their transmission versus voltage
(birefringence) response at 623, 545, and 466 nm between 0 and
7.5 V (rms at 10 kHz). Initially, measurements were taken daily
for about a week and then every two to four days thereafter. Cell
failure was manifested in the transmission versus voltage curves
through a decrease of transmission and a lack of resolution of the
peaks in the 2-4 V region. Additionally, formation of a cloudy
spot or streak caused by reorientation of the LC alignment was
concurrently noticed. Figures 6 and 7 are examples of transmission versus voltage curves at the start and end of a lifetime
test.

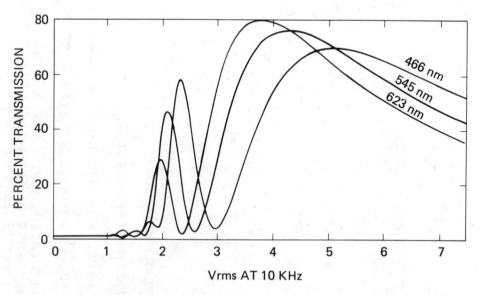

Figure 6.   Transmission versus voltage response of a 6.4 μm, 45°
             twist cell before exposure (cell 2, Table 1).

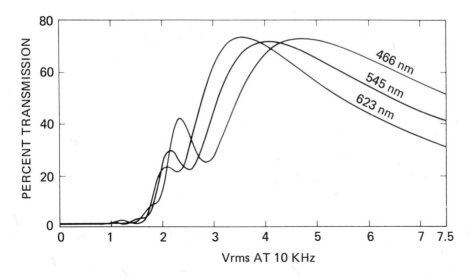

Figure 7. Transmission versus voltage response of a 6.4 μm, 45°
           twist cell at the failure point following exposure
           (cell 2, Table 1).

Analytical data were obtained on a Waters Associates Model
ALC-202 High Performance Liquid Chromatograph equipped with a 4 mm
x 30 cm μ-Porasil (10 μm particle size silica) column using a solvent mixture of hexane/chloroform/acetonitrile (19:4:1), respectively. This solvent elutes E7 as one major peak but allows
separation of the more polar decomposition products. The four main
components of E7 may be separated by diluting the above mixture
with an equal portion of hexane. All solvents were Burdick and
Jackson "Distilled in Glass" spectrograde. They were filtered
through Millipore 0.45 μm Teflon filters and degassed in an ultrasonic generator prior to use. Quantification was done by calibration with pure known components and measuring peak heights.
Figure 8 shows a chromatogram of E7 and decomposition products
following exposure.

Nuclear magnetic resonance (nmr) spectra were obtained on a
Varian Associates A-60 spectrometer in deuterochloroform. Chemical
shifts are reported in parts per million (δ) downfield from internal
tetramethylsilane. Infrared (IR) spectra were taken on a Beckman
IR-12 Spectrophotometer as 1% solutions in chloroform. Thin layer
chromatography was done on Analtech, Inc. silica gel GF precoated
plates (250 μm) using a 254 nm lamp for spot detection. Mass
spectra were taken by the analytical chemistry laboratory at
California Institute of Technology, Pasadena, California. Melting
points taken on a Thomas-Hoover Uni-Melt apparatus are reported in
degrees Celsius and are uncorrected.

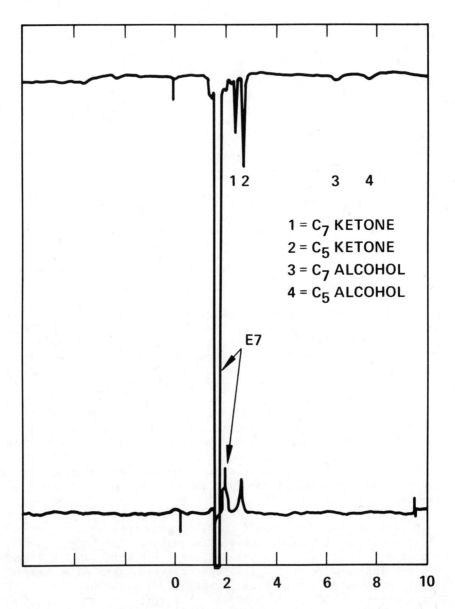

Figure 8.   Liquid chromatogram of E7 following cell failure due to
              photochemical decomposition.

4-Cyano-4'-heptanoylbiphenyl (C_7 Ketone). Into 75 ml of freshly distilled dimethylformamide was added 4-bromo-4'-heptanoyl-biphenyl[6] (4 g, 11.6 mmol) and cuprous cyanide (3.11 g, 34.8 mmol). The reaction was protected with a calcium sulfate drying tube and heated to reflux for 17.5 hr. After cooling, the dark green slurry was poured into a solution composed of $FeCl_3 \cdot 6H_2O$ (6.6 g), concd HCl (1.5 ml), and water (75 ml). This was heated to 65° for 20 min, cooled and extracted with two 150-ml portions of $CHCl_3$. The organic extracts were combined and washed with two 150-ml portions of 5N HCl, H_2O, 10% NaOH, H_2O, and brine. The solution was dried over anhyd Na_2SO_4 and the solvent removed in vacuo to give a brown residue. This was chromatographed on 500 g of silica gel (benzene eluent, 300-ml fractions). Fractions 6 to 12 contained the product (1.4 g, 41.4%). Recrystallization from absolute ethanol gave a white solid, mp 79.5 to 82.5°. Its IR spectrum showed absorptions at 1684 cm^{-1} (C=O) and 2225 cm^{-1} (C≡N). Its nmr spectrum shows a multiplet at 7.59 to 8.20 δ (aromatic, 8H), a triplet centered at 3.01 δ (-CH_2-C-, 2H), and a broad multiplet at 0.72 to 2.1 δ (aliphatic, 13H). Its high resolution mass spectrum showed a molecular ion at m/e = 291.165 (Calcd 291.162).

4-Cyano-4'-(1-hydroxyheptyl)biphenyl (C_7 Alcohol). The ketone (0.3 g, 1.03 mmol) prepared above was dissolved in 30 ml of abs ethanol and 3 ml of dry benzene. Sodium borohydride (0.06 g, 1.59 mmol), in 5 ml of 2N NaOH, was added dropwise over 15 min. The reaction was stirred at 25° under a blanket of nitrogen for 1 hr after which the solvent was removed in vacuo and the residue diluted with 25 ml of water. This was extracted with two 25 ml portions of ether which was washed with water, 10% HCl and brine. After drying over anhyd Na_2SO_4, the solvent was removed to give an off-white oil which slowly solidified. The product was isolated by preparative thin layer chromatography (silica gel GF, 2000 μm, $CHCl_3$, three passes), 71 mg (23.5%), mp 51-55°. Its nmr spectrum showed a multiplet 7.40 to 7.77 δ (aromatic, 8H), a triplet centered at 4.74 δ (-CH-OH, 1H), a singlet at 2.51 δ (-OH, 1H), and a broad multiplet at 0.63 to 1.92 δ (aliphatic, 13H). Its high-resolution mass spectrum showed a parent ion at m/e = 275.167 (-H_2O) (Calcd 293.178).

4-Cyano-4'-valerylbiphenyl (C_5 Ketone). The product was prepared as above in 30.4% yield, mp 99-100°. Its IR spectrum showed absorptions at 1686 cm^{-1} (C=O) and 2225 cm^{-1} (C≡N). Its nmr showed a multiplet at 7.59 to 8.20 δ (aromatic, 8H), a triplet centered at 3.0 δ (2H), and a broad multiplet at 0.71 to 2.1 δ (aliphatic, 7H). Its high-resolution mass spectrum showed a molecular ion at m/e = 263.134 (Calcd 263.131).

4-Cyano-4'-(1-hydroxypentyl)biphenyl (C_5 Alcohol). Sodium borohydride reduction of 4-cyano-4'-valerylbiphenyl produced the product as a clear oil. The IR and nmr spectra are consistent with

the structure. Its high resolution mass spectrum showed a molecular
ion at m/e = 265.149 (Calcd 265.147).

Identification of 4-cyano-4'-biphenylcarboxylic Acid. The
solid isolated from the thermal decomposition of K21 had a mp at
260-267° (lit.[6] 263-265°) and an IR and nmr spectrum consistent
with the structure.   Its mass spectrum gave a molecular ion at
m/e = 223.064 (Calcd 223.063).

RESULTS AND DISCUSSION

Photochemical Stability

Table 1 presents the results of six photolyzed cells with
their corresponding experimental conditions and lifetimes. Lifetimes expressed as W-hr/cm^2 allow direct comparison of cells since
this takes into account varying light intensities for each test.
Furthermore, an additional column has been included for calculated
lifetimes of a practical LC light valve projection system operating
under conditions listed for each cell.

Cells were monitored by their transmission versus voltage
response. Figure 9 shows a diagram of the measurement apparatus.
It was expected that any photodegradation of the LC would be manifested by an alteration of the alignment. This method is very
sensitive to alignment changes.

Figures 3 and 8 show HPLC chromatograms of E7 before (used as
received) and after irradiation. Figure 3 shows traces from a uv
detector (254 nm) and a refractive index detector. E7 is eluted, in
this case (uv), as one large peak since the column was heavily loaded
so that impurities such as C and D could be observed. Peaks C
and D were later identified as 4-cyano-4'-heptanoylbiphenyl and
4-cyano-4'-valerylbiphenyl, respectively. They are present from
0.03 to 0.06%. The refractive index detector trace shows the separation of E7 into its major components.

A cell was deemed to have failed when the transmission versus
voltage curves were significantly altered from the one taken before
exposure. This change showed up as a loss of peak sharpness in
the 2-4 V range, an overall lowering of transmission throughout the measured voltage range, and occasionally by differing
transmissions of the three wavelengths through the 0 to 2 V range.
Figures 6 and 7 are typical examples of a set of curves before and
after exposure. Accompanying the change in transmission versus
voltage was the development of a cloudy spot or streak in the
cell. Observation of the interface of the streak with the bulk

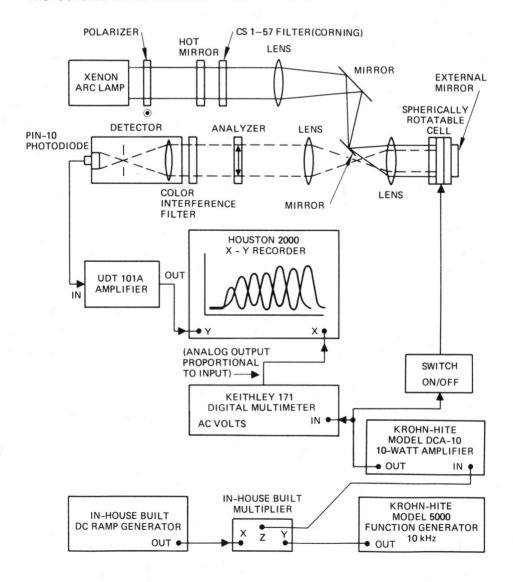

Figure 9. Diagram of the transmission versus voltage measurement apparatus.

LC through a polarizing microscope showed that E7 was reoriented toward a homeotropic alignment. One can therefore conclude that the photodegradation products are surface-perpendicular aligning agents and are probably fairly long chain groups which likely contain a surface-active functional group. It is known that an SiO_2 surface can be chemically treated with alcohols[7] to produce homeotropic alignment.

The effect of the near uv incident upon the test cell is shown by comparison of cells 1 and 5 in Table 1. These cells were exposed under nearly the same conditions but at different cutoff wavelengths. Exposure of cell 5 for $\lambda > 385$ nm caused a decrease in the lifetime of the cell by a factor of 13 over cell 1 which was exposed for $\lambda > 420$ nm. Lowering of the wavelength causes a greater absorption by either E7 or its photodegradation products in their absorption tails. The choice of a 420 nm filter was to eliminate as much near uv light as possible while retaining the blue region of the visible. However, although utilization of a longer wavelength cut-off would lead to a longer lifetime because of decreased absorption, the output color characteristics of the display would suffer.

The increase of light intensity has a nonlinear effect on cell lifetime. Comparison of the lifetimes of cells 1 and 2 show that increasing the light intensity by a factor of 1.9 decreases the lifetime by a factor of 1.6. Additionally, increasing the intensity during an experiment, as in cell 3, causes a dramatic decrease in lifetime. As an approximation, it would appear that the decomposition of the LC and the subsequent realignment is caused by a combination of primary and secondary chemical reactions. The effect of increasing the intensity serves to increase the concentration of reactive free radicals. Thus, one would expect that more products arising from radical-radical interaction would be formed which may lead to shorter lifetimes.

Temperature measurements were taken by a thermocouple attached to the external mirror. Readings taken in this manner are no more than 1^O lower than the internal temperature. Fan cooling controlled the temperature to 30 to 33° whereas the temperature reached as high as 47° without cooling. We feel that heating alone at this temperature range does not play a direct role in the decomposition of the LC. Two cells were heated at 50 to 52° in the absence of any light for more than 4000 hr and showed no alteration of alignment. Therefore, heating of the LC at temperatures below 50° is not detrimental to the performance of the device, but it may be an important factor in conjunction with high intensity light. It is not clear whether heating has a synergistic effect or if it merely lowers the viscosity of the LC allowing a greater mobility to the free radicals formed by irradiation. Experiments are under way to determine the dependence of cell lifetime with heating.

As mentioned earlier, peak C in Figure 3 was identified as 4-cyano-4'-heptanoylbiphenyl. This material was synthesized and doped into E7 to the extent of 0.37% for the runs with cells 4 and 6. This doped LC mixture greatly reduced the cell lifetimes. The

effect of added ketone is greatest for $\lambda > 360$ nm, but comparison
of cells 4 and 1 run under similar conditions show that the increased
concentration of ketone also dramatically increases the decomposition rate of the LC exposed with a 420 nm cutoff filter. With the
occurrence of a ketone in the LC, a photochemically active chromo-
phore is present and it is likely that it is important in the
mechanism of degradation. The possible mechanism of photochemical
decomposition will be discussed below.

A projection device using the hybrid field effect can be utilized with incident polarization of the projection light either
parallel or perpendicular to the alignment direction at the first
surface. One would expect that the pleochroic effect (dependence
of absorption of radiation with polarization direction with respect
to the long axis of the liquid crystal) would be important to the
lifetime of the cell. We have tested this using E7 containing
0.37% 4-cyano-4'-heptanoylbiphenyl for $\lambda > 360$ nm using Polacoat
polarizers at 83 mW/cm^2. When the polarization and alignment
directions were parallel, the cell failed after 162 hr; when the
two were perpendicular, failure occurred after 298 hr.   The amount
of ketone found in the parallel case (1.01%) was greater than the
amount found in the perpendicular (0.94%). The rate of ketone
formation per hour was 2.1 times greater for the parallel case than
the perpendicular case. These preliminary results indicate that
polarization effects are important, and further experiments are
being planned to more fully characterize the relationship between
the surface alignment, the incident light polarization, and the
photochemical degradations.

Thermal Stability

Our initial studies of the thermal stability of E7 were carried
out in an oven at about 100° in air. After four days, the LC turned
yellow, and after seven days a precipitate separated from the LC.
A concurrent run carried out under a nitrogen atmosphere caused no
change in the LC over the same time period. It is believed that
oxygen is an important factor in the decomposition of E7. As pointed
out earlier in the photochemical decomposition of E7, two ketones
were identified as well as their respective alcohols. The source
of oxygen comes from air to which the LC is exposed.

To simplify analytical complexities, samples of K21 were heated
at 90° and 100° in air protected by a calcium sulfate drying tube.
Samples were periodically removed and quantitatively analyzed by
HPLC. Figures 10 and 11 show the decrease of K21 with heating time
at 90° and 100°, respectively. In both cases there is a rapid
decomposition of K21 over the first three days of heating which
represents a 20 to 25% loss. From that time the decrease of K21 is

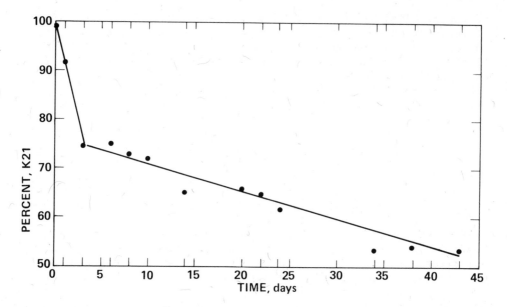

Figure 10.   Decrease in K21 concentration when heated at 90°.

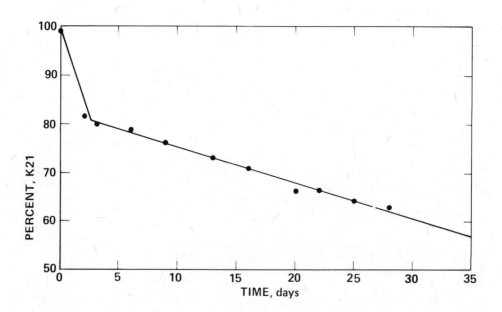

Figure 11.   Decrease in K21 concentration when heated at 100°.

nearly linear for the remainder of the experiment. The initial decrease of K21 does not correspond to the increased concentration of ketone and alcohol, as shown in Table 2. It should also be noticed that the alcohol concentration increases at a greater rate than the ketone.

The rapid initial decomposition could be explained by the presence of a small concentration of an impurity that could act as a thermal radical initiator such as peroxide. Propagation of a radical chain mechanism would ensue until the initiator had been consumed, after which generation of radicals would occur by another pathway at a slower rate. This mechanism would dictate, however, that the initiator be present in the LC prior to its being heated or that it be rapidly formed upon heating. We have found no evidence, as yet, of a peroxide prior to using the LC in experiments.

A more likely explanation is that an impurity in K21 reacts more rapidly with oxygen than the bulk and that a radical chain mechanism would proceed until the impurity was consumed. One possible candidate is the olefin 8 which is known to be present as a minor impurity.[8] Allylic hydrocarbons are known to autoxidize more rapidly than benzyl moieties.

$$NC-\langle\bigcirc\rangle-\langle\bigcirc\rangle-CH=CH-(CH_2)_4-CH_3$$

8

which is known to be present as a minor impurity.[8] Allylic hydrocarbons are known to autoxidize more rapidly than benzyl moieties.

As in the case of E7, a solid separated from K21 after prolonged heating. This material was isolated and identified as 4-cyano-4'-biphenylcarboxylic acid. This should be partially soluble in E7 and would be a surface-active, homeotropic aligning agent. Our initial photolyzed cells were exposed past the now measured failure point and an insoluble solid, which we now assume is this acid, separated from the bulk LC.

Mechanism

We feel that the major decomposition pathway for E7, both photochemically and thermally, is autoxidation. It is well documented that autoxidation occurs at benzylic centers and E7 has three of these in K15, K21, and T15, although no oxygenated compounds derived from the latter have been isolated. A mechanism is outlined below in Scheme 1.

Table 2. Thermal Decomposition of K21 at 90° and 100°

DAYS HEATED		KETONE, %		ALCOHOL, %		K21, %	
90°	100°	90°	100°	90°	100°	90°	100°
1	2	0.26	0.02	1.91	2.93	92.05	83.11
3	3	0.59	0.04	8.77	5.76	74.80	80.08
6	6	1.30	0.20	13.49	11.54	75.14	79.22
10	9	2.57	0.41	19.02	14.06	72.30	76.44
14	13	5.12	0.89	23.62	17.27	65.48	73.32
–	15	–	1.46	–	17.36	–	71.90
20	20	6.79	2.10	20.90	–	63.59	66.50
22	22	8.95	3.04	27.80	22.39	61.77	66.41
24	25	9.00	3.28	28.38	–	61.05	66.34
34	28	13.17	4.68	28.02	–	53.67	63.16

INITIAL K21: 99.80 %; KETONE: 0.2 %

SCHEME 1

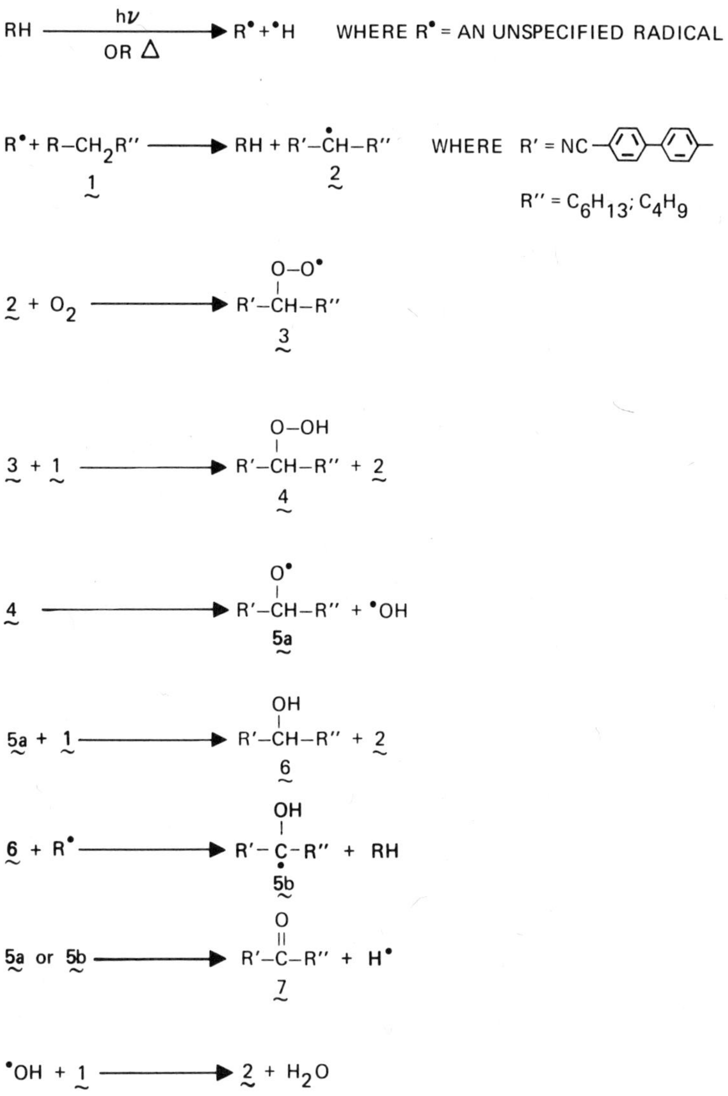

$$RH \xrightarrow[\text{OR } \Delta]{h\nu} R^{\bullet} + {}^{\bullet}H \qquad \text{WHERE } R^{\bullet} = \text{AN UNSPECIFIED RADICAL}$$

$$R^{\bullet} + \underset{\underset{\widetilde{}}{1}}{R-CH_2R''} \longrightarrow RH + \underset{\underset{\widetilde{}}{2}}{R'-\overset{\bullet}{C}H-R''} \qquad \text{WHERE } R' = NC\text{—}\langle \rangle\text{—}\langle \rangle\text{—}$$

$$R'' = C_6H_{13}; C_4H_9$$

$$\underset{\widetilde{}}{2} + O_2 \longrightarrow \underset{\underset{\widetilde{}}{3}}{R'-\overset{\overset{O-O^{\bullet}}{|}}{C}H-R''}$$

$$\underset{\widetilde{}}{3} + \underset{\widetilde{}}{1} \longrightarrow \underset{\underset{\widetilde{}}{4}}{R'-\overset{\overset{O-OH}{|}}{C}H-R''} + \underset{\widetilde{}}{2}$$

$$\underset{\widetilde{}}{4} \longrightarrow \underset{\underset{\widetilde{}}{5a}}{R'-\overset{\overset{O^{\bullet}}{|}}{C}H-R''} + {}^{\bullet}OH$$

$$\underset{\widetilde{}}{5a} + \underset{\widetilde{}}{1} \longrightarrow \underset{\underset{\widetilde{}}{6}}{R'-\overset{\overset{OH}{|}}{C}H-R''} + \underset{\widetilde{}}{2}$$

$$\underset{\widetilde{}}{6} + R^{\bullet} \longrightarrow \underset{\underset{\widetilde{}}{5b}}{R'-\overset{\overset{OH}{|}}{\underset{\bullet}{C}}-R''} + RH$$

$$\underset{\widetilde{}}{5a} \text{ or } \underset{\widetilde{}}{5b} \longrightarrow \underset{\underset{\widetilde{}}{7}}{R'-\overset{\overset{O}{\|}}{C}-R''} + H^{\bullet}$$

$${}^{\bullet}OH + \underset{\widetilde{}}{1} \longrightarrow \underset{\widetilde{}}{2} + H_2O$$

A similar mechanism for autoxidation of the olefin 8 can be drawn as in Scheme 2.

SCHEME 2

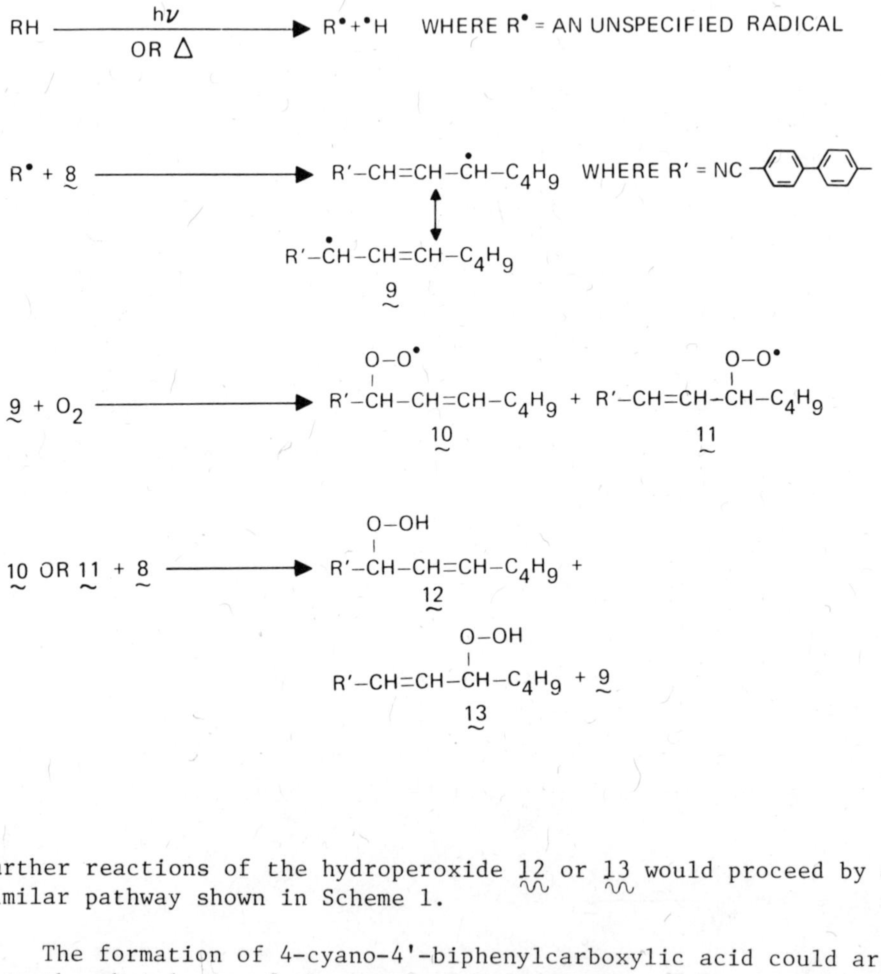

$$RH \xrightarrow[\text{OR } \Delta]{h\nu} R^\bullet + {}^\bullet H \quad \text{WHERE } R^\bullet = \text{AN UNSPECIFIED RADICAL}$$

$$R^\bullet + 8 \longrightarrow R'-CH=CH-\overset{\bullet}{C}H-C_4H_9 \quad \text{WHERE } R' = NC-\!\!\!\bigcirc\!\!\!-\!\!\!\bigcirc\!\!\!-$$

$$R'-\overset{\bullet}{C}H-CH=CH-C_4H_9$$
$$9$$

$$9 + O_2 \longrightarrow \underset{10}{R'-CH-CH=CH-C_4H_9} + \underset{11}{R'-CH=CH-CH-C_4H_9}$$
with $O-O^\bullet$ groups

$$10 \text{ OR } 11 + 8 \longrightarrow \underset{12}{R'-CH-CH=CH-C_4H_9} +$$
with $O-OH$ group

$$\underset{13}{R'-CH=CH-CH-C_4H_9} + 9$$
with $O-OH$ group

Further reactions of the hydroperoxide 12 or 13 would proceed by a similar pathway shown in Scheme 1.

The formation of 4-cyano-4'-biphenylcarboxylic acid could arise from the photolysis of the ketone 7 and reaction of the resulting cyanobiphenylacyl radical with hydroxy radical or oxygen.

After a period of irradiation, the concentration of ketones present in E7 could become significant enough to allow them to enter the mechanisms as photosensitizers. The ketones do have an

absorption ·tail at 420 nm. We suggest that the ketones participate
by the mechanism shown in Scheme 3.

SCHEME 3

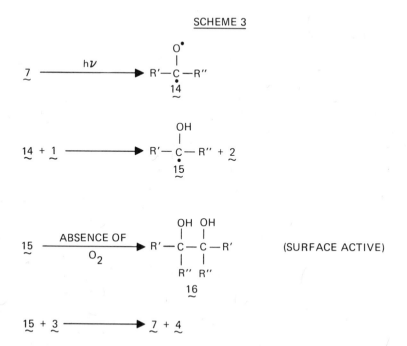

The glycol 16 has not yet been isolated. Experiments in which
oxygen is excluded are being planned. In this case the formation
of 16 would cause alignment degradation since 16 would be surface
active.

SUMMARY

We have shown that the commercial LC E7 can be used in high-
intensity light, large-screen projection light valves with lifetimes approaching 2000 hr at $\lambda > 420$ nm, and light intensities about
400 mW/cm^2. In practical projection systems where the cell is operated at lower light intensity, longer lifetimes are expected. A
lifetime of >3100 hr can be expected at 250 mW/cm^2. With a wavelength
cutoff higher than 420 nm, lifetimes are expected to be even substantially higher. Decomposition of E7, and subsequent realignment, which
leads to cell failure, is accelerated for $\lambda > 420$ nm, by use of
greater light intensities and by ketone impurities present or formed
in the LC. We have attributed the major mechanism of photochemical
and thermal decomposition to autoxidation of the alkyl cyanobiphenyl
components of E7.

We expect longer lifetimes can be achieved by control of the parameters listed above, particularly LC purity and reduced exposure to oxygen. We are continuing our efforts to achieve longer lifetimes for these devices.

ACKNOWLEDGMENTS

We are grateful to Dr. Hugh L. Garvin for assistance with the ion-beam etching, to Gary D. Myer for making the transmission versus voltage measurements, and to Donald E. Sprotbery and Ernst J. Scipio for their help in some of the photolysis work. This work was partially supported by the Naval Sea Systems Command under Contract N00024-76-C-5366.

REFERENCES

1.   L.A. Goodman, J. Vac. Sci. Technol. 10, 804 (1973).

2.   J.D. Margerum and L.J. Miller, J. Colloid Interface Sci. 58, 599 (1977).

3.   J. Grinberg, W.P. Bleha, A.D. Jacobson, A.M. Lackner, G.D. Myer, L.J. Miller, J.D. Margerum, L.M. Fraas, and D.D. Boswell, IEEE Trans. Electron. Devices ED-22, 775 (1975); and references cited therein.

4.   J. Grinberg, A.D. Jacobson, W.P. Bleha, L.J. Miller, L.M. Fraas, D.D. Boswell, and G.D. Myer, Opt. Eng. 14, 217 (1975).

5.   M.J. Little, H.L. Garvin, and L.J. Miller, Abstracts, 174th National Meeting of the American Chemical Society, Chicago, Ill., August 1977, No. COLL-53; M.J. Little, H.L. Garvin, and L.J. Miller, in Liquid Crystals and Ordered Fluids, J.F. Johnson and R.S. Porter, Eds (Plenum Press, New York, N.Y.), Vol. 3.

6.   G.W. Gray, K.J. Harrison, J.A. Nash, J. Constant, D.S. Hulme, J. Kirton, and E.P. Raynes, Liquid Crystals and Ordered Fluids, J.F. Johnson and R.S. Porter, Eds (Plenum Press, New York, N.Y.), Vol. 2, p. 617.

7.   L.J. Miller, J. Grinberg, G.D. Myer, D.S. Smythe, and W.H. Smith, Jr., Abstracts, 174th National Meeting of the American Chemical Society, Chicago, Ill., August 1977, No. COLL-52; L.J. Miller, J. Grinberg, G.D. Myer, D.S. Smythe, and W.H. Smith, Jr., in Liquid Crystals and Ordered Fluids, J.F. Johnson and R.S. Porter, Eds (Plenum Press, New York, N.Y.), Vol. 3.

8.   C.R.P. Wilcox, BDH Chemicals, Ltd., private communication. He reports that the olefin has been removed from more recent batches of K21.

A NEW METHOD FOR INDUCING HOMOGENEOUS ALIGNMENT OF NEMATIC LIQUID CRYSTALS

Michael J. Little, Hugh L. Garvin, and Leroy J. Miller

Hughes Research Laboratories

3011 Malibu Canyon Road, Malibu, CA. 90265

ABSTRACT

A uniform homogeneous alignment of nematic liquid crystals can be induced by sputtering a silica coating onto the substrate and by ion-beam etching the silica surface at a shallow angle. The resulting liquid crystal alignment is parallel to the etch direction with a very small tilt angle out of the surface toward the ion beam source. Orientational control is believed to be derived from the presence of extremely minute grooves in the etched silica surface. One can measure a splay in the alignment caused by a slight divergence of the ion beam. Small defects in the alignment occasionally appear but are metastable and can be eliminated by heating or applying an ac field.

INTRODUCTION

For most liquid crystal electro-optic devices it is necessary to have the liquid crystal oriented uniformly at the walls that confine it. In the usual case, these walls are the surfaces of two parallel electrodes. Two methods have been widely used for inducing an orientation of nematics in which the director is parallel or almost parallel to these electrode surfaces. The first, and simplest, is to rub or groove the surface with any of a variety of suggested materials.[1] Rubbing, however, inevitably produces striations that are readily visible if the display is viewed with polarizers and magnified with a microscope or a projection display system. In recent years the industry has widely adopted the Janning method[2] of inducing alignment by vapor-depositing a material

497

such as SiO or MgF$_2$ onto each of the electrode surfaces at an
oblique angle.[3] Several years ago (prior to the Janning publication) we developed a method for inducing homogeneous alignment by
using surfaces that were shallow-angle ion-beam etched (SAIBE).
This technique has produced a very uniform and stable liquid crystal
alignment suitable for the demanding requirements of projection
displays. In this paper we are describing the ion-beam etching
technique as well as some of the characteristics of the alignment
that can be achieved.

EXPERIMENTAL SECTION

Ion Beam Etching

The electrodes used to evaluate the effect of shallow-angle
ion-beam etching on liquid crystal alignment were typically 3 mm
(1/8 in.) thick glass substrates upon which were deposited various
thin films. The uppermost coating was then ion-beam etched. Subsequently, the electrodes were used to make a liquid crystal cell
whose alignment was critically evaluated.

In our early studies, argon ions accelerated to 3 keV were
used to bombard glass surfaces coated with transparent electrode
materials including indium-tin oxide (ITO) and tin-antimony oxide.
In the duoplasmatron type of system shown schematically in Figure 1,
the ions were made to impinge on the surface at a shallow angle of
10° to 30° from parallel. Although various etching depths were
produced (from 100 to 600 Å), uniform and reproducible alignment
was not observed.

Further studies showed that liquid crystal alignment was much
better controlled by first sputter depositing a thin semitransparent
film (about 100 Å thick) of carbon onto the ITO and then shallow-
angle etching this surface. This procedure has two disadvantages,
however: (1) because of poor adhesion of the carbon film, the
resulting surface is very fragile and easily damaged and (2) the
transparency of the electrode is reduced by the carbon film.

The duoplasmatron system (Figure 1) was also used in our first
studies of other types of SAIBE coating materials. Substrates were
first coated with an overlay material by directing the ion beam at
a target made of the overlay material. Material sputtered from this
target was deposited onto the substrate. The coated substrate was
then repositioned so that the ions impinged on the surface directly
at a shallow angle. In this system the collimated argon ion beam
is relatively narrow (about 2 cm diam); thus surfaces larger than

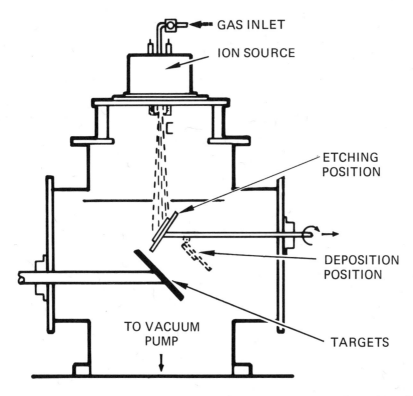

Figure 1.   Duoplasmatron ion beam sputtering system for thin film
deposition and etching.

2 cm in diameter had to be coated and then shallow-angle etched by
translating them back and forth under the beam.

The overlay coatings for the majority of our electrode samples
were deposited in a commercially available rf plasma sputtering
system* (Figure 2). Argon gas at 3 to 10 x 10^{-3} Torr is excited
and ionized by rf power applied to parallel plates, one of which
serves as a cathode and the other (holding the sample plates) as an
anode. The ions are accelerated by the rf voltage to bombard the
cathode, causing material to be sputter-deposited onto the opposing
samples. Under this type of normal "straight sputtering" operation,
the ions bombard only the cathode to cause deposition onto the
samples. In an alternative mode of operation known as "bias

---

*Model 8802 by Materials Research Corp., Orangeburg, N.Y.

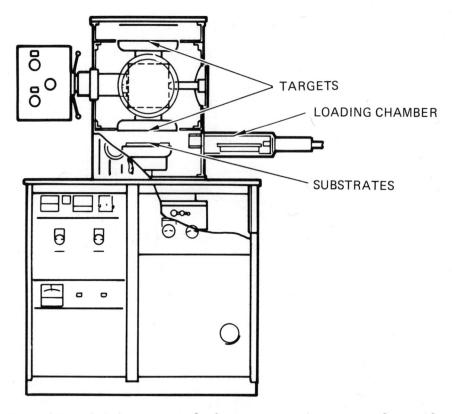

Figure 2. Multiple target rf plasma sputtering system for uniform
           thin-film deposition over 20 cm in diameter.

sputtering" the rf power is divided (about 90% to the cathode and
10% to the anode) so that a small amount of material is continually
being deposited and etched off the samples during the overcoating. This bombardment of the freshly deposited material generally
causes densification of the materials. The outermost layer in most
of our test samples was SiO_2. Much better alignment was obtained
when this coating was deposited by straight sputtering than by bias
sputtering.

    The ion-beam etching of the coated electrode was typically
carried out in a commercially available ion-beam etching apparatus*
shown diagramatically in Figure 3. The ion source is an electron
bombardment type, which uses a multi-aperture electrode structure

---

Microetch model by Veeco Inst., Plainview, N.Y.

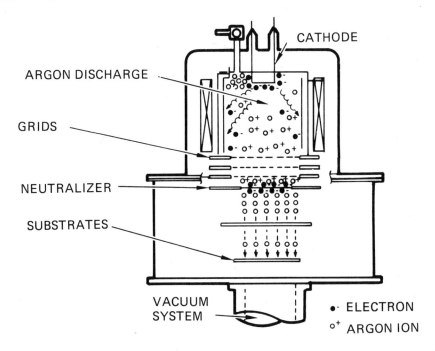

Figure 3. Microetch ion-beam sputtering system for uniform etching
           over 7 cm in diameter.

to accelerate an argon ion beam that is 5 to 7 cm in diameter. The
ions are accelerated to an energy of 500 to 2500 eV and then are
passed by a filamentary source of neutralizing electrons. This
beam neutralization prevents charge buildup on insulating coatings,
such as SiO_2, from repelling ions arriving later. The ion beam
produced by the Microetch is reasonably well collimated with a
uniform intensity across the central 5 cm of the beam at a distance
of 8 cm from the source. The beam intensity decreases beyond this
diameter and the fringing ions have a divergence of about 8°. Under
typical shallow-angle etching conditions the samples are held in
the central portion of the beam with the ions incident at 10 to 30°
from the surface. An ion density of 0.1 mA/cm^2 impinges on the
surface which is located at 7 to 10 cm from the ion source. With
2.5 keV ion energy, about 100 Å of SiO_2 are removed per minute at a
20° etching angle. Coating thicknesses before and after etching
were measured with a mechanical stylus.*

---

*
  Sloan Dektak.

Alignment Evaluation

Typical sandwich-type test cells were assembled with two SAIBE electrodes and either 6 or 13 μm (1/4 or 1/2 mil) Mylar spacers. Usually, the commercial biphenyl liquid crystal mixture E7* or an ester mixture HRL-2N10 (Ref. 4) doped with tetrabutylammonium perchlorate was used in testing, although other nematic materials were aligned similarly.

The quality of the surface alignment was evaluated by visual observations made with a Zeiss WL polarizing microscope at a magnification of 125x. The alignment was either smooth and uniform at one extreme, as inferred from a uniform grey color across the device aperture, or at the other extreme was totally erratic and dependent only on the manner in which the liquid crystal flowed into the cell. Cases were also noted in which the alignment was streaky or grainy, but generally enforced by the etched surface, or in which there was a smooth background alignment with embedded, small isolated defect areas ("assembly marks") with a different type of alignment.

Alignment directions were measured on each electrode surface relative to one edge of the cell. Such measurements were made by keeping the polarizer in a fixed position while adjusting the analyzer and the rotating stage on which the cell was mounted until the positions producing a minimum of light transmission were located. The alignment direction on the bottom surface was then obtained from the angular difference between the polarizer (P) and the stage setting (S). The twist angle (T) was equal to the analyzer setting (A) minus the analyzer setting at the experimentally crossed position in the absence of a birefringent sample (A_0). The alignment direction at the top surface was the sum of S and T, while S + T/2 located the bisector of the twist angle. A SAIBE electrode was paired with a rubbed electrode to determine the alignment direction by reference to the direction of rubbing.

Twist angles were measured with white light. In 6 μm-thick twisted nematic cells the rotation of polarized light is dependent on the wavelength, and misleading results can be obtained with monochromatic light. With white light, however, the eye integrates all of the wavelengths and the results are usually less than 1^0 different from the results obtained for the same electrodes in thicker cells (∿25 μm or more), in which the wavelength dependence is negligible.

---

*BDH, Ltd.

The tilt angle of the director with respect to the SAIBE
surfaces was measured by monitoring the effect of a 10 kG magnetic
field on the capacitance of a sandwich-type liquid crystal cell
as a function of the angle the cell formed with the field.   The
liquid crystal tilt angle was equal to this angle when there was no
effect on the cell capacitance as the magnetic field was turned on
and off.[5]

RESULTS AND DISCUSSION

When a silica coating is deposited by the rf straight sputtering technique and etched under the preferred conditions, it is
capable of inducing a very smooth, uniform homogeneous alignment.
This capacity to align a liquid crystal is very durable; it is not
destroyed by heating the substrate to $500^{\circ}C$ or by washing it with
solvents.   The nematic director lies in the plane of incidence of
the ion beam.   It is tilted slightly out of the surface, the
raised end pointing in the direction of the ion beam source.   A
tilt angle of 4° was measured for the liquid crystal E7.

The angle that the ion beam formed with the silica surface had
no apparent affect on the quality of homogeneous alignment in the
range of 10° to 30°.   When this angle was 40°, the quality of alignment was somewhat degraded by the presence of assembly marks, and
when it was 50°, the alignment was extremely poor and unsatisfac-
tory.   Since the etching rate decreases rapidly at angles below
10°, these extremely shallow angles were not explored.   An etch
angle of 20° was generally used in experiments designed to study
these parameters.

When a 2700 Å thick coating of SiO_2 was ion-beam etched, good
alignment was obtained after as little as 2 Å was removed.   Further
etching to remove up to 900 Å of SiO_2 produced little or no change
in the quality of alignment.   Removal of 1000 to 1500 Å caused
increasing streakiness in the resulting alignment while leaving
some areas unaligned; with a removal of 2000 Å there was no induced
alignment.   These etching depths were not studied on thicker coatings, and it is possible that some areas of the coating may have
been completely removed by the etching.   Etching depths of about
500 to 700 Å were used routinely in the study of other variables.

A major limitation of the ion-beam etching method is the small
number of samples that can be placed under the beam at one time in
the equipment that is currently available.   For this reason we
investigated the use of movable carriages that could be used to
place several samples under the beam at one time and then advanced
to a new position when etching of the first samples was completed.
Unfortunately, this approach was not satisfactory because scattered

material from the etching of subsequent substrates reached the
previously etched surfaces and ruined the quality of alignment.
Moreover, ions were found to be scattered surprisingly far from the
center of the beam, and these low-intensity scattered ions were
sufficient to alter the alignment direction of portions of previously etched surfaces. Glass shields placed between samples
failed to provide adequate protection from scattered material, and
in some cases, scattering from the shields themselves reduced the
uniformity of alignment. Repeatedly passing the substrate under
the beam emerging through a 0.5 cm slit between glass plates
appeared to suffer from the same problems, and there was no uniform
alignment on these substrates. All of these results demonstrated
the astonishing sensitivity of the liquid crystal in detecting the
condition of the etched silica surface.

Although the direction of liquid crystal flow over the SAIBE
surface does not affect the fundamental alignment direction, flow
parallel to the etch direction is preferred. Flow perpendicular to
the etch direction increases the tendency to form small isolated
defects or assembly marks.

Splay in alignment is a persistent problem caused by the imper-
fect collimation of the ion beam. (Splay also occurs in the alignment of nematics on surfaces treated by the vapor deposition
techniques of Janning.) In all cases observed thus far, the splay
is consistent with the divergence of the ion beam, which behaves as
if it originated from a virtual point source (Figure 4). Data for
each electrode of two typical cells are given in Figure 5, along
with the resulting variations in the twist angle across the nematic
layer. Under the conditions we usually employed, the splay was
about $2.2^{\circ} \pm 0.2^{\circ}$/cm across the surface perpendicular to the ion
beam. While the splay is obviously a function of the geometry of
the etching apparatus, it can also be altered electronically. For
example, increasing the magnet current in the discharge chamber of
the ion source from 0.18 to 0.4 Å decreased the splay to about
1.3°/cm. Decreasing the ion acceleration voltage, on the other
hand, increased the splay and caused some deterioration in uniformity of alignment.

We were particularly interested in the nature of the assembly
marks that occasionally marred the alignment. These alignment
defects can appear in any carelessly assembled cell, but they are
formed more readily when the liquid crystal flow during cell assembly is perpendicular rather than parallel to the etch direction.
Sometimes they are left in the wake of an air bubble that moves
across the cell surface during assembly. They consist of small
defect areas separated from the surrounding background by a disclination. Within the defect area the alignment is optically smooth
and uniform. We studied these defects extensively in homogeneously

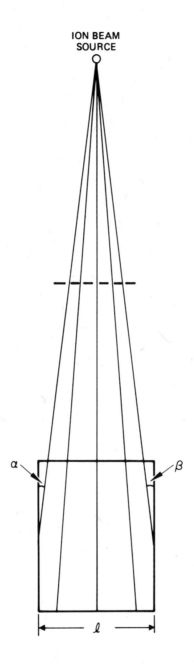

Figure 4.   Diagram of the splay in alignment on a SAIBE electrode surface.   Splay is given by $(\alpha + \beta)/\ell$.

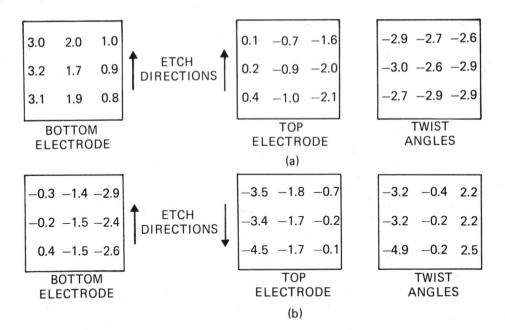

Figure 5.   Alignment variations at each electrode and the resulting
            twist angles in two cells with (a) parallel and
            (b) antiparallel SAIBE electrodes. Data were taken at
            points across a 1.6 cm square and are given as angles
            between the director and the right-hand edge of the cell.

aligned cells that were assembled with antiparallel etching of the
SAIBE electrodes so that the liquid crystal would be oriented with
a uniform tilt throughout the cell. In these cells there were
always two distinct, predominant types of defects, although three
or four types could be found occasionally. For a given cell, each
type had its own color when viewed between polarizers at any given
setting (Figure 6). Clearly the alignment in these defects is
controlled by the structure of the SAIBE silica surfaces, just as
the surface structure controls the normal homogeneous alignment of
the surrounding background.

    Within the defect areas the liquid crystal had the character-
istics of a twisted nematic alignment. However, the rotation of
polarized light passing through the defects was strongly dependent
on the wavelength, much more so than when it passed through an
equivalent twisted-nematic cell with the same thickness and a 90°
twist. This indicated that, if the twist was less than 90°, it

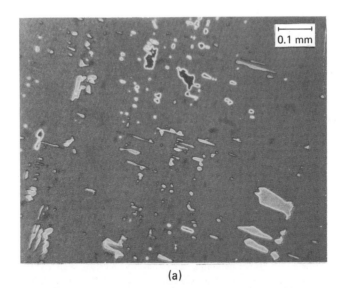

(a)

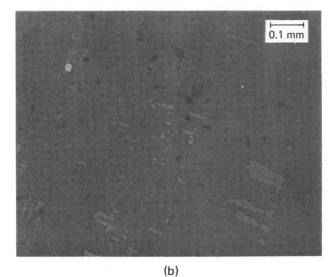

(b)

Figure 6. Assembly marks in a 6 μm cell; (a) before and (b) after heating to the isotropic state and cooling. Between polarizers set to show maximum contrast prior to heating.

was accompanied by considerable tilt,* or, alternatively, that the
twist was considerably more than 90°. Values could be obtained for
an apparent twist angle, however, using the minimum transmission
settings with the color of the transmitted light being midway
between blue and magenta. These apparent twist angles in the two
main types of assembly marks were always opposite in sign, and
usually were roughly equal in magnitude (relative to the background,
which could also have a twist). Moreover, the bisectors of the
twist angles in the three areas (background and two types of defects)
were all parallel within experimental error. This was usually true
for the third and fourth types of assembly marks as well. Therefore, at least in the two major types of assembly marks, the director at one electrode surface appeared to be rotated clockwise while
the director at the opposite surface appeared to be rotated counter-
clockwise simultaneously through approximately the same angle.

When an ac voltage was applied to establish Williams domains,
the domains within the defect areas ran in directions that were
distinctly different from those of the uniform background. In
defects large enough to minimize the distortion caused by the surrounding disclination, the Williams domains were approximately perpendicular to those of the normal alignment areas (Figure 7(a)).
This observation provides strong evidence that the alignment has a
twist of about 180° and -180° in the two main types of defects.
Similar alignment defects were observed by Spruijt[6] in cells with
rubbed electrodes.

These assembly marks are metastable alignment states, although
they have persisted unchanged in some cells for more than three
years at room temperature. Heating above the clear point of the
nematic and recooling causes them to convert to a stable alignment
state that is very similar to, but still perceptibly different from,
that of the background. This is illustrated in Figure 6(b) and by
the data for one cell shown in Table 1. Although most of the
assembly marks can survive dynamic scattering unchanged, they are
converted to the stable alignment by secondary dynamic scattering[7]
(Figure 7). A sudden change in pressure, such as that encountered
by an electrode cracking, can also induce realignment.

---

*This twisted tilted alignment can be simulated by applying an
electrical field across a 6 μm thick cell with a nematic liquid
crystal having a positive dielectric anisotropy and a twist of
about 45°. When the apparent twist of such a cell is measured,
the result is strongly dependent on the field strength, i.e., the
tilt. However, the bisector of the apparent twist angle remains
constant.

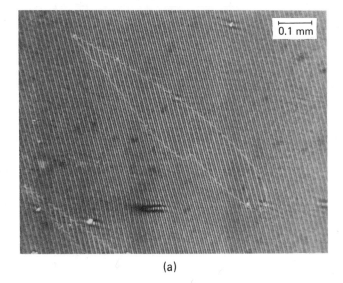

(a)

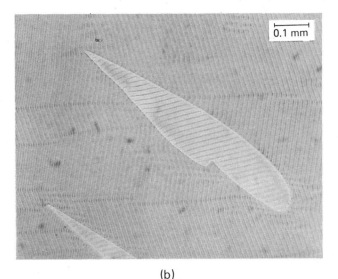

(b)

Figure 7. Williams domains in a 6 μm cell with assembly marks.
(a) Before secondary scattering. (b) After secondary
scattering.

    An examination of the SAIBE surfaces with a scanning electron
microscope at magnifications ranging from 11,000x to 65,000x failed
to reveal any surface features that could be construed to induce
alignment. A study with transmission electron microscopy is being
pursued.[8]

Table 1. Alignment Data for a 6 µm Cell with Antiparallel SAIBE Electrodes*

	Number of Measurements	Bottom Electrode (S-P)	Top Electrode (S+T)	Apparent Twist Angle (T or $A-A_o$)	Bisector (S + T/2)	Apparent Twist Angle Minus Background Twist ($T-T_B$)
Before Heating						
Background	3	-13.4 ± 0.4	-0.1 ± 0.4	+13.3 ± 0.3	-6.7 ± 0.3	—
Type 1 Assembly Mark	5	+3.7 ± 1.0	-14.2 ± 1.2	-18.9 ± 0.5	-5.7 ± 1.1	-32.2
Type 2 Assembly Mark	8	-27.8 ± 2.0	+7.1 ± 1.6	+34.9 ± 1.0	-10.4 ± 1.7	-21.6
After Heating						
Background	3	-14.6 ± 0.1	-2.0 ± 0.1	+12.6 ± 0.1	-8.3 ± 0.1	—
Type 1 Assembly Mark	5	-11.3 ± 0.4	-2.8 ± 0.4	+8.5 ± 0.1	-7.1 ± 0.4	-4.1
Type 2 Assembly Mark	5	-17.9 ± 0.2	-0.7 ± 0.2	+17.2 ± 0.1	-9.3 ± 0.2	+4.6

*Values are given in degrees relative to one edge of the electrode, ± the standard deviation.

ACKNOWLEDGMENT

We wish to thank Dr. Y.S. Lee and Dr. J.D. Margerum for advice and suggestions, and Mr. Klaus Robinson and Mr. G.D. Myer for technical assistance. We also appreciate the valuable comments of Dr. J.L. Fergason.

REFERENCES

1. D.W. Berreman, Phys. Rev. Lett. $\underline{28}$, 1683 (1972); D.W. Berreman, Mol. Cryst. Liq. Cryst. $\underline{23}$, 215 (1973); U. Wolff, W. Greubel, and H. Krüger, Mol. Cryst. Liq. Cryst. $\underline{23}$, 187 (1973); L.T. Creagh and A.R. Kmetz, Mol. Cryst. Liq. Cryst. $\underline{24}$, 59 (1973).

2. J.L. Janning, Appl. Phys. Lett. $\underline{21}$, 173 (1972).

3. E. Guyon, P. Pieranski, and M. Boix, Lett. Appl. Eng. Sci. $\underline{1}$, 19 (1973); W. Urbach, M. Boix, and E. Guyon, Appl. Phys. Lett. $\underline{25}$, 479 (1974); G.D. Dixon, T.P. Brody, and W.A. Hester, Appl. Phys. Lett. $\underline{24}$, 47 (1974); D. Meyerhofer, Appl. Phys. Lett. $\underline{29}$, 691 (1976); M.R. Johnson and P.A. Penz, IEEE Trans. Electron Devices $\underline{ED-24}$, 805 (1977).

4. H.S. Lim and J.D. Margerum, Appl. Phys. Lett. $\underline{28}$, 478 (1976).

5. K. Toriyama and T. Ishibashi, Nonemissive Electrooptic Displays, A.R. Kmetz and F.K. von Willisen, eds. (Plenum Press, New York, 1976), p. 145.

6. A.M.J. Spruijt, Solid State Commun. $\underline{13}$, 1919 (1973).

7. A. Sussman, Appl. Phys. Lett. $\underline{21}$, 269 (1972); R. Chang, J. Appl. Phys. $\underline{44}$, 1885 (1973).

8. J.C. Potosky and H.L. Garvin, to be submitted for publication.

A NEW METHOD FOR INDUCING HOMEOTROPIC AND TILTED ALIGNMENTS OF

NEMATIC LIQUID CRYSTALS ON SILICA SURFACES

Leroy J. Miller, Jan Grinberg, Gary D. Myer,
Deborah S. Smythe, and Willis H. Smith

Hughes Research Laboratories

3011 Malibu Canyon Road, Malibu, CA. 90265

ABSTRACT

Homeotropic alignment of a nematic liquid crystal on a silica
surface can be induced by means of a prior treatment of the surface
with a long-chain aliphatic alcohol or a mixture of such an alcohol
and an amine. The treatment is presumed to replace some or all of
the hydroxy groups normally attached to silicon atoms at the surface with alkoxy groups derived from the alcohol, thereby reducing
the critical surface tension of the surface. As the chain length
of the alcohol is decreased, the director of the nematic tilts away
from the perpendicular orientation. The direction of tilt can be
controlled by shallow angle ion beam etching the silica surface
prior to the alcohol treatment, in which case the nematic molecules
are uniformly tilted toward the ion beam source. The tilt angle is
a function of both the length of the alcohol chain and the composition of the liquid crystal. Tilted homeotropic alignment is essential for obtaining optimum performance of field effect devices that
use nematics with a negative dielectric anisotropy.

INTRODUCTION

Typically, nematogenic compounds have a rod-shaped molecular
structure in which the central portion is more polar and/or more
readily polarizable than the ends. When they come in contact with
a solid having a high surface energy (e.g., most metals and metal
oxides) this central portion of the molecule becomes adsorbed on
the surface, causing the surface layer of the nematic to be oriented
parallel to the surface. An orientation of the nematic

513

perpendicular to the surface is generally induced by substantially reducing the surface energy of the solid. This induces the nematic molecules to orient their polar or polarizable central groups toward each other leaving their nonpolar molecular ends (usually saturated alkyl groups) in contact with the low energy surface.

This reduction in the surface energy has been accomplished in three ways. The first is to dissolve an aligning agent with surfactant properties in the liquid crystal[1] and rely on its tendency to concentrate at the solid-nematic interface with the polar end adsorbed on the solid and the nonpolar end facing the liquid crystal. Alternatively the surfactant is applied directly to the surface prior to contact with the nematic.[1a,2] The second method is to overcoat the surface with polymer having a low surface energy, usually with polymerization being carried out on the surface.[3] The third method, which includes the technique reported here, involves the formation of covalent bonds between relatively low molecular weight compounds and the surface, thus modifying the surface chemically to produce the desired effect.[4] This last method has advantages over the soluble aligning agents, which may alter the bulk properties of the liquid crystal or may exert an aligning effect that varies as the ratio of the surface area to the volume is varied. It may also have advantages over polymeric coatings in that it is extremely thin and uniform, and therefore it is able to transmit the influence of the contour of the underlying surface to the nematic.

A relationship between the critical surface tension (γ_C) of the solid, the surface tension of the liquid crystal (γ_L), and the alignment was proposed by Creagh and Kmetz[5] and elaborated on by Kahn, Taylor, and Schonhorn.[1a] These investigators reported that when γ_C was greater than γ_L, the nematic aligned parallel to the surface, while when γ_L was greater, a homeotropic alignment was obtained. The validity of this rule was soon questioned, however, as exceptions were found.[3c,d] Moreover, the proposed relationship made no provision for tilted alignments in which the director was oriented close to, but tilted slightly away from, the position perpendicular to the surface.[6] A uniform tilted homeotropic or tilted perpendicular alignment is especially desirable for field effect cells with tunable birefringence,[6a] where the off-state tilt predetermines the manner in which the nematic will respond to an applied field.

We first became interested in the use of alcohols for chemically modifying silica surfaces when we needed homeotropic alignment in dynamic scattering devices. Uncontrolled tilting away from the normal was observed in some cells, which were considered to have a poor quality homeotropic alignment. Subsequently our attention became focused on tunable birefringence devices,[7] and we were able to use the alcohol treatment of shallow angle ion beam etched (SAIBE) silica surfaces[8] to control both the tilt direction and the tilt angle.

EXPERIMENTAL SECTION

Substrates

The substrates were prepared by overcoating Nesatron (from PPG) with silica and ion beam etching at a shallow angle as described in the previous paper.[8] Samples were routinely etched to a depth of 700 Å.

Treatment of the Silica Surface with Alcohols

For alcohols with 7 to 22 carbon atoms per molecule, the substrates were placed face-up in a Pyrex container and covered with a melt of the alcohol or alcohol-amine mixture. The container was covered with aluminum foil and heated for 1 to 24 hours in an oven at temperatures between 100 and 150°C. In the usual case the material was heated overnight at about 120°. A stream of nitrogen was used to purge vapors from the oven when the alcohol had 7 to 9 carbons. For alcohols with 4 to 6 carbons, the substrates were placed on a glass rack in a refluxing mixture of the alcohol and an amine. The alcohols and amines were all commercial materials used without purification. Although occasionally there appeared to be small differences between batches of alcohol, the selection of the desired alcohol treatment was empirical and automatically compensated for the presence of any reactive impurities. Generally, 1-hexadecylamine and 1-octadecylamine were used as the amine because of their low volatility, but shorter chain amines were also used in some cases. When amines were used, they were mixed with an equal weight of alcohol.

After the treatment, the substrates were washed thoroughly with reagent grade or electronic grade solvents to remove all traces of unreacted material. Hexane and methanol sufficed for removing the shorter alcohols, while a solvent such as chloroform or methylene chloride was necessary to remove alcohols and amines with 18 or more carbons. Ordinarily combinations of two or three solvents were used, and no systematic difference due to the choice of solvents was detected. Treated electrodes were stored in desiccators, but the aligning effect was at least qualitatively unchanged after storage in the laboratory atmosphere for several months.

Surface Evaluation

Test cells were constructed with a variety of Schiff base, ester, and tolane liquid crystal mixtures with negative dielectric anisotropies. They included N-(p-methoxybenzylidene)-p-butylaniline (MBBA) and a 1:2 mixture by weight of MBBA and N-(p-ethoxybenzyli-dene)-p-butylaniline (EBBA); HRL-2N10[9] and a variety of related proprietary phenyl benzoate ester mixtures with nematic properties at room temperature; and a nematic mixture of tolanes and esters. The exact liquid crystal composition is of little consequence since the surface treatment must be optimized empirically for each new mixture. The test cells ranged in thickness from less than 2 μm to 13 μm, and the alignment was found to be independent of thickness. Thickness was controlled with Mylar spacers or evaporated SiO_x spacer pads.

Initial evaluation of alignment quality was made with a Zeiss WL polarizing microscope. Tilt directions could be evaluated quickly for even small domains by manually tilting the cell between crossed polarizers to locate the optic axis, or by applying an ac field and noting the direction from which the typical progression of colors[7] appeared. Twist angles were measured in the manner pre-viously described[8] while the field was applied.

Tilt angles were measured with apparatus shown schematically in Figure 1. The cell was rotated between crossed polarizers with the nematic director lying in a plane that is perpendicular to the axis of rotation and that forms an angle of 45° with the plane of vibration of the incident polarized light.[10] When the light passes through the cell parallel to the optic axis, the transmission is at a minimum. This minimum transmission angle was located most accu-rately by plotting the transmission as a function of the rotation angle and determining a series of midpoints between paired positions of equivalent transmission on both sides of the minimum; extrapo-lating a line connecting these midpoints to intersect the trans-mission curve gave the minimum transmission angle. The tilt angle was defined as the angle formed by the nematic director and the normal to the cell surfaces. It was equal to the angle between the collimated light and the cell normal at minimum transmission, divided by the ordinary refractive index of the nematic. At tilt angles below 5° the result was reproducible within 0.2^O. While the data could be taken manually, the acquisition was accelerated by placing the process under the control of a minicomputer, which was pro-grammed to analyze the data and determine the tilt angle auto-matically. Measurements were made either at room temperature (measured next to the cell) or at elevated temperatures controlled to $\pm 1^O$ within a compartment surrounding the test cell and measured at the cell surface. The maximum tilt angle that could be deter-mined by this method was about 17°.

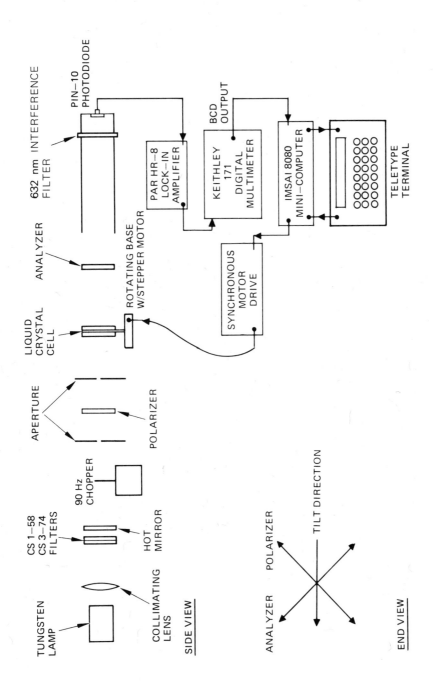

Figure 1. Apparatus for measuring tilt angles.

The critical surface tension (γ_C) for treated surfaces was determined with measurements of the advancing contact angle (θ) for a series of suitable liquids using a Rame-Hart telescopic goniometer. Plots of the literature values for the surface tension of the liquids against cos θ were extrapolated to cos θ = 1 to obtain γ_C. Water, glycerol, ethylene glycol, formamide, 1-bromonaphthalene, tricresyl phosphate, bromobenzene, and squalane were used as test liquids.

RESULTS AND DISCUSSION

The treatment of silica surfaces with long-chain alcohols produced low energy surfaces that induced spontaneous homeotropic alignment of nematics. The alignment characteristics were not changed by prolonged or repeated washing with organic solvents. Homeotropic alignment was not induced by merely coating the surface with the alcohol or alcohol-amine mixture at room temperature and then washing. Clearly the aligning influence was due to a chemical reaction at the surface and not to a physically adsorbed surface layer.

Generally, the only two types of groups at the silica surface are silanol and siloxane groups.[11] The silanol groups can be categorized as isolated, vicinal, and geminal types. The free or isolated OH groups are probably the most reactive and may be the only groups to react with the alcohols to form alkoxy groups.

$$
\begin{array}{ccccccc}
\text{OH} & & & & \text{OR} & & \\
| & & & & | & & \\
\text{Si} & + & \text{ROH} & \longrightarrow & \text{Si} & + & \text{H}_2\text{O} \\
/\,|\,\backslash & & & & /\,|\,\backslash & &
\end{array}
$$

Amines have been shown to catalyze the reaction,[12] and therefore amines were mixed with the alcohols in many of our experiments. Equivalent aligning effects were achieved, however, by using the alcohol without the amine.[13] The amines by themselves had no effect on alignment.

When unetched silica surfaces were given the alcohol treatment, a uniform homeotropic alignment was obtained for long chain alcohols, but the quality of this alignment deteriorated as the chain length was decreased. The nematic director in these cells was tilted away from the perpendicular orientation in directions that appeared to be controlled only by the vagaries of the surface structure (Figure 2).

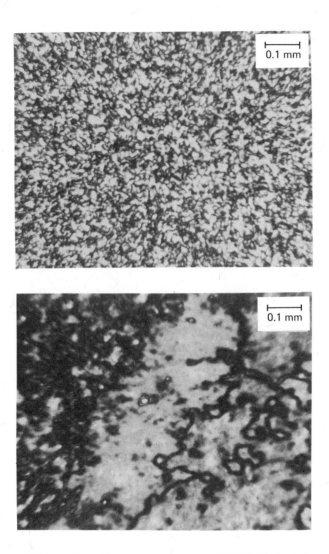

Figure 2. The appearance of tilted homeo-
           tropic alignment on unetched
           silica surfaces viewed between
           crossed polarizers.

When the SAIBE silica was treated with the alcohols, the tilting direction was controlled by the topography of the underlying silica surface. The nematic director always tilted toward the position of the ion beam source. A uniformly tilted homeotropic alignment was obtained only when the two parallel surfaces were etched in opposite directions (Figure 3).

For any given liquid crystal, the tilt angle was a function of the chain length of the alcohol used. It was also a function of the composition of the liquid crystal. The empirically determined relationship is shown for two liquid crystals in Figures 4 and 5. Opposing electrode surfaces can be treated with different alcohols to obtain intermediate tilt angles. If this does not provide adequate control, alcohols can also be mixed. The reproducibility is approximately ±2° over the linear portions of the curves.

The tilt angle obtained with a given alcohol can vary from batch to batch of the liquid crystal if the purity is not rigidly maintained. It is also dependent on the temperature, and even the sign of the temperature coefficient is a function of the liquid crystal composition (Figures 6 and 7).

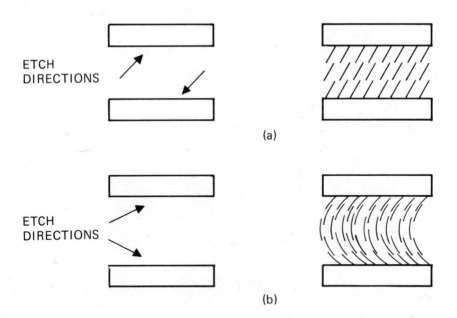

Figure 3.   Correct (a) and incorrect (b) method of cell assembly
            for obtaining a uniformly tilted homeotropic alignment.
            Ion beam etch directions are indicated with arrows.

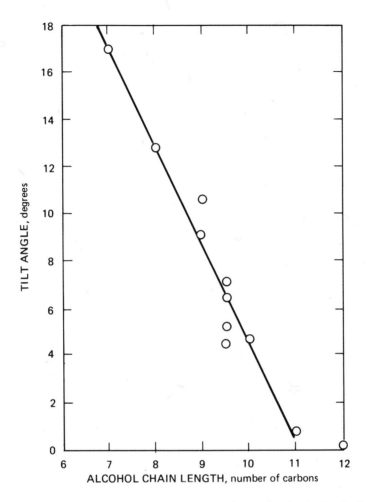

Figure 4.   Relationship between tilt angle and alcohol chain length for a 1:2 mixture of MBBA and EBBA at 22-23°C.

Misaligned areas are very commonly encountered in preparing tilted homeotropic cells, particularly when the normal tilt angle is higher than 6-8°. The misaligned areas are usually metastable. Sometimes they will vanish when a field is applied, but it is generally more effective to heat the cell above the clearpoint of the liquid crystal.   If the misaligned areas are resistant to realignment by this technique, they may realign if a field is applied while the cell is cooled through the clearpoint.   However, in some cases the expected tilted alignment simply cannot be obtained.

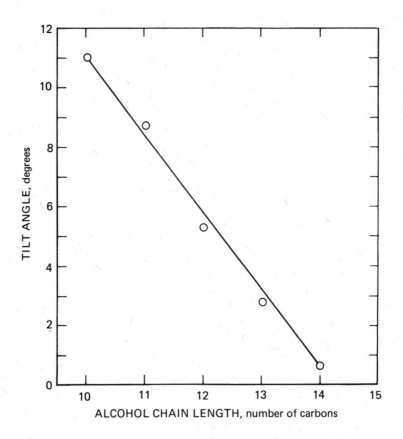

Figure 5.   Relationship between tilt angle and alcohol chain length
            for a proprietary nematic mixture of phenyl benzoates,
            HRL-2N13, at 22°.

        The misaligned areas frequently have a peculiar set of proper-
ties reminiscent of the characteristics of the assembly marks formed
in cells with homogeneous alignment on SAIBE electrodes.[8]   Between
crossed polarizers the misaligned area is darker than the stable
alignment area, indicating a lower tilt angle in the misaligned
area.   When an ac field is applied across the nematic to orient the
nematic director in the bulk parallel to the electrode surfaces,
there is no twist in the stable alignment area.   However, in the
misaligned area there is a twist, with the bisector of the twist
angle lying parallel to the director in the stable alignment area.
If the field is applied suddenly, the abnormally aligned area forms
two twisted domains, with the apparent twist angles being approximately equal in magnitude but opposite in sign.   This is illustrated

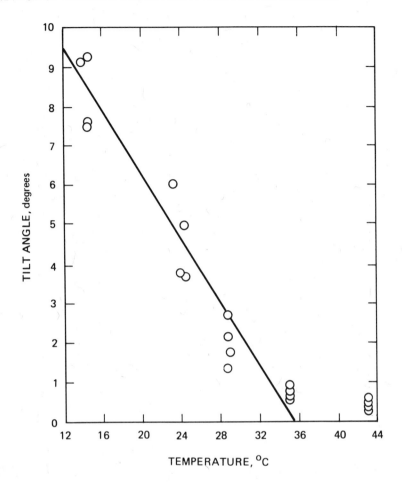

Figure 6.   The effect of temperature on the tilt angle of a 1:2
            mixture of MBBA:EBBA. One surface was treated with
            1-nonanol and the other with 1-decanol.

with the data for one cell in Table 1. One twisted area is unstable
relative to the other and slowly shrinks in size and merges into the
domain with opposite twist. If the liquid crystal is capable of
forming a domain pattern due to conduction, the domains are roughly
perpendicular to each other in the normal and abnormal alignment
areas (Figure 8). The abnormal alignment apparently represents a
metastable alignment state that is derived from the topography of
the SAIBE surface and has almost the same energy as the normal tilted
homeotropic alignment. The tendency to form these misaligned areas
decreases as the ion beam etch angle (the angle between the ion beam
and the surface) increases from 10° to 30°.

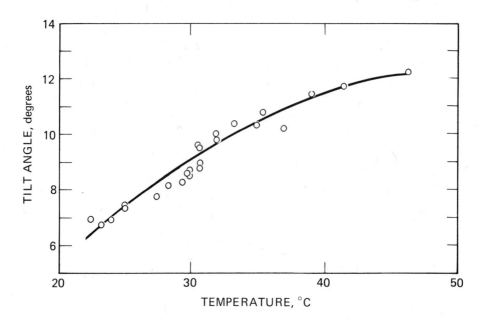

Figure 7.   The effect of temperature on the tilt angle of a
            proprietary nematic mixture of phenyl benzoates, HRL-
            2N20.   Both surfaces were treated with a 4:1:5 mixture
            by weight of 1-octadecanol, 1-eicosanol, and
            1-octadecylamine.

     In addition to the abnormal "twisted tilted" alignment described
above, some cells have misaligned areas in which the molecules are
tilted in the direction opposite to that of the normal alignment
pattern.   Efforts to convert this type of misaligned area to the
normal alignment have always failed.

     Tilted homeotropic alignment is reasonably stable for long
periods of time, but unfortunately the tilt angle is very sensitive
to chemical changes in the liquid crystal.   The stability of the
tilt angle is also affected by the angle of ion beam etching.   Figure 9 illustrates the temporal stability of the tilt angle for four
cells with a 1:2 mixture of MBBA and EBBA that were stored in a
desiccator at a low level of illumination.   The electrodes used
were SAIBE at an angle of 20°.   Increasing the etch angle to 30°
caused the tilt angles to increase with time as shown by the curves
in Figure 10.   Irradiating the cells with visible or near ultraviolet light caused the tilt angles to decrease fairly rapidly,
undoubtedly because of photochemical changes in the liquid crystal
composition.[14]

Table 1. Alignment Data for a Cell with Tilted Homeotropic Alignment in an Applied AC Field[a]

Alignment Area	Bottom Electrode	Top Electrode	Apparent Twist Angle	Bisector
Normal alignment area	0.0	0.0	0.0	0.0
More stable of abnormal alignment areas	+18.3	-25.1	-43.4	-3.4
Less stable of abnormal alignment areas	-22.9	+25.3	+48.2	+1.2

[a] SAIBE electrodes with antiparallel etching, treated with 1-octanol; 13 μm thick cell with a 1:2 mixture of MBBA and EBBA by weight; 35 V, 10 kHz. Values are given in degrees relative to one edge of the cell.

Hydrolytic changes in the liquid crystal can also cause changes in the tilt angle, but the surface itself is remarkably stable toward hydrolysis. One cell was disassembled and the electrodes were soaked in water for 17 hours. When the cell was reassembled, the original tilt angle was reproduced almost exactly.

In speculating on the forces controlling the tilt angle, we considered an extension of the hypothesis propounded by Creagh and Kmetz[5] and Kahn, Taylor and Schonhorn.[1a] These authors considered the surface tension of the liquid crystal to be a single value, γ_L, which was measured experimentally at the nematic-air interface. However, because of the anisotropy of the liquid crystal, in theory there should exist two characteristic values for the surface tension: $\gamma_{L\perp}$ when the molecules are oriented perpendicular to the nematic-air interface, and $\gamma_{L\parallel}$ when the orientation is parallel. We considered the possibility that when $\gamma_{L\parallel}$ and $\gamma_{L\perp} < \gamma_C$, the alignment is parallel; when $\gamma_C < \gamma_{L\parallel}$ and $\gamma_{L\perp}$, the alignment is perpendicular; and when $\gamma_{L\parallel} < \gamma_C < \gamma_{L\perp}$, the alignment is tilted, with the tilt angle being determined by the relationship of these three values. Our observations do not support this hypothesis, however. Since the longer alcohols give rise to a more nearly perpendicular alignment, this hypothesis would predict that γ_C should decrease as the chain length increases. Our data shown in Figure 11 indicate that in general γ_C increases as the chain length

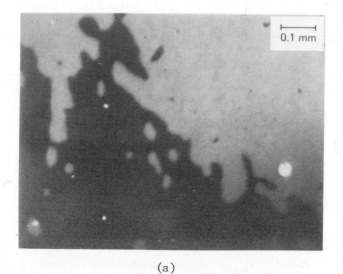

(a)

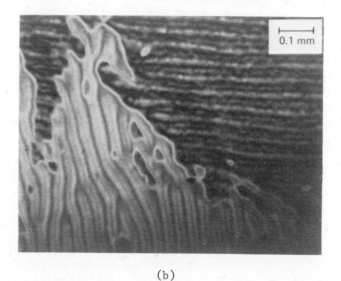

(b)

Figure 8.   Normal and abnormal alignment areas in a cell with
            tilted homeotropic alignment in (a) the absence of
            an applied field and (b) with a 37 V, 100 Hz field
            applied to induce a domain pattern due to conduction.
            Normal alignment area corresponds to the lighter area
            in (a).   (13 µm cell with liquid crystal HRL–2N10 and
            SAIBE electrodes treated with a mixture of 1–dodecanol,
            1–tetradecanol, and 1–hexadecylamine.)

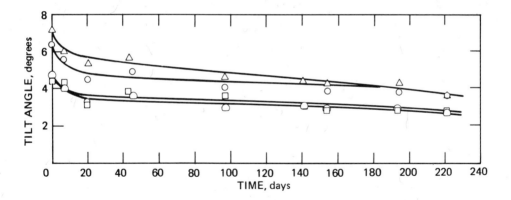

Figure 9. Temporal stability of the tilt angle in four cells
          prepared in the same manner and stored in a desic-
          cator. (Etch angle, $20°$; alcohol treatment, $C_9H_{19}OH$
          and $C_{10}H_{21}OH$; 1:2 mixture of MBBA and EBBA by weight).

of the alcohol used in the surface treatment increases.[15] Moreover,
surfaces that had equal γ_C values sometimes had widely different
alignment effects. For example, a surface treatment with behenyl
alcohol yielded a surface with the same γ_C as one treated with
cyclohexanol, yet the tilt angle of a proprietary nematic mixture
was $0°$ on the former surface and too high to measure on the latter.

    An alternative view is that the tilted alignment is caused by
the combined influence of a minutely grooved surface and the steric
effects of the attached alkoxy groups. In the absence of the
alkoxy groups the molecules align parallel to the grooves, whereas
on the alcohol-treated surface they align parallel to the alkoxy
chains. As these chains are shortened, the underlying grooves
exert an increasing influence on the nematic orientation. The
density of alkoxy chains on the surface would also be expected to
play a role, but in our studies we made every effort to eliminate
this as a variable by carrying the reaction of the surface with the
alcohol to completion.

    Since alcohols are available in almost infinite structural
variety, the silica surface could conceivably be modified to produce a wide range of nematic orientations and anchoring energies.
Our investigation is proceeding along these lines.

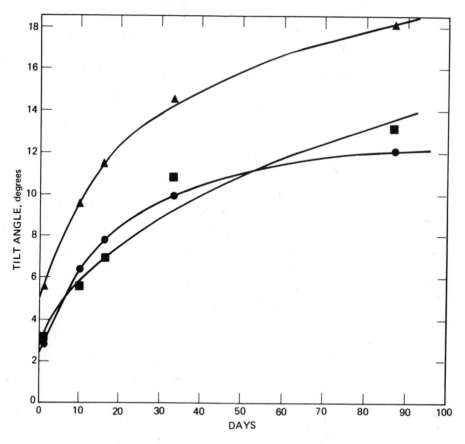

Figure 10.   Temporal stability of the tilt angle in three cells
prepared in the same manner and stored in a desiccator, when the etch angle is 30° (alcohol treatments,
$C_9H_{19}OH$ and $C_{10}H_{21}OH$; 1:2 mixture of MBBA and EBBA by
weight).

ACKNOWLEDGMENT

We are pleased to acknowledge the support of Naval Sea Systems
Command, U.S. Department of the Navy, for part of this work. We
also thank Prof. A.W. Adamson and Dr. J.D. Margerum for helpful
discussions, Dr. P. Hu for ellipsometric measurements, and Mr.
M.J. Little, Dr. H.L. Garvin, and Mr. K. Barnett for valuable
assistance.

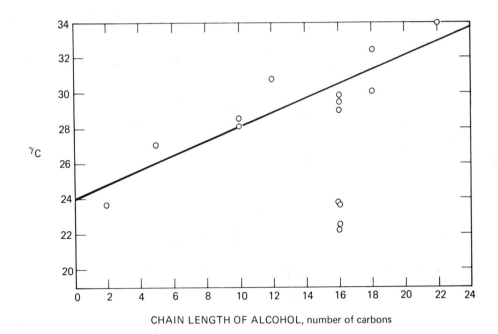

CHAIN LENGTH OF ALCOHOL, number of carbons

Figure 11. Effect of the chain length of the alcohol used in the silica surface treatment on the critical surface tension of the resulting surface.

REFERENCES

1. (a) F.J. Kahn, G.N. Taylor, and H. Schonhorn, Proc. IEEE 61, 823 (1973); (b) I. Haller and H.A. Huggins, U.S. Pat. 3,656,834, April 18, 1972; J.E. Goldmacher and M.G. Tayag, U.S. Pat. 3,809,456, May 7, 1974; G.W. Smith and D.B. Hayden, U.S. Pat. 3,848,966, Nov. 19, 1974; M. Fukai, K. Asai, S. Nagata, H. Tatsuta, and K. Mori, U.S. Pat. 3,979,319, Sept. 7, 1976.

2. (a) H. Sorkin and R.I. Klein, U.S. Pat. 3,698,449, Oct. 17, 1972; (b) J.E. Proust, L. Ter-Minassian-Saraga, and E. Guyon, Solid State Commun. 11, 1227 (1972); J.E. Proust and L. Ter-Minassian-Saraga, C.R. Acad. Sci. Paris, Ser. C, 276, 1731 (1973); J.E. Proust and L. Ter-Minassian-Saraga, J. Phys. Colloq. C1, 36, C1-77 (1975).

3. (a) F.J. Kahn, Appl. Phys. Lett. 22, 386 (1973); (b) J.C. Dubois, M. Gazard, and A. Zann, Appl. Phys. Lett. 24, 297 (1974); (c) J.C. Dubois, M. Gazard, and A. Zann, J. Appl.

Phys. 47, 1270 (1976); (d) I. Haller, Appl. Phys. Lett. 24, 349 (1974); (e) G.J. Sprokel and R.M. Gibson, Electrochemical Soc. Mtg., Las Vegas, Oct. 17-22, 1976, Abst. No. 214.

4.   S. Matsumoto, M. Kawamoto, and N. Kaneko, Appl. Phys. Lett. 27, 268 (1975).

5.   L.T. Creagh and A.R. Kmetz, Mol. Cryst. Liq. Cryst. 24, 59 (1973).

6.   (a) F.J. Kahn, Appl. Phys. Lett. 20, 199 (1972); (b) F.J. Kahn, U.S. Pat. 3,694,053, Sept. 26, 1972; (c) G. Ryschenkow and M. Klemen, J. Chem. Phys. 64, 404 (1976); M. Klemen and G. Ryschenkow, J. Chem. Phys. 64, 413 (1976); (d) K. Fahrenschon, H. Gruler, and M.F. Schiekel, Appl. Phys. 11, 67 (1976); (e) W. Urbach, M. Boix, and E. Guyon, Appl. Phys. Lett. 25, 479 (1974).

7.   J. Grinberg, W.P. Bleha, A.D. Jacobson, A.M. Lackner, G.D. Myer, L.J. Miller, J.D. Margerum, L.M. Fraas, and D.D. Boswell, IEEE Trans. Elect. Dev. ED-22, 775 (1975).

8.   M.J. Little, H.A. Garvin, and L.J. Miller, Liquid Crystals and Ordered Fluids, J.F. Johnson and R.S. Porter, Eds (Plenum Press, New York, N.Y.), Vol 3, p. 497.

9.   H.S. Lim and J.D. Margerum, Appl. Phys. Lett. 28, 478 (1976).

10.   G. Baur, V. Wittwer, and D.W. Berreman, Phys. Lett. 56A, 142 (1976).

11.   V.L. Snoeyink and W.J. Weber, Jr., "Surface Functional Groups on Carbon and Silica," in Progress in Surface and Membrane Science, ed. by J.F. Danielli, M.D. Rosenberg, D.A. Cadenhead, Vol. 5, Academic Press, New York, 1972, pp.63-119.

12.   R.G. Azrak and C.L. Angell, J. Phys. Chem. 77, 3048 (1973).

13.   We used the alcohols alone only in the case of heptanol and the higher homologues. There is some recent evidence that low boiling alcohols require an amine catalyst when the treatment is carried out at reflux.

14.   F.G. Yamagishi, unpublished data.

15.   We believe that the surface coverage is more nearly complete with shorter alcohols. This view was supported by ellipsometric measurements made by Prof. Arthur Adamson and Dr. Patrick Hu on quartz surfaces treated with 1-decanol and 1-octadecanol.

CONTRIBUTORS

S. Aftergut. General Electric Corporate Resech & Development,
          Schenectady, New York.

S. I. Ahmad. Technicon Instruments Corporation, Tarrytown, New York.

B. L. Bales. Centre de Fisich I.V.I.C. Caracas, Venezuela.

E. M. Barrall, II. IBM Research Laboratory, San Jose, California.

K. Brezinsky. Hunter College, CUNY, New York, New York.

T. R. Britt. University of Southern Mississippi, Wattiesburg, Mississippi.

N. W. Buckley. University of Southern Mississippi, Hattiesburg, Mississippi.

B. J. Bulkin. Polytechnic Institute of New York, Brooklyn, New York.

E. F. Carr. University of Maine, Orono, Maine.

R. Chang. Rockwell International, Thousand Oaks, California.

L. L. Chapoy. The Technical University of Denmark, Lyngby, Denmark.

N. J. Clecak. IBM Research Laboratory, San Jose, California.

H. S. Cole, Jr. General Electric Corporate Research & Development,
          Schenectady, New York.

R. J. Cox. IBM Research Laboratory, San Jose, California.

A. G. De Rocc. The University of Maryland, College Park, Maryland.

D. B. DuPre. University of Louisville, Louisville, Kentucky.

A. C. Eringen. Princeton University, Princeton, New Jersey.

L. R. Farmer. Gillette Research Institute, Rockville, Maryland.

R. F. Fisher. University of Southern Mississippi, Hattiesburg, Mississippi.

531

J. B. Flannery. Webster Research Center, Xerox Corporation, Rochester, New York.

B. M. Fung. University of Oklahoma, Norman, Oklahoma.

H. L. Garvin. Hughes Research Laboratories, Malibu, California.

J. H. Goldstein. Emory University, Atlanta, Georgia.

D. W. Goodman. National Bureau of Standards, Gaitnersburg, Maryland.

B. Grant. IBM Research Laboratory, San Jose, California.

A. F. Gregges. IBM Research Laboratory, San Jose, California.

J. Grinberg. Hughes Research Laboratories, Malibu, California.

A. C. Griffin. University of Southern Mississippi, Wattiesburg, Mississippi.

E. A. Grula. Oklahoma State University, Stillwater, Oklahoma.

S. J. Havens. University of Southern Mississippi, Wattiesburg, Mississippi.

C. F. Hayes. University of Hawaii, Honolulu, Hawaii.

E. Iizuka. Shinshu University, Ueda, Japan.

M. J. Janiak. Boston University School of Medicine, Boston, Massachusetts.

J. H. Johnson. Oklahoma State University, Stillwater, Oklahoma.

J. F. Johnson. IBM Research Laboratory, San Jose, California.

L. L. Jones. University of Illinois at Chicago Circle, Chicago, Illinois.

T. E. Kelley. Oregon State University, Coruallis, Oregon.

M. J. Little. Hughes Research Laboratories, Malibu, California.

R. C. Long, Jr. Emory University, Atlanta, Georgia.

C. R. Loomis. Boston University School of Medicine, Boston, Massachusetts.

R. A. Mackay. Drexel University, Philadelphia, Pennsylvania.

J. D. Margerum. Hughes Research Laboratories, Malibu, California.

D. L. Melchior. Brown University, Providence, Rhode Island.

L. J. Miller. Hughes Research Laboratories, Malibu, California.

W. G. Miller. University of Minnesota, Minneapolis, Minnesota.

G. D. Myer. Hughes Research Laboratories, Malibu, California.

M. N. L. Narasimhan. Oregon State University, Corvallis, Oregon.

T. J. Novak. Aberdeen Proving Ground, Aberdeen, Maryland.

C. S. Oh. Beckman Instruments, Inc., Fullerton, California.

J. M. Pollack. Webster Research Center, Xerox Corporation, Rochester,
          New York.

E. J. Poziomek. Aberdeen Proving Ground, Aberdeen, Maryland.

R. G. Priest. Naval Research Laboratory, Washington, D. C.

M. S. Rapport. The University of Maryland, College Park, Maryland.

E. L. Samulski. University of Connecticut, Storrs, Connecticut.

T. Sarada. The American University, Washington, D. C.

J. M. Schnur. Naval Research Laboratory, Washington, D. C.

R. N. Schwartz. University of Illinois at Chicago Circle, Chicago, Illinois.

J. P. Sheridan. Naval Research Laboratory, Washington, D. C.

W. H. Smith. Hughes Research Laboratories, Malibu, California.

D. S. Smythe. Hughes Research Laboratories, Malibu, California.

M. K. Stanfield. Tulane Medical School, New Orleans, Louisiana.

J. M. Steim. Brown University, Providence, Rhode Island.

A. E. Stillman. University of Illinois at Chicago Circle, Chicago, Illinois.

P. H. Von Dreele. Northwestern University, Evanston, Illinois.

B. L. Warren. Tulane Medical School, New Orleans, Louisiana.

P. F. Waters. The American University, Washington, D. C.

P. K. Watson. Webster Research Center Xerox Corporation, Rochester, New York.

F. H. Wilson. Tulane Medical School, New Orleans, Louisiana.

E. L. Wee. University of Minnesota, Minneapolis, Minnesota.

F. G. Yamagishi. Hughes Research Laobratories, Malibu, California.

J. T. Yang. University of California, San Francisco, California.

INDEX